27th International Symposium on Space Terahertz Technology (ISSTT 2016)

Nanjing, China
12 – 15 April 2016

ISBN: 978-1-5108-3890-1

Printed from e-media with permission by:

Curran Associates, Inc.
57 Morehouse Lane
Red Hook, NY 12571

Some format issues inherent in the e-media version may also appear in this print version.

Printed by Curran Associates, Inc. (2017)

International Symposium on Space Terahertz Technology
c/o NRAO Headquarters
520 Edgemont Road
Charlottesville, VA, USA 22903-2475

Phone: (434) 296-0254
Fax: (434) 296-0278

mbishop@nrao.edu

Additional copies of this publication are available from:

Curran Associates, Inc.
57 Morehouse Lane
Red Hook, NY 12571 USA
Phone: 845-758-0400
Fax: 845-758-2633
Email: curran@proceedings.com
Web: www.proceedings.com

Organizers and Sponsors

Organizers

Purple Mountain Observatory

Key Lab of Radio Astronomy

Funding agency

National Nature Science Foundation of China (NSFC)

Chinese Academy of Sciences (CAS)

Sponsors

Sumitomo Heavy Industries

Rohde & Schwarz

Oxford Instruments

Welcome

Purple Mountain Observatory and Key Lab of Radio astronomy, Chinese Academy of Sciences, welcome you to the 27th International Symposium on Space TerahertzTechnology (ISSTT2016), held from April 12 to 15, 2016, in Nanjing, China.

A total of 106 abstracts were accepted, with49 abstracts scheduled for oralpresentation and 57 for poster presentation. There are 3 invited contributions. Wewould like to thank the Scientific Organizing Committee for the abstract review.

Scientific Organizing Committee

Andrey Baryshev	SRON, the Netherlands
Jochem Baselmans	SRON/TUDelft, the Netherlands
Victor Belitsky	Chalmers University of Technology, Sweden
Raymond Blundell	Smithsonian Astrophysical Observatory, USA
Tom Crowe	Virginia Diodes Inc., USA
Herbert Eisele	University of Leeds, UK
Brian Ellison	Rutherford Appleton Lab, UK
Neal Erickson	University of Massachusetts - Amherst, USA
Gregory Goltsman	Moscow State Pedagogical University, Russia
Cornelia E. Honingh	University of Cologne, Germany
Anthony Kerr	National Radio Astronomy Observatory, USA
Arthur Lichtenberger	University of Virginia, USA
Alain Maestrini	Paris Observatory, France
Doris Maier	IRAM, France
Imran Mehdi	Jet Propulsion Lab, USA
Yutaro Sekimoto	National Astronomical Observatory of Japan, Japan
Shengcai Shi	Purple Mountain Observatory, China
Jan Stake	Chalmers University of Technology, Sweden
Edward Tong	Smithsonian Astrophysical Observatory, USA
Christopher K. Walker	University of Arizona, USA
Wolfgang Wild	European Southern Observatory, Germany
Stafford Withington	University of Cambridge, UK
Ghassan Yassin	University of Oxford, UK

Local Organizing Committee

Sheng-Cai Shi, Rui-Qing Mao, Jing Li, Wen Zhang, Wei Miao, Xue-Mei Chen

The 27th International Symposium on Space Terahertz Technology

April 12-15, 2016, Nanjing, China

Program

Tuesday, April 12, 2016

14:00-20:30 Registration

18:30-20:30 Reception

===

Wednesday, April 13, 2016

8:00-8:30 Registration

8:30-8:40 Welcome & Opening Remarks

8:40-10:10 W1 Session: THz Projects & Instruments (I)

(Chair: Heinz-Wilhelm Hübers)

10:10-10:40 Coffee Break

10:40-12:20 W2 Session: THz Mixers& Detectors (I)

(Chair: Jonathan Kawamura)

12:20-14:00 Lunch Break

14:00-16:00 W3 Session: THz Receivers

(Chair: Valery Koshelets)

==

Thursday, April 14, 2016

8:30-10:20 T1 Session: THz Projects & Instruments (II)

(Chair: Tetsuo Hasegawa)

10:20-10:50 Coffee Break

10:50-12:30 T2 Session: Quantum Cascade Lasers

(Chair: JianRongGao)

12:30-14:00 Lunch Break&SOC Meeting

14:00-16:40 T3 Session: THz Sources &Optics

(Chair: Scott Paine)

9:20 Fast Terahertz Imaging using a Quantum Cascade Amplifier up to 20,000 pps 82
Yuan Ren (*University of Cambridge*)

9:40 (4) The Sardinia Radio Telescope Front-Ends 83
AlessandroNavarrini (*INAF-Radio Astronomy Observatory*)

10:00 (3) Multi-Gbit/s Data Transmission in Sub-Terahertz Range 87
Zhe Chen (*University of Electronic Science and Technology of China*)

10:20-10:50 Coffee Break

10:50-12:30 F2 Session: THz Mixers & Detectors (III)

(Chair: Sergey Cherednichenko)

10:50 Study of Image Rejection Ratio of 2SB SIS receiver 91
AndreyKhudchenko (*SRON Netherlands Institute for Space Research*)

11:10 A Zero-Bias Ultrasensitive THz Hot-Electron Direct Detector with Large Dynamic Range 92
BorisKarasik (*Jet Propulsion Laboratory*)

11:30 Room-temperature Direct and Heterodyne Detectors Based on Field-effect Transistors 93
Hua Qin (*Suzhou Institute of Nano-tech and Nano-bionics*)

11:50 Photon Counting Detector as a Mixer with Picowatt Local Oscillator Power Requirement 94
MichaelShcherbatenko (*Moscow State Pedagogical University*)

12:10 Development of a 2 THz Solid-state Radiometer for Atmospheric Sounding 95
JeanneTreuttel (*Jet Propulsion Laboratory*)

12:30-12:40 Closing

12:40-14:00 Lunch

14:00-17:00 Tour to Purple Mountain Observatory & SMLab

===

T4: Poster Session

W1 Session: THz Projects & Instruments (I)

ALMA - Scientific Results and Future Developments

Tetsuo Hasegawa*

NAOJ Chile Observatory, National Astronomical Observatory of Japan, Osawa, Mitaka, Tokyo 181-8588 Japan
*Contact:tetsuo.hasegawa@nao.ac.jp

Atacama Large Millimeter/submillimeter Array (ALMA) has been built and operated at 5,000-m altitude site in the Andean plateau of northern Chile, as an international collaboration between North America, Europe, East Asia and Chile. It is an aperture synthesis radio telescope with 66 high precision antennas with maximum baseline of 15 km to observe astronomical objects at 30 - 950 GHz with unprecedented sensitivity and angular resolution that represents a state of art of the remote sensing technology at these frequencies. Since its start of early science operations in 2011, it has been providing data that are revolutionizing our view of the universe from the formation of galaxies in the early universe to formation of planetary systems in the Solar neighborhood. In this talk, I present some of these exciting scientific results, as well as the plans of future upgrades of the ALMA system.

ALMA isa partnership of ESO (representing its member states), NSF (USA), and NINS (Japan),together with NRC (Canada), NSC, and ASIAA (Taiwan), in cooperation with the Republic of Chile. The Joint ALMA Observatory is operated by ESO, AUI/NRAO, and NAOJ.

The Far Infrared Spectroscopic Explorer (FIRSPEX)

D. Rigopoulou[1*], B. K. Tan[1], G. Yassin[1]

On behalf of the FIRSPEX Consortium

[1] Department of Physics, University of Oxford, Keble Road, Oxford, OX1 3RH, UK

*Contact: dimitra.rigopoulou@physics.ox.ac.uk, phone +44 1865 273296

Abstract— **The Far InfraRed Spectroscopic EXplorer (FIRSPEX) is a novel concept for an astronomy satellite mission that will revolutionise our understanding of the properties of the Interstellar Medium (ISM) and star formation through velocity resolved spectroscopic observations at multi-terahertz frequencies. FIRSPEX comprises a fully cryogenic (~4K) heterodyne payload and a ~1.2 m primary antenna to scan the sky in a number of discreet spectroscopic channels delivering 3-dimensional spectral information. The spectral range selected contains important molecular, atomic and ionic species; the majority of which cannot be observed from the ground.**

FIRSPEX is UK led with additional contributions from partners throughout Europe. FIRSPEX opens up a relatively unexplored parameter space that will produce an enormously significant scientific legacy by focusing on the properties of the multi-phase ISM, the assembly of molecular clouds in our Galaxy and the onset of star formation topics which are fundamental to our understanding of galaxy evolution.

INTRODUCTION

The far-infrared (FIR) to submillimetre (submm) window is one of the least-studied regions of the electromagnetic spectrum. Yet, this wavelength range is absolutely crucial to our understanding of star formation and stellar evolution in the Universe. These complex physical processes leave their imprint on the Interstellar medium (ISM) of our Galaxy and that of external galaxies. By studying the phase structure of the ISM we can begin to unravel the processes that control the heating and cooling of the clouds that regulate star formation. FIR/submm spectroscopy is essential to answer these questions since this regime contains important cooling lines of the different phases of the ISM. The fine-structure line of singly ionised carbon [CII] at 158μm is the most important cooling line of the neutral ISM. FIRAS/COBE maps showed that this is the strongest cooling line in the ISM at about 0.3% of the continuum infrared emission (Fixsen et al. 1999). Other important atomic fine-structure lines include the atomic oxygen lines [OI] at 63 and 145 μm, and the atomic carbon [CI] lines at 370 and 609 μm.

The FIR/submm window also contains a number of atomic fine-structure lines that trace ionized gas, such as the singly ionized nitrogen [NII] 205μm. Finally, it contains a large number of high- J molecular rotational transitions. The mid- to high-J transitions, which are either difficult to observe or completely inaccessible from the ground, have a large span in critical densities, making them excellent tracers of the physical conditions of gas over a wide range in temperatures and densities. The Herschel Space Observatory (Pilbratt et al. 2010) highlighted the immense potential of far-infrared spectroscopy in understanding the complex physics of the ISM. While the superb imaging capabilities of Herschel resulted in panoramic views of the Galactic Plane that provided a full census of the dust reservoir available (e.g. Molinari et al. 2010), there is a distinct lack of similar information for the gaseous component of the ISM. Only then, will be able to establish the gas-to-dust ratio and how this affects star formation in our own Galaxy and in external galaxies. But Herschel spectroscopic observations that allow us to probe the gas phase structure of the ISM were limited to less than 1% of the whole sky. Our mission concept, the Far-Infrared Spectroscopic Explorer (FIRSPEX) comes to fill this gap: FIRSPEX will carry out large area surveys in four discreet spectral channels centered on key FIR lines: [CI] 370 μm (809 GHz), [NII] 205μm (1460 GHz), [CII] 158μm (1900 GHz), and [OI] 63μm (4700 GHz). The choice of the spectral bands has been motivated by the need to study the dynamics of the multi-phase ISM in the Universe. In what follows we present the FIRSPEX scientific drivers and give an overview of the payload and it capabilities.

FIRSPEX SCIENCE DRIVERS

Observations in the four FIRSPEX bands have been designed to serve three intertwined science themes. Firstly, a fully sampled velocity resolved map of the Galactic Plane (~3600 sq. deg) targets the physical and dynamical properties of the ISM in our Galaxy. Secondly, pointed observations of a volume-limited sample of nearby galaxies will establish the role of environment on the ISM, and how the properties of the ISM affect star formation. Thirdly, a deep blind FIR spectroscopic survey (~400 sq. deg.) will probe the physics of distant galaxies and establish the properties of the gas and how it impacts on galaxy evolution from ``cosmic noon'' to the galaxies we see today.

A. *FIRSPEX deep Galactic Plane Survey*

How do molecular clouds form?

Theoretical studies (Goldbaum et al 2011) have suggested that molecular clouds are formed by large-scale accretion of new material onto existing dense clouds. This scenario is consistent with recent Herschel observations that suggest the global filamentary structure of molecular clouds is created by large scale flows of atomic material at earlier times (e.g. Molinari 2010, Peretto 2012). Nonetheless, mass accretion has not been convincingly demonstrated observationally. As the accreted material undergoes a transition from atomic hydrogen to low-density molecular hydrogen and finally denser molecular gas, also visible in CO, we need to disentangle the various phases of the ISM to probe molecular cloud formation. Resolving the velocities of the different components that determine the accretion time scales is a pre-requisite to our understanding of star-formation on galactic scales. Without the ability to separate the different ISM phases proper assessment of the atomic hydrogen emission is impossible.

What fraction of the baryonic matter is in CO dark clouds?

High spectral resolution observations of C+ will enable us to quantify the fraction of CO-dark gas, which is likely to contain a significant fraction of the baryonic matter in the Galaxy, but is invisible in most other tracers. By comparing the phase distribution with the known distribution of energetic sources in the Galaxy we can quantify the effects of the Galactic metallicity gradient and measure the global trapping efficiency of the ISM for the different heating sources.

What heats the ISM?

Radiative feedback from young stellar sources is one of the main heating sources for the Galactic ISM. FIRSPEX observations of the two main cooling lines of the ISM [CII] and [OI] will provide an complete census of the UV heating of the Galactic ISM from UV radiation in terms of photon-dominated regions (PDRs). This allows to globally access the impact of stellar feedback on the evolution of the ISM structure and it allows us to ``calibrate" the contribution to the different phases to the integrated [OI], [NII], and [CII] emission observed in other galaxies.

To provide answers to these fundamental question requires surveys with reasonable angular resolution (< 1') together with large spatial coverage, with the velocity resolution to detect the predicted velocity shifts of less than or equal to 1 km/s. The FIRSPEX Galactic plane survey (Fig.1) will unravel the physics controlling the evolution of these structures and their role in the process of star formation.

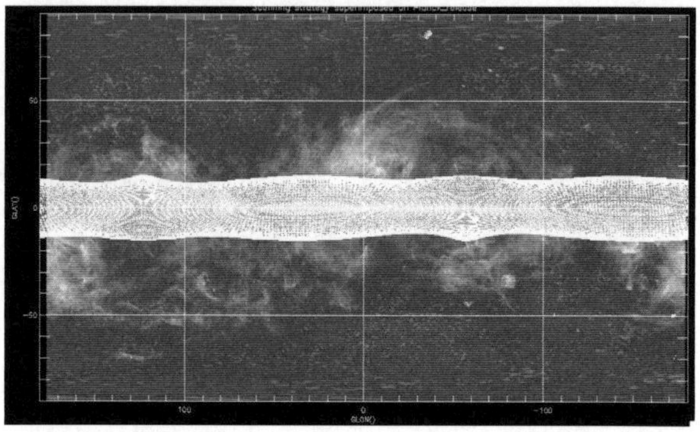

Fig. 1: The planned FIRSPEX Galactic Plane Survey superimposed on Planck all sky map (credit: ESA/HFI/LFI).

B. *FIRSPEX Observations of Nearby Galaxies*

Some of the fundamental science questions in the study of local galaxies are: ***What regulates star formation in galaxies? How does the ISM vary with environment?*** These questions have important cosmological implications and, as such, it is imperative to answer them in the local Universe, to be able also to understand the distant Universe. We can answer these questions with a detailed study of the gas components and properties in external galaxies of different types (starburst, ellipticals, mergers etc). These properties seem to be quite different from those seen in galactic star forming regions and this may impact the extragalactic star formation efficiencies and rates (e.g. Magdis et al. 2012). It is now well established that massive galaxies show large variations in the physical and chemical properties of their star forming gas, which in turn regulates the evolution of galaxies. Existing atomic and molecular data, for at least the nearest galaxies, show a chemical diversity and complexity that cannot be explained by a one-component, steady-state chemical model, and indicates how relative abundances between atoms and molecules may be able to provide insights into the physical distribution of the gas and the energetics of these galaxies.

FIRSPEX offers the unique opportunity to obtain velocity resolved maps of the major cooling lines of the ISM in galaxies. High spectral resolution is needed to disentangle the contributions of the various ISM phases along the same line of sight, such observarions will shed unprecedented light on the interplay of the ISM phases and their specific role in the cycle of matter inside galaxies as well as in their evolution.

C. *Probing the ISM of Distant Galaxies with FIRSPEX*

FIR fine structure (FS) lines provide an extremely powerful tool to probe the physical conditions of star formation in galaxies. Several studies have examined their reliability as Star Formation Rate (SFR) indicators (e.g. de Looze et al. 2014, Rigopoulou et al. 2014, Magdis et al. 2014). However, there is a considerable gap in redshifts between observations of FIR FS lines in relatively nearby galaxies with Herschel and detections of FIR FS lines in distant (z>3) galaxies routinely reported by ALMA. If we are to establish the use of

FIR FS lines as SFR indicators in the very distant Universe then it is essential to trace their evolution from present day till about redshift 3. We need to understand the physics driving the emission of cooling lines at an epoch where the Universe was very energetic, the cosmic `noon'.

FIRSPEX deep `blind spectroscopic surveys' will detect galaxies based on their line emission. Using the four FIRSPEX Bands we will detect [OI], [OIII], [CII], [NII] and [NIII] in various redshifts bins between $0.3<z<3$. The FIRSPEX deep fields are located in prime strategic fields with deep ancillary data in other wavelengths. Such observations will allow us to characterize the ISM of distant galaxies and how the properties of the gas affect star formation efficiencies and subsequent galaxy evolution.

FIRSPEX PAYLOAD

The two prime scientific requirements for the FIRSPEX mission are high detection sensitivity and high spectral resolving power (resolution order > 10^6) within the Terahertz frequency domain. The sensitivity requirement is satisfied with the use of advanced heterodyne receivers based upon superconducting mixer technology used in conjunction with frequency stable local oscillators (LOs) and advanced digital sampling and analysis techniques.

The FIRSPEX payload comprises four parallel receiver channels that can operate simultaneously and therefore independently sample neighbouring regions of the sky. Each receiver channel is located within the focal plane of a 1.2m diameter primary antenna, and offers angular resolution on the scale of 1 arcmin. A schematic view of the payload is shown in Fig. 2.

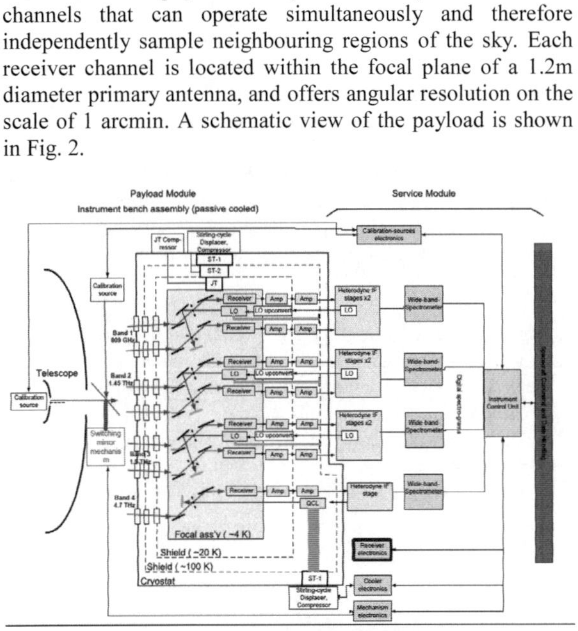

Fig.2: A schematic view of the FIRSPEX payload

In addition, each receiver channel comprises a superconducting mixer, LO chain, stages of intermediate frequency (IF) amplification followed by digital sampling and signal processing. The need for state-of-the-art sensitivity defines the use of superconducting mixers configured either as tunnel junctions or hot electron bolometers. Such technology requires cooling to a low temperature, and approaching 4K, in order to operate. Achieving such low temperatures in space is non-trivial, but has been demonstrated by various past missions e.g. Planck used active closed cycle coolers. Our baseline configuration proposes to passively cool the instrument to 50K (L2 orbit) and with active cooler technology providing sub-stages with necessary heat lift at 4K and 15K. The 4K stage cools the sensitive mixers and the 15K and 50K stages, in addition to reducing the thermal conductive and radiation load on the 4K stage, also cool vital elements of the receiver chain such as the final LO stage and IF low noise amplifiers (LNAs).

A total of four receiver channels are planned with each operating in a double sideband configuration and providing a total of 7 sampling pixels on the sky. Fig. 3 shows the receiver optics for a two receiver channel.

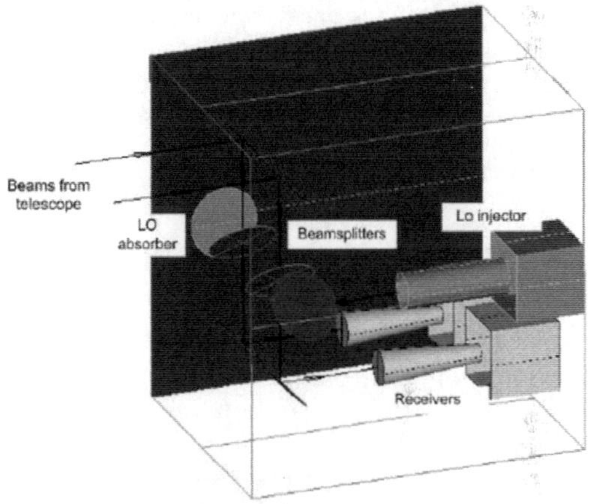

Fig. 3: Receiver optics for double pixel channel

The frequency band allocations are described within Table 1 along with estimated system sensitivities. Bands 1 through 3 use two independent mixers per frequency band giving multiple pixel sampling to compensate for the smaller beams. Coupling to the primary quasi-optical focal plane is accomplished via relay optics comprising a series of re-imaging mirrors. For Band 2, 3 and 4, conventional harmonic frequency up-convertors provide suitable sources of LO power injected into the mixer using simple beam splitter. For band 1, the LO source is provided by a quantum cascade laser (QCL) cooled to ~50K. Each mixer is followed by a cooled LNA and a further stage of room temperature amplification while the IF final output is processed by a dedicated fast Fourier Transform spectrometer (FFTS). Each receiver system is calibrated through the use of blackbody targets of known brightness temperature that are sequentially introduced into the optical path.

Table 1: FIRSPEX receiver channel frequency, species and performance designations

Designation	Frequency (THz)	Primary Species	Secondary Species	No of Pixels	System Noise (K)
Band 1	4.7	OI	-	2	800
Band 2	1.9	CII	$^{13}C+$	2	500
Band 3	1.45	NII	SH^+,SO,CF^+,H^2O+	2	350
Band 4	0.81	CI	CO(7-6)	1	180

The spacecraft (SC) design consists of a cold Payload Module (PM) that includes the telescope and instrument box. Positioned at the Lagrangian obit L2, the PM is located on the anti-sun side of a sunshade and using a multiple V-groove thermal shield arrangement as per other missions at L2 (Planck, NGCryoIRTel) is passively cooled to ~50K. The cold PM provides a low temperature starting point for the closed- cycle cooler system, with use of instrument radiators, allowing the coolers to achieve the 15K, 4K stages, The warm Service Module (SM) is located on the sun-facing side.

The proposed structural mounting scheme is shown in Fig. 4.

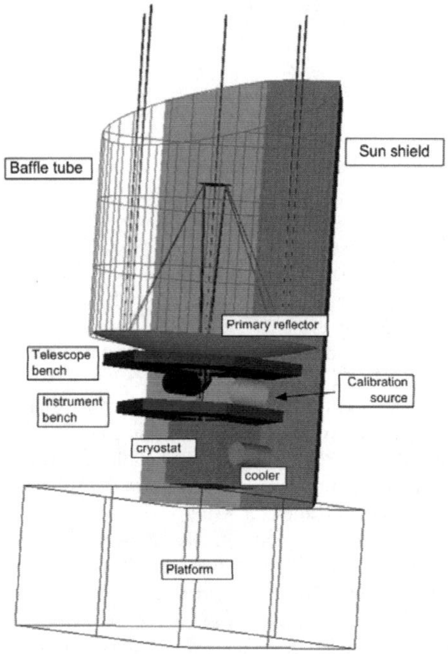

Fig.4: Outline model of the FIRSPEX payload showing the relationship between the various payload elements.

The payload harnesses, and cryo-coolers for the instrument, have to penetrate through the V-groove shields, and they make use of the progressively colder stages of the shields to achieve required cooling. The cooling scheme uses a 2-stage Stirling cryo-cooler in the shields, and following ESA's NG-CryoIRTel design, and cools to ~15K. This latter stage cools a 15K enclosure for the receiver and precools the Joule-Thompson cooler which provides heat lift at ~4K. A design driver is the need to minimise the power dissipation in the cold stages (4K and 15K), due to the limited capacity of the cryo-coolers. This is accomplished through careful thermal and harnesses design.

The receiver back-end electronics, e.g. LO generation and IF components and FFTS, is mostly located within the SM. Also located within the SM are the compressors required for the Stirling and Joule-Thompson cryo-coolers, including a gas-preparation system for the latter; the cooler and on-board calibration target and target deployment electronics, and instrument control functions. A schematic of the proposed payload structure is shown in Fig. 5.

This mission will be operated in two distinct modes. Mode 1 allows large area surveys around the galactic plane to be performed by continuously varying the SC pointing. Mode 2 extends the spatial sampling range to higher out-of-plane positions and allows for period of deep integration on extragalactic sources.

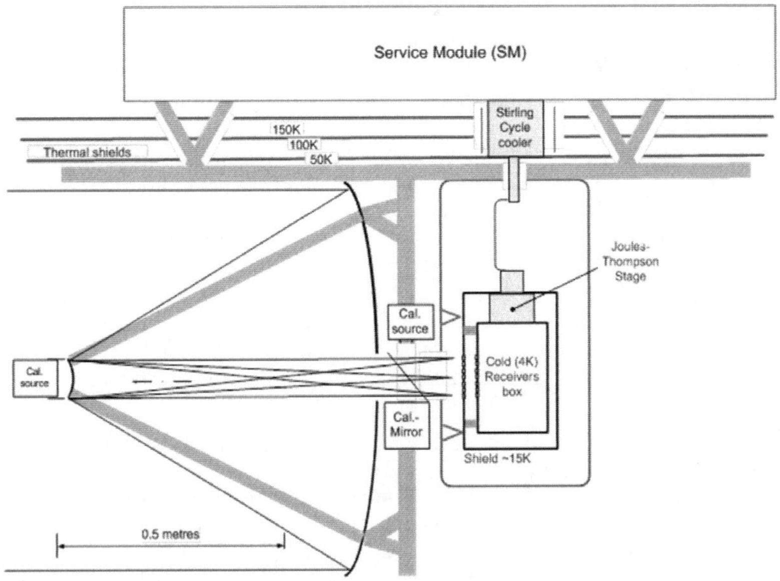

Fig. 5: A side view of the proposed FIRSPEX payload

REFERENCES

Pilbratt, G., L., Riedinger, J. R., Passvogel, T., et al., 2010, A&A 518, 1

Molinari S., Swinyard, B., Bally, J., et al. 2010, A&A, 518, 100

Goldbaum, N.J., Krumholtz, M., Matzner, C.D., McKee, C., 2011, ApJ, 738, 101

Perreto, N., Andre, Ph., Konvyes, V., et al. 2012, A&A,514,63

Pineda J. L., Langer, W. D., Velusamy, T., et al. 2013, A&A, 554, 103

Fixsen, D. J., Bennett, C. L., Mather, J., 1999, ApJ, 526, 207

Nakagawa, T., et al. 1998, ApJS 115, 259

Velusamy T., Langer, W. D., Pineda, J. L., Goldsmith, P. F., 2012, A&A 541, 10

Bisbas T. G., Bell, T. A., Viti, S., et al. 2014, MNRAS, 443, 1111

de Looze I., Fritz, J, Baes, M, et al 2014, A&A, 571, 69

Rigopoulou, D., Hopwood, R, Magdis, GE, et al, 2014, ApJLett, 781, 15

Magdis, G.E., Rigopoulou, D, Hopwood, R, et al, 2014, ApJ 796, 63

CMB Polarization Experiment "GroundBIRD"

C. Otani[1,2*], O. Tajima[3], S. Oguri[3], S. Mima[1], J. Choi[3,4], T. Damayanthi[1], N. Furukawa[1], M. Hattori[5], M. Hazumi[3,6], H. Ishitsuka[3], R. Koyano[7], M. Minowa[8], M. Nagai[3], T. Nagasaki[3], Y. Sekimoto[9], M. Semoto[7],T. Taino[7], N. Tomita[3,8], T. Uchida[3], E. Won[4], M. Yoshida[3],and GroundBIRD project

[1]*RIKEN Center for Advanced Photonics, RIKEN, Sendai 980-0845 and Wako 351-0198, Japan*
[2]*Department of Physics, Tohoku University, Sendai 980-8577, Japan*
[3]*Institute of Particle and Nuclear Physics, KEK, Tsukuba 305-0801, Japan*
[4]*Department of Physics, Korea University, Seoul 02841, Republic of Korea*
[5]*Department of Astronomy, Tohoku University, Sendai980-8577, Japan*
[6]*Kavli Institute for the Physics and Mathematics of the Universe, Kashiwa 277-8583, Japan*
[7]*Department of Electrical and Electronic Systems, Saitama University,Saitama 338-8570, Japan*
[8]*Department of Physics, University of Tokyo, Tokyo 113-0033, Japan*
[9]*Advanced Technology Center, National Astronomical Observatory of Japan, Mitaka 181-8588, Japan*
* Contact:otani@riken.jp

Cosmic microwave background radiation (CMB) is the oldest electromagnetic radiation we can observe in the universe. The observations of its 2.7 K blackbody spectrum and its anisotropy strongly support the Big Bang cosmology, and also strongly imply the "inflation" before the bang. In order to prove its presence, the best way is to find the sign of the primordial gravitational waves. The waves are expected to imprint the rotational polarization patterns named "B-mode" on the CMB.The precise measurement of the spatial spectrum of the polarization distribution is strongly required. For this purpose, we are developing the ground-based experiment, GroundBIRD. The telescope is planned to be situated in Canary Islands and its operation will start in 2017. In this presentation, we will show the current status of the development.

GroundBIRD introduces several keytechnologies to realize the low detection limit for the CMB polarization observations such asa high-speed rotatingtelescope, cooled optics, and microwave kinetic inductance detectors (MKIDs). The high-speed rotating telescope is realized by amechanical cryostat on a rotating table. We have newly developed a rotary jointto pass through the high-pressure helium gas as well as the electrical lines.The joint enables usto connect the head of the cryostat to the compressor placed on the ground. By combining the mechanical cooler with a ^{3}He depressurized stage (He-10), we have achieved the temperature of 0.23 K with a holding time of more than 24 hrswith the continuous rotation at 20 rpm.The rotation enables us to scan a wide range of sky continuously and repeatedly. The expectedmultipolecoverageis $6 < l < 300$. The rotation is also advantageous to reduce the $1/f$ noise of the atmospheric perturbation.

We have adopted a cold optics below 4K tosuppress the radiation noise. The size of the aperture window is 30 cm in diameter. To minimize theincident thermal radiation, the metal-meshed filters (QMC Instruments Ltd.) and the radio-transform multi-layer insulations (RT-MLI) areutilized. The latter consists of a set of stacked formed-polystyrene layers. Each layeris transparent to radio waves andopaque to infrared radiation. Because the RT-MLIis cooleddown by itself by radiating heat, no thermal link is required.Its performance does not change even if the aperture size changes. We have achievedthe mirror temperature at 3.4 K andthe cold stage at 0.21 K.

As the focal plane detector, we have adopted MKIDsbecause of its fast response time (~100 μs),a large number of pixels, and high sensitivity. The fast time response is crucial for the high-speed scanning observations. The readout of MKIDsis based on the frequency domain multiplexing, and it enable us to realize a large-format detector array.The number of pixels is 109 and 330 pixels at 220and 145 GHz, respectively.

First Flight of the PILOT Balloon Borne Experiment

F. Pajot[1*], J.Aumont[2], J.P. Bernard[1] and the PILOT collaboration

[1]Institut de Recherche en Astrophysique et Planétologie, Toulouse,France
[2]Institut d'Astrophysique Spatiale, Orsay, France
* Contact: francois.pajot@irap.omp.eu

The PILOT experiment is dedicated to the measurement of the polarized emission of the galactic interstellar dust in the submillimeter wavelength range. The instrument payload consists of a bolometer camera at the focus of a 1 m diameter telescope, and its associated readout and control electronics. The detection is based on silicon bolometer arrays of the same technology as the Herschel/PACS instrument. The 2048 pixels are cooled to 320 mK by a ^{3}He sorption refrigerator inside a ^{4}He cryostat. The optics splits the incoming radiation on two orthogonal polarization directions, making this instrument the first polarized camera operated at 240 μm. The first flight of PILOT was launched from the Canadian balloon facility in Timmins, operated by the CSA (Canadian Space Agency) and the CNES (Centre National d'Etudes Spatiales, France), on September 20th 2015.The operation of the instrument was nominal. It allowed 20 hours of astrophysical observations from the float altitude of 39.5 km. The data are being analysed at IAS and IRAP.

W2 Session: THz Mixers & Detectors (I)

HEB Waveguide Mixers for the upGREAT 4.7 THz Heterodyne Receiver Array

P. Pütz*, K. Jacobs, M. Schultz,M. Justen, R. Higgins, S. Fathi, C. E. Honingh and J. Stutzki
Kölner Observatorium für Submm-Astronomie (KOSMA),
I. Physikalisches Institut, Universität zu Köln, Zülpicher Strasse 77, 50937 Köln, Germany
*Contact: puetz@ph1.uni-koeln.de

***Abstract*— We report on integration and testing of HEB waveguide mixers for the upcoming upGREAT 7-pixel 4.7 THz high frequency array (HFA). upGREAT is the array extension of the modular GREAT (German Receiver for Astronomy at Terahertz frequencies) heterodyne receiver platform.The mixers are based on similar technology to the mixer of the single pixel 4.7 THz receiver in operation on SOFIA and use waveguide feedhorn coupling. Forthe array application the mixers are optimized tolowerlocal oscillator powerby reduction of the thickness of the NbN superconducting microbridge layer. Preliminary laboratory measurement results indicate performance comparable to the single pixel mixers.**

INTRODUCTION

Above 1 THz superconducting hot electron bolometer (HEB) mixers are employed in the lowest noise heterodyne receivers([1]-[4]). TheHEB mixers for the GREAT (German Receiver for Astronomy at Terahertz frequencies)single-pixel receivers operated on SOFIA have been continuously improved in performance (Trec, intermediate frequency bandwidth), reliability and higher operation frequency, ultimately leading to the 4.7 THz mixerin operation since 2014 ([5]-[7]).

All mixers are based on waveguide membrane devices and use a feedhorn antenna coupling. The main motivationof this approach is that the center axis of the feedhorn defines a precise locationand orientation of the optical axis with respect to a mechanical reference on the mixer unit. Forfocal plane array applications, i.e. receiver front-endswith several mixer units (pixels),this should favor aregularly spaced and well-defined beam pattern on sky, and therewith an optimum in mapping efficiency of the instrument.

The mixers we report on here are being developed for the upGREAT high frequency array (HFA) receiver, which is the focal plane array extension of GREAT. After successful commissioning of the 14-pixel low frequency array (LFA) receiver in 2015 ([8], [9]), with frequency range centered on the 1900 GHz [CII] fine structure transition,we are now focusing on integration of the second upGREAT receiver, the 7-pixel HFA, which solely targets the [OI] fine structure transition at 4745 GHz (63 μm) ([10], [11]).

MIXER DESIGN

The 4.7 THz HEB mixer design and hardware for the HFA is similar to the 4.7 THz operated in the single-pixel receiver, for full details see [7]. In the following we will only give a brief summary.

Device Fabrication

Each HFA mixer is based on a 3–4 nm thick by 200 nm long and 4000 nm wide NbN microbridge embedded into an on-chip matching circuit and integrated probe antenna, precisely aligned to a 24 μm x 48 μm waveguide. Beamleads are used for DC and RF contacting of the device and they also serve as mounting points. The 2 μm thick Si membrane substrateis fabricated by means of backside processing of a thick SOI wafer after the front side circuit definition has been performed. All critical circuit layers (HEB microbridge and contacts, RF circuit elements) are defined by e-beam lithography.

Waveguide Fabrication

The operation frequency leads to the technically very challenging fabrication of 48 μm x 24 μm sized waveguides. The waveguide backshort and device channel features are fabricated by direct-machining into a Cu-Te block by means of a sophisticated mix of precision stamping and milling techniques in our in-house workshop. For the HFA the mixer blocks have a reduced footprint of 16 mm x 16 mm that is compatible with the array optics. The footprint currently is determined by the use of a standard Huber-Suhner SMA flange type female connector that we kept due to its long heritage in operation and because the SiGe LNA units have SMA input connectors [12].

Feedhorn

The feedhorns are designed and fabricated at Radiometer Physics GmbH using a mandrel / electroforming process, see [13] for details. We press-fit each feedhorn into a T6061 Al clamp that is mounted to the top of the mixer block. The waveguide output is put into direct contact to the device. For the HFA we dropped to usage of M1 dowel pins for alignment to the receiver optics and for improved accuracy now use machined reference stops on the sides. Our machine shop readily achieves the +/- 10 µm specification on the horn center position.

Mixer Qualification

The main challenge for array mixer development is to make all mixers sufficiently similar in performance. For the HFA a single local oscillator (LO) sourcewill be used [14] and for interchangeability of the array mixers an even split for the LO power by means of a Fourier grating mirror is chosen [15]. Therefore the mixers need to be sufficiently similar in LO power consumption for optimum array sensitivity. For the HFA we selected devices from a batch that that haveapprox. 1/3of the LO power requirement of the older single-pixel devices. For upGREAT this will ensurea safeLO power margin whilst permitting maximum signal couplingthrough the diplexer.

DC Qualification

We pre-characterize the RT and IV curves each device on diced wafer pieces with up 18 devices each using a liquid Helium dipstick setup. TheIV and RT curve characteristicsare used for selection of the devices to be mounted into a waveguide mixer block after the final backside (membrane) fabrication process. The typical device yield at DC testing is 90%.

RF Qualification

For the RF heterodyne characterization of the HFA mixers we have developed a new setup that includes a completely evacuated signal path that ensures measurements independent from atmospheric influences in particular near the rest frequency of [OI]. It has a selectable signal path to hot (RT) and cold (LN2) loads as well as agas cell. This methanol gas cell is used for in-line frequency determination of the LO. For simplifying the procedures with have opted to keep the cryostat on its separate vacuum using a Parylene coated Si dewar window with 92% transmission procured from Tydex.

The LO is based on a quantum-cascade laser (QCL) from the ETH Zürich and is an in-house developed prototype for the HFA. We have the option to select from two different QCLs in-situ, one emitting at 4785 GHz and the other close to 4745 GHz. Signal/LO diplexing is achieved with a 5 µm thick Mylar beamsplitter and we remotely attenuate the LO with a rotatable wire grid. Our new setup features 0–5 GHz intermediate frequency (IF) processing with two 2.5 GHz digital FTS spectrometers kindly on loan from the MPIfR. We combine the spectra for single-shot spectrometer measurements over the full IFrange [16].

Due to the QCL's intrinsic tunability limitation we can only measure a very small frequency range of the mixer RF bandpass in heterodyne mode. In order to measure the broadband response we therefore use a commercial FIR FTS from Bruker and operate the HEB mixer as a direct detector. The FTS measurements are important as they can feedback observed changes in the RF bandpass to our 3D-EM circuit modeler (CST Microwave Studio^{TM}), and e.g. allow more precise modeling of the on-chip THz circuit.

Conclusions

We are in the integration and testing phase for the upGREAT HFA 4.7 THz array scheduled for commissioning on SOFIA in late 2016. We are measuring the HFA flight mixers and preliminary Trec(IF) results are similar to our initial results for the single-pixel receiver mixers (Fig. 1). Optimization is still ongoing and based on our experience with the LFA arrays we will need to carefully select HEB mixers not only based on best Trec(IF) but also on LO power consumption and uniformity.

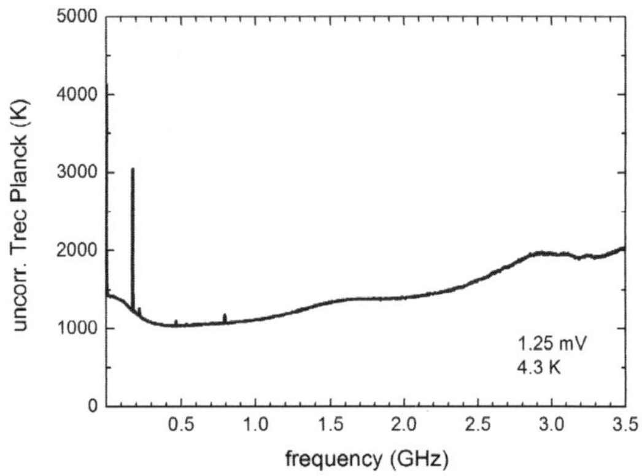

Fig. 1Uncorrected DSB Trec(IF) (Planck) for a HFA flight mixer candidate at a LO frequency of 4785 GHz. A 5 µm thick Mylar beamsplitter in vertical polarization was used as diplexer with a fully evacuated signal path. No corrections for signal path losses were applied (beamsplitter T=70%, dewar window T=92%) were applied. The spurs at low IForiginate from the QCL local oscillator.

Acknowledgment

We like to further thank our mechanical workshop for their persistent and skillful work in fabricating the waveguide mixer blocks.

The GREAT receivers are developed, built and operated as a PI instrument by a consortium of German institutes, the MPI fü r Radioastronomie (PI R. Gü sten) and the Universitä t zu Kö ln (KOSMA), in cooperation with the MPI fü r Sonnensystemforschung and the DLR Institut für Planetenforschung.

Our HEB mixer work currently is supported by the Federal Ministry of Economics and Technology via the German Space Agency (DLR), grant 50 OK 1103 and by Collaborative Research Council 956, sub-project area D3, funded by the Deutsche Forschungsgemeinschaft (DFG).

References

[1] A. Shurakov, Y. Lobanov, G. Goltsman, "Superconducting hot-electron bolometer: from the discovery of hot-electron phenomena to practical applications," *Supercond. Sci. Technol.* 29, 023001, 2016

[2] P. Khosropanah, J. R. Gao, W. M. Laauwen, M. Hajenius, and T. M. Klapwijk, "Low noise NbN hot electron bolometer mixer at 4.3THz," *Appl. Phys. Lett.*, 91, p. 221111, 2007

[3] S. Cherednichenko, P. Khosropanah, E. Kollberg, M. Kroug, H. Merkel, "Terahertz superconducting hot-electron bolometer mixers," *Physica C: Superconductivity*, vol. 372–376, Part 1, pp. 407-415, Aug. 2002.

[4] W. Zhang, P. Khosropanah, J. R. Gao, T. Bansal, T. M. Klapwijk, W. Miao, S. C. Shi, "Noise temperature and beam pattern of an NbN hot electron bolometer mixer at 5.25 THz", *J. Appl. Phys.*, 108, 093102, 2010.

[5] S. Heyminck *et al.*, "GREAT: the SOFIA high-frequency heterodyne instrument," *Astron. Astrophys.*, vol. 542, L1, June 2012

[6] P. Pütz, C. E. Honingh, K. Jacobs, M. Justen, M. Schultz and J. Stutzki, "Terahertz hot electron bolometer waveguide mixers for GREAT," *Astron. Astrophys.*, vol. 542, L2, 2012.

[7] D. Büchel *et al.*, "4.7 THz Superconducting Hot Electron Bolometer Waveguide Mixer," *IEEE THz Sci. Techn.*, vol.5, no.2, pp. 207–214, March 2015.

[8] C. Risacher *et al.*, "First supra-THz Heterodyne Array Receivers for Astronomy with the SOFIA Observatory," *IEEE Trans. THz Sci. Techn.*, vol. 6, no. 2, pp. 199–211, Mar. 2016.

[9] C. Risacher *et al.*, "The upGREAT 1.9THz multi-pixel high resolution spectrometer for the SOFIA Observatory," *Astron. Astrophys.*, submitted for publication.

[10] R. T. Boreiko and A. L. Betz, "Heterodyne Spectroscopy of the 63 μm O I Line in M42", *Astrophys. J.*, 464, pp. L83–L86, 1996.

[11] C. Risacher*et al.*, "The upGREAT THz arrays for SOFIA: successful commissioning at 1.9 THz," in *Proc. 27th Int. Symp. Space THz Technol.*, Nanjing, China, Apr. 13–15, 2016.

[12] S. Weinreb, "Design of cryogenic SiGe low-noise amplifiers," *IEEE Trans. Microw. Theory Techn.*, vol. 55, no. 11, pp. 2306–2312, Nov. 2007.

[13] B. Thomas *et al.*, "1.9–2.5 THz and 4.7 THz electroformed smooth-wall spline feedhorns for the HEB mixers of the upGREAT instrument onboard SOFIA aircraft," in *Proc. 25th Int. Symp. Space THz Technol.*, Moscow, Russia, Apr. 27–30, 2014.

[14] H. Richter, M. Wienold, L. Schrottke, K. Biermann, H. T. Grahn, and H.-W. Hu¨bers "4.7-THz local oscillator for GREAT," IEEE Trans. THz Sci. Technol., vol. 5, no. 4, pp. 539–545, Jul. 2015.

[15] U. U. Graf and S. Heyminck, "Fourier gratings as submillimeter beam splitters," *IEEE Trans. Antennas Propag.*, vol. 49, no. 4, pp. 542–546, Apr. 2001.

[16] B. Klein, S. Hochgürtel, I. Krämer, A. Bell, and R. Güsten, "High-resolution wide-band fast-Fourier transform spectrometers," *Astron. Astrophys*, vol. 542, p. L3, Jun. 2012.

Study of IF bandwidth of NbN hot electron bolometers on GaN buffer layer using a direct measurement method

S. Krause[1*], V. Mityashkin[2], S. Antipov[2], G. Gol'tsman[2], D. Meledin[1], V. Desmaris[1], V. Belitsky[1], M. Rudzinski[3]
[1]*Group for Advanced Receiver Development, Chalmers University of Technology, SE-41296, Gothenburg, Sweden*
[2]*Moscow State Pedagogical University, Moscow, Russia*
[3]*Institute of Electronic Materials Technology (ITME), 01-919 Warsaw, Poland*
*Contact: sascha.krause@chalmers.se, phone +46-73-303 36 72

Abstract— **In this paper, we present a reliable measurement method to study the influence of the GaN buffer layer on phonon-escape time in comparison with commonly used Si substrates and, in consequence, on the IF bandwidth of HEBs. One of the key aspects is to operate the HEB mixer at elevated bath temperatures close to the critical temperature of the NbN ultra-thin film, where contributions from electron-phonon processes and self-heating effects are relatively small, therefore IF roll-off will be governed by the phonon-escape.**

Two independent experiments were performed at GARD and MSPU on a similar experimental setup at frequencies of approximately 180 and 140 GHz, respectively, and have shown reproducible and consistent results. The entire IF chain was characterized by S-parameter measurements. We compared the measurement results of epitaxial NbN grown onto GaN buffer-layer with Tc of 12.5 K (4.5 nm) with high quality polycrystalline NbN films on Si substrate with Tc of 10.5 K (5 nm) and observed a strong indication of an enhancement of phonon escape to the substrate by a factor of two for the NbN/GaN material combination.

INTRODUCTION

Utilizing the heterodyne receiver scheme at submillimetre and terahertz frequencies with high spectral resolution offers a great opportunity to study objects in the distant universe and helps understanding the processes taking place in the cold universe [1]. In order to observe these weak signals from Earth, highly sensitive receivers are required. Beyond approximately 1.2 THz, only hot electron bolometers based on NbN material provide ultimate noise performance and are the technology of choice where lowest noise is required. However, due to their physical working principle, they inherently possess a roll-off of their IF bandwidth, which can be associated with the cooling rate of "hot" electrons [2]. Two major relaxation processes have been identified in phonon cooled HEBs, namely the electron-phonon interaction and the subsequent phonon escape to the underlying substrate [2], [3]. Buffer-layers to promote the growth of high quality ultrathin–NbN films as well as to improve the acoustic matching have been considered essential to overcome their limited IF bandwidth, which only amounts to typically 3-4 GHz for recent operational receivers [4] [5] [6]. Even though improvements at the thin-film level have already been shown with the use of GaN buffer layers [7], the impact of the improved thin-films on mixer- and especially IF performance of HEBs made out of them has not been demonstrated yet. This study is aiming at determining the influence of the GaN substrate on escape time in comparison to conventional Si substrates, used to grow the NbN films onto.

EXPERIMENT

The measurement technique applied here is based on recording the IF output power of the HEB mixer, which is operated at elevated bath temperature close to the critical transition temperature (Tc) of the NbN ultra-thin film. In this operation mode, the inherent complex energy relaxation process of electrons is dominated by the escape of phonons to the substrate [2], [3]. Moreover, the raise of bath temperature has further implications such as the reduction of the energy gap of the superconducting material, thus being able to operate the mixer at lower LO frequencies, where measurement equipment and sufficiently high RF power is easily available. The characteristic escape time is proportional to the thickness of the NbN film and inversely proportional to the acoustic matching between the film and the substrate as well as the mean sound velocity of the film [8].

A. Ultra-thin NbN films

For this experiment, we prepared two high quality NbN films on Si substrate and GaN (0001) buffer-layer by means of reactive DC magnetron sputtering at elevated substrate holder temperatures. The film grown onto silicon at MSPU exhibited a poly-crystalline structure and a Tc of 10.5 K for 5 nm thickness. In contrast, the NbN film, which was deposited onto the GaN featured a single crystal structure due to the small lattice mismatch with high Tc of 12.5 K for 4.5 nm thickness.

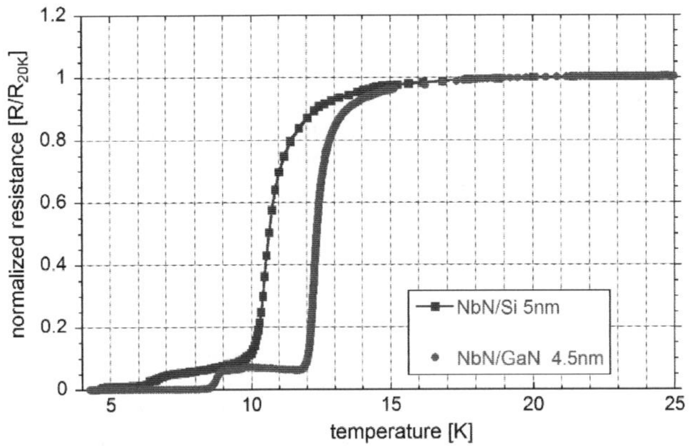

Fig. 1 Resistance versus temperature curve of the HEB bridges made of NbN, which was grown onto Si substrate (Tc=10.5 K) and GaN buffer-layer (Tc=12.5 K). The double transition is caused by the weakened superconductivity under the contact pads.

The electrical resistance versus temperature behaviour is shown in Fig. 1 and reflects the high quality of the NbN ultra-thin films, used to make the HEB from. The bolometer bridges with dimensions of 0.3x3 μm (fabricated at MSPU) and 0.2x2 μm (fabricated at GARD) and the log-spiral antenna was defined by e-beam lithography and subsequent dry etching.

B. Experimental setup

Two similar measurement systems were established at MSPU and GARD in order to cross-verify the measurement data of both laboratories.

The RF signal frequency of the HEB was chosen at 180 GHz (GARD) and 140 GHz (MSPU) and the signal was combined with the local oscillator signal (LO) through a waveguide hybrid (GARD) and beam splitter (at MSPU), respectively. The patterned chip was directly placed onto the waveguide opening, despite the resulting poor impedance match, the pumping of the mixer was easily achieved due to the raised bath temperature. The cooling of the mixer was achieved in a cryostat (GARD) to 4 K and subsequent temperature control of the mixer block by a resistive heater. At MSPU the device was cooled in a liquid helium Dewar and positioned carefully above the helium surface in order to control the temperature of the mixer. The current-voltage (IV) characteristic was recorded frequently during the measurement to exclude effects of drift of the operational point and to ensure constant conditions throughout the experiment. While sweeping the LO frequency, the IF output power level was measured with a spectrum analyzer in the frequency range of 0.2 to 7 GHz.

The entire IF chain including the LNA, bias T, coaxial cables, the mixer block and the HEB in its operating point was characterized by a S-parameter measurement and calibrated with gating and time domain reflectometry (TDR) techniques [9]. The recorded IF data was corrected with the deduced transfer response of the connected IF chain in the frequency band of interest.

Results and Discussion

The measurement systems were verified by characterizing one particular HEB (NbN on Si with 0.3x3μm) in both laboratories. The recorded IF response, after de-embedding the IF transmission chain, are in very good accordance, as depicted in Fig. 2.

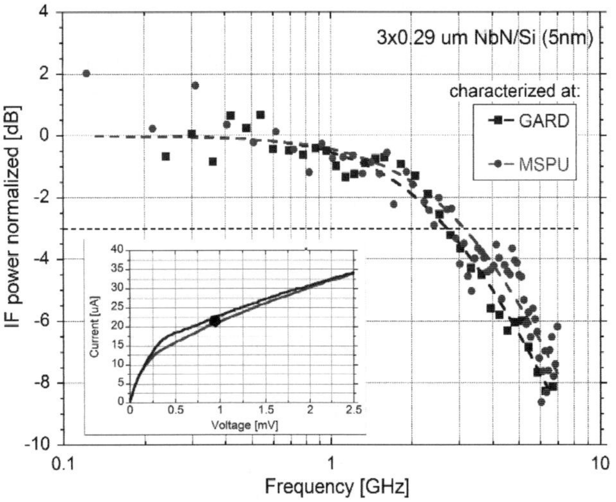

Fig. 2 Normalized recorded IF response for one particular NbN/Si HEB, which was characterized at MSPU (red) and at GARD (blue). The IF power versus frequency is similar and shows a roll-off at approximately 3 GHz. The operating point is indicated in the IVC in the lower left section.

Moreover, the effects on the response by changing bias point and LO power level were investigated and do not show a significant change in the IF roll-off (drop of power by 3 dB). This supports the assumption that the self-heating effect is small in this operation mode of the mixer and therefore, particularly suitable to investigate the influence of film and substrate on the phonon escape.

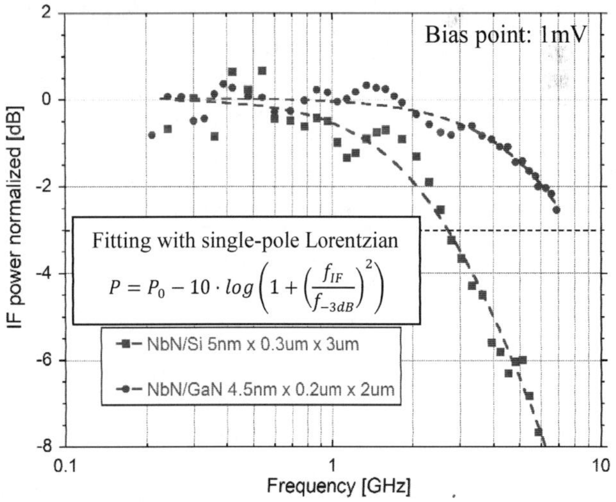

Fig. 3 Normalized recorded IF response for the HEB made of NbN/Si and NbN/GaN.

The 3-dB roll-off as indicated by the dashed line amounts to approximately 3 GHz for the NbN/Si HEB. However, the roll-off frequency, observed for two separately fabricated NbN/GaN HEBs, was significantly increased and amounts to 7.0-7.5 GHz at pumping and bias conditions similarly to the investigated NbN/Si as seen in Fig. 3.

CONCLUSION

It was presented a measurement technique, which allowed to study the effect of buffer-layers on the escape of phonons in phonon-cooled HEBs by operating the mixer at elevated bath temperatures close to the critical temperature of the NbN film, where the contribution of the electron-phonon interaction process and self-heating effect is comparably small. The HEB utilizing a GaN buffer-layer to promote high quality growth of NbN shows a significant enhancement of IF roll-off frequency compared to commonly used Si substrate. Thus, employing the GaN buffer-layer in future THz phonon-cooled HEB mixers based on NbN material may eventually help to overcome the limitation of their small IF bandwidth.

Further studies using this measurement setup will focus on differently sized bolometers to investigate the impact of bolometer area on IF bandwidth and more extensive studies on the effect of underlying buffer-layers.

REFERENCES

[1] D. Leisawitz, “Scientific motivationand technology requirements for the SPIRIT and SPECS far-infrared/submillimeter space interferometers,” *Proc. SPIE,* vol. 4013, pp. 36-46, 2000.

[2] R. Nebosis, A. Semenov, Y. Gousev and K. Renk, “Analysis of a Superconducting Hot-Electron Bolometer Mixer: Theory and Comparison with Experiment,” *7th International Symposium on Space and Terahertz Technology,* 1996.

[3] K. Il'in, I. Milostnaya, A. Verevkin, G. Gol'tsman, E. Gershenzon and R. Sobolewski, “Picosecond hot-electron energy relaxation in NbN superconducting photodetectors,” *Applied Physics Letter,* vol. 76, no. 19, 2000.

[4] D. Meledin, A. Pavolotsky, V. Desmaris, I. Lapkin, C. Risacher, V. Perez, D. Henke, O. Nyström, E. Sundin, D. Dochev, M. Pantaleev, M. Fredrixon, M. Strandberg, B. Voronov, G. Gol'tsman and V. Belitsky, “A 1.3-THz Balanced Waveguide HEB Mixer for the APEX Telescope,” *IEEE Transactions on Microwave Theory and Techniques, vol. 57, no. 1,* 2009.

[5] P. Pütz, D. Büchel, K. Jacobs, M. Schultz, C. E. Honingh and J. Stutzki, “Waveguide hot electron bolometer mixer development for upGREAT,” in *39th International Conference on Infrared, Millimeter, and Terahertz waves (IRMMW-THz),* Tucson, AZ, 2014.

[6] S. Cherednichenko, V. Drakinskiy, T. Berg, P. Khosropanah and E. Kollberg, “Hot-electron bolometer terahertz mixers for the Herschel Space Observatory,” *Review of Scientific Instruments,* vol. 79, p. 034501, 2008.

[7] S. Krause, D. Meledin, V. Desmaris, A. Pavolotsky, V. Belitsky, M. Rudzinski and E. Pippel, “Epitaxial growth of ulta-thin NbN films on AlxGa1-xN buffer-layers,” *Supercond. Sci. Technol.,* pp. vol. 27, no. 6, Apr 2014.

[8] S. B. Kaplan, “Acoustic Matching of Superconducting Films to Substrates,” *Journal of Low Temperature Physics,* vol. 37, pp. 343-364, 1979.

[9] H. Rashid, D. Meledin, V. Desmaris, A. Pavolotsky and V. Belitsky, “Superconducting 4-8 GHz Hybrid Assembly for 2SB Cryogenic THz Receivers,” *IEEE Transactionson Terahertz Science and Technology,* vol. 4, no. 2, pp. 193-200, 2014.

[10] D. Meledin, E. Tong, R. Blundell, N. Kaurova, K. Smirnov, B. Voronov and G. Gol'tsman, “Study of the IF Bandwidth of NbN HEB Mixers,” *Applied Superconductivity,* vol. 13, no. 2, pp. 164-167, 2003.

[11] K. Ilin, R. Schneider, D. Gerthsen, A. Engel, H. Bartolf, A. Schilling, A. Semenov, H.-W. Huebers, B. Freitag and M. Siegel, “Ultra-thin NbN films on Si: crystalline and superconducting,” *Journal of Physics: Conference Series,* vol. 97, p. 012045, 2008.

[12] J. Gao, M. Hajenius, F. Tichelaar, T. Klapwijk, B. Voronov, E. Grishin, G. Gol´tsman, C. Zorman and M. Mehregany, “Monocrystalline NbN nanofilms on a 3C-SiC/Si substrate,” *Applied Physics Letters,* vol. 91, p. 062504, 2007.

[13] D. Dochev, V. Desmaris, A. Pavolotsky, D. Meledin, Z. Lai, A. Henry, E. Janzen, E. Pippel, J. Woltersdorf and V. Belitsky, “Growth and characterization of epitaxial ultra-thin NbN films on 3C-SiC/Si substrates for terahertz applications,” *Supercond. Sci. Technol.,* 2011.

THzSensors Based on Superconducting MgB_2

B. S. Karasik[1*], D. P. Cunnane[1], N. Acharya[2], M. A. Wolak[2],
T. Tan[2], J. H. Kawamura[1], and X. X. Xi[2]

[1]Jet Propulsion Laboratory, California Institute of Technology *, Pasadena, CA 91109, USA
[2]Temple University, Philadelphia, PA 19122, USA
* Contact: boris.s.karasik@jpl.nasa.gov

Terahertz range is rich with molecular lines important for understanding of the chemistry associated with evolution of star-forming molecular clouds. High-resolution spectroscopy of such clouds greatly relies on hot-electron superconducting bolometric (HEB) mixers. Current state-of-the-art receivers use mixer devices made from ultrathin (~ 3-5 nm) films of NbN with critical temperature ~ 9-11 K. Such mixers have been deployed on a number of ground based and suborbital (SOFIA, STO, TELIS) platforms, and have been used in HIFI instrument on the Hershel Space Observatory. Despite its good sensitivity and well-established fabrication process, the NbN HEB mixer suffers from the narrow intermediate frequency (IF) bandwidth ~ 2-4 GHz and is limited to operation at liquid Helium temperature. As an interest in high-resolution spectroscopy of high THz frequency lines (e.g., [HD] 2.68 THz, [OIII] 3.39 & 5.79 THz, [OI] 4.75 THz, [NIII] 5.23 THz, etc.) is growing the need in larger IF bandwidth becomes more pressing.

A possibility to increase both the operating temperature and the IF bandwidth of HEB mixers lies with the use of superconducting MgB_2 with critical temperature of 39 K. Realization of a receiver operating at 20 K would allow for the use of relatively low-cost mechanical cryocooling in space. This would be a big impact on the cost reduction and lifetime increase of an associated space mission.

Recently, thin films of this superconductor have become achievable, which opened a door for development of various detectors. Our current work focuses on the development of practical HEB mixers using ultrathin (8-20 nm) MgB_2 films. We prepare films using the Hybrid Physical-Chemical Vapor Deposition (HPCVD) process in combination with ion mill yielding high-quality ultrathin films with critical temperature ~ 37-39 K on THz-transparent SiC substrates.The combination of small film thickness, large sound velocity, high acoustic phonon transparency at the interface with the substrate, and short electron-phonon relaxation time results in an IF bandwidth ≈ 9 GHz, which has been measured in 15-nm thick HEB devices. Even larger IF bandwidth is expected in thinner (5-10 nm) films, which have been already achieved. Micron- and submicron-sized spiral-antenna coupled HEB mixers have been fabricated in order to minimize the local oscillator (LO) power requirements. Preliminary measurements yielded a double-sideband noise temperature of≈ 2,500 K weakly dependent on temperature between 4 and 20 K. This indicates that, indeed, mixer operation may be possible using a cryogen-free cooling system. An on-going material development work focuses on achieving disordered films (but still with high critical temperature) where the intrinsic quantum efficiency is expected to be high. An additional benefit of the high resistivity will be better rf and IF impedance match of HEB devices.

We will report on experimental results to date as well as on the progress in development of a cryocooler based 4.7 THz HEB receiver using a Quantum Cascade Laser (QCL) as an LO source. We will also discuss other potential applications of MgB_2 films to THz sensors. Promising devices include tunnel-junction mixers, kinetic inductance detectors and parametric amplifiers, and flux-flow oscillators.

MgB_2 HEB Mixers at Operation Temperatures above Liquid Helium Temperature

E. Novoselov[1*], S. Bevilacqua[1], N. Zhang[1], S. Cherednichenko[1],H. Shibata[2], and Y. Tokura[2]

[1]*Chalmers University of Technology, Gothenburg 41296,Sweden*
[2]*NTT Basic Research laboratories, Atsugi 243-0198, Japan*
* Contact: evgenii@chalmers.se

Superconducting HEB mixers are widely used in heterodyne instruments for sub-mm wave astronomy observations at frequencies of 1.3-5.3THz providing high sensitivity and high spectral resolution. The limiting factors for the currently used NbN HEB mixers are a low operation temperature (4K) and a narrow intermediate frequency (IF) bandwidth of 3-4 GHz. It has been demonstrated that MgB_2 HEB mixers can provide an operation temperature above 10-20K and an IF bandwidth up to 10 GHz due toa high T_c and short electron-phonon interaction time. The demonstrated sensitivity is comparable to NbN HEBs.Fabrication of MgB_2 HEB mixers with a critical temperature (T_c) above 30K will allow for the use of compact cryocoolers instead of liquid helium for the device cooling, which will lead to the increase of spaceborne mission lifetimes. Several methods for thin film growth were reported, among them co-evaporation, molecular beam epitaxy (MBE) and hybrid physical-chemical vapor deposition (HPCVD). HPCVD has been demonstrated as a very effective MgB_2 thin films fabrication method with excellent crystallinity and superconducting properties.

In this presentation we will discuss results of MgB_2 thin films deposition for the HEB fabrication using a custom made HPCVD system. The parameters of achieved 10-40nm thick MgB_2 films (such as aT_c and sheet resistance) and morphologywill be analyzed. Results of sensitivity characterization and IF gain measurements at 1.6THz local oscillator of MgB_2 HEB mixers fabricated using these films and 40nm thick MBE films with a T_c of about 30K atbath temperatures above 20K will be presented and compared.

Experimental Studies of IF Impedance of MgB_2 HEB Mixers at Various Bias Conditions and Operation Temperatures

S. Bevilacqua, E. Novoselov, S. Cherednichenko

Chalmers University of Technology, Gothenburg 41296,Sweden
* Contact: serguei@chalmers.se

Accurate measurements of IF impedance for hot-electron bolometers (HEB) mixers is very important for designing matching circuits between the HEBs and the first stage Low Noise Amplifiers (LNA). Apart of giving rise to the receivers noise temperature, IF impedance mismatch leads to ripples in the receiver gain, which reduces accuracy of the base line subtraction during the receiver calibration.

IF impedance can be obtained from vector measurements of the reflection coefficient (S11) in the frequency range of interest. It is 0.1-10GHz in our case. Accuracy of the S11 measurements depends on the calibration precision. The main problem here is that HEB mixers operate at cryogenic temperatures, 4K-20K. Therefore, they are frequently packaged in a mixer block. There have been several ideas explored in the literature for how to deduce the HEB mixer IF impedance from the measurements. Furthermore, IF impedance measurements on HEB mixers has also a more fundamental importance. It is one of the methods to explore the physics behind the HEB operation. Furthermore, it is one of the methods to measure the HEB mixer gain bandwidth (GBW).

In our work we will present and experimentally compare several methods for S11 calibration procedures for packaged HEB mixers. We have used magnesium diboride (MgB_2) HEB mixers with critical temperatures of 22K and 30K. Different calibration procedures have been used. Furthermore, HFSS modeling of the complete mixer block was done and experimental results will be compared to the simulations. As a final step, calibration loads were made of (non-superconducting) NiCr thin films in the exact layouts of MgB_2 HEB mixers. Their frequency and temperature independent microwave impedance allow for verification of the HEB mixers measurements procedure.

W3 Session: THz Receivers

1.9 THz 4-Pixel Heterodyne Array Receiver

Jonathan Kawamura[1], Jenna Kloosterman[1], Robert Lin[1], Faouzi Boussaha[2],
Jose Siles[1], Imran Mehdi[1], Choonsup Lee[1], and Bruce Bumble[1]

[1]*Jet Propulsion Laboratory, California Institute of Technology*
[2]*Observatoire de Paris, 75014 Paris, France*
* Contact: kawamura@jpl.nasa.gov

We are developing a 16-pixel 1.9 THz superconducting receiver module for future balloon and aircraft instruments. Towards achieving this goal, we have developed a 4x1-pixel heterodyne receiver system. This array uses a 4x1-pixel multiplier chain for the local oscillator (LO) and a matching 4x1-pixel feedhorn array with hot electron bolometer (HEB) mixers, coupled weakly through a free-standing wire-grid. Simultaneous pumping of mixers is verified through I-V curves with sensitivities of ~850 K. This design can be extended into larger arrays, ultimately 4x4 or possibly larger by stacking the 4x1 LO and mixer modules, and the approach can be scaled in size to create arrays at other THz frequencies.

The receiver consists of a 4x1-pixel multiplier chain and a corresponding 4x1-pixel feedhorn focal plane unit (FPU) housing the HEB mixers. In the laboratory setup, the 4x1-pixel LO [1] is injected quasioptically via a polarizer grid and reimaged with two parabolic mirrors onto the FPU. The first mirror sits outside a liquid cryostat and the second mirror is bolted to the 4 K plate inside the cryostat. The HEB mixer block is also bolted to the 4 K plate and connected to low noise amplifiers (LNAs) via semi-rigid coax with GPO connectors. The first stage low-noise amplifiers are also in thermal contact with the 4 K plate. Each HEB is constructed from a niobium nitrate (NbN) bridge and a bowtie antenna on a silicon-on-insulator (SOI) chip [2]. These chips are inserted into individual waveguide circuit and then placed into the 4x1-pixel mixer block.

The local oscillator [1] takes input from a Ka-band synthesizer and through a series of amplification and frequency multiplication stages ultimately produces power from four output horns at 1.9 THz. Each LO pixel is designed to output a minimum of 10 μW of power, which provides sufficient power to pump each of the 4 HEB mixers with a 10% beam splitter. The output power from each output can be controlled electronically by using the diode bias.

The backend electronics have a 1-2 GHz IF passband and fed to digital spectrometers and microwave detectors, which are used to complete Y-factor measurements. A computer-controlled bias system scheme has the capability to both set the bias voltages on the LO and HEBs and read back the bias voltages and currents from the HEBs, thus enabling active local oscillator level trimming for each pixel critical to the operation of stable HEB-based receivers.

This work was carried out at the Jet Propulsion Laboratory, California Institute of Technology, under a contract with the National Aeronautics and Space Administration.

REFERENCES

[1].J. V. Siles, I. Mehdi, C. Lee, R. Lin, J. Kawamura, E. Schlecht, P. Bruneau, and P. F. Goldsmith, "A multi-pixel room-temperature local oscillator subsystem for array receivers at 1.9 THz," in Proc. SPIE 9147, Ground-based and Airborne Instrumentation for Astronomy V, 2014, paper 914777.

[2]. F. M. Boussaha, J. H. Kawamura, J. A. Stern, A. Skalare, and V. White, "A Low Noise 2.7 THz Waveguide-Based Superconducting Mixer," Terahertz Science and Technology, IEEE Transactions on, vol. 2, no. 3, pp. 284-289, May 2012.

The upGREAT THz Arrays for SOFIA: Successful Commissioning at 1.9 THz

C. Risacher[1*], R. Güsten[1], J. Stutzki[2], H.-W. Hübers[3], P. Pütz[2], A. Bell[1], D. Büchel[2], I. Camara[1], R. Castenholz[1], M. Choi[1], U.U. Graf[2], S. Heyminck[1], C. E. Honingh[2], K. Jacobs[2], M. Justen[2], B. Klein[1,4], T. Klein[1], C. Leinz[1], N. Reyes[1,5], H. Richter[3], O. Ricken[1], A. Wunsch[1]

[1]*Max Planck Institut für Radioastronomie, Auf dem Hügel 69, 53121 Bonn, Germany*
[2]*KOSMA, I. Physikalisches Institut, Universität zu Köln, Zülpicher Strasse 77, 50937 Köln, Germany*
[3]*German Aerospace Center (DLR), Institute of Planetary Research, Rutherfordstr. 2, Berlin, Germany*
[4]*European Southern Observatories, Vitacura, Santiago de Chile, 19001, Chile*
[5]*Universidad de Chile, Santiago, Chile*
* Contact: crisache@mpifr.de

We present the upGREAT heterodyne arrays for astronomy, used with the SOFIA airborne observatory, a 2.5-m telescope flying on a NASA/DLR Boeing 747. The upGREAT array receivers operate in two different frequency ranges, the low frequency array (LFA) covering the 1.9-2.5 THz band with 14 pixels, and the high frequency array (HFA)targeting the 4.745 THz line of atomic oxygen [O I] with 7pixels. The frontend uses superconducting Hot Electron Bolometers (HEB) waveguide devices as mixers. The local oscillators are based on commercial synthesizer driven solid-state multiplier chains for the LFA and a quantum cascade laser for the HFA. Both receivers are cooled using closed cycle pulse tube refrigerators, reaching temperatures below 4K.

The upGREAT LFA receiver, with its 14 channels, was successfully commissioned in May and December 2015. We will present the main results of the commissioning flights. At 1.9 THz the array performed nominally on sky, with state of the art performance for 12 out of 14 pixels, reaching about 600-800 K DSB uncorrected receiver noise temperature at 0.5 GHz IF with an IF noise bandwidth of ~3.5 GHz. The stability was excellent and allowed efficient large scale mapping, demonstrating already that a factor of 10 was gained in time efficiency compared to the previous single pixel receiver at 1.9 THz. We are currently integrating the HFA for the planned commissioning flights in October-November 2016 and will present preliminary performance data measured during integration at the MPIfR.

This work was supported in part by the Federal Ministry of Economics and Technology via the German Space Agency (DLR) under Grants 50 OK 1102, 50 OK 1103 and 50 OK 1104 and by the Collaborative Research Council 956, sub-projects D3 and S, funded by the Deutsche Forschungsgemeinschaft (DFG).

7-pixels prototype for a 230 GHz multi-beam receiver

D. Maier, A.L. Fontana, M. Parioleau, Q. Moutote, J. Reverdy, A. Barbier,D. Billon-Pierron, E. Driessen, and J.M. Daneel
Institut de RadioAstronomie Millimétrique, St. Martin d'Hères, France
Contact: maier@iram.fr, phone +33 76 89 49 16

Abstract— **IRAM is currently working on a 230 GHz multi-beam receiver to replace the Heterodyne Receiver Array installed at its 30 m telescope at Pico Veleta, Spain. The new instrument features more pixels and an increased RFobserving bandwidth. It will employ state-of-the-art sideband-separating mixers with wide IF band. The receiver design is based on 7 adjacent modules of seven pixels. A fully integrated prototype module featuring seven pixels and complete LO distribution has been designed and evaluated. A feasibility study of the unit has been carried out, in particular considering the needs of fabrication reproducibility, resulting in a less integrated, but also less risky design.**

INTRODUCTION

The HEterodyne Receiver Array (HERA) installed at the IRAM 30 m telescope in Spain is a key instrument for large-scale mapping in the 230 GHz (1 mm) atmospheric window[1]. It will be replaced in the near future by a new multi-beam receiver, which will not only provide a larger RF observing range, but will also increase the instantaneous bandwidth by a factor of 16 by employing state-of-the-art wide IF band sideband-separating (2SB) SIS mixers, similar to those currently deployed in the EMIR or NOEMA receivers[2], [3]. In addition, the number of pixels will be increased from 3x3 to 7x7, each pixel delivering two polarizations.

SINGLE-PIXEL DESIGN

In a first step a 230 GHz 2SB SIS mixer suitable to serve as the base unit for a multi-pixel mixer array has been designed. Apart from requirements concerning the RF and IF operating ranges of 200 to 280 GHz and 4 to 12 GHz, respectively, the mixer had therefore to comply with several mechanical constraints. First of all the size of the mixer block had to be kept reasonably small to achieve a suitable distance between the individual pixels. Additionally, in order that the mixer could be stacked horizontally and vertically, RF input and IF outputs had to be placed at the front and the rear of the block, respectively. And finally, the mixer should achieve state-of-the-art performances regarding noise temperatures and image rejections.

In order to obtain a very small footprint for the 2SB SIS mixer prototype for the 7 pixels array, it has been designed by integrating all components of the 2SB mixer, i.e. the RF hybrid coupler, two DSB mixer blocks, the LO splitter, two LO couplers and the IF coupler, into one E-plane splitblock[4]. In view of the size constraints for the multi-pixel receiver the Josephson effects are suppressed by employing permanent micromagnets. A photo of one half of this splitblock with mounted mixer chips, IF coupler chip, micromagnets, and waveguide loads is shown in Fig. 1.

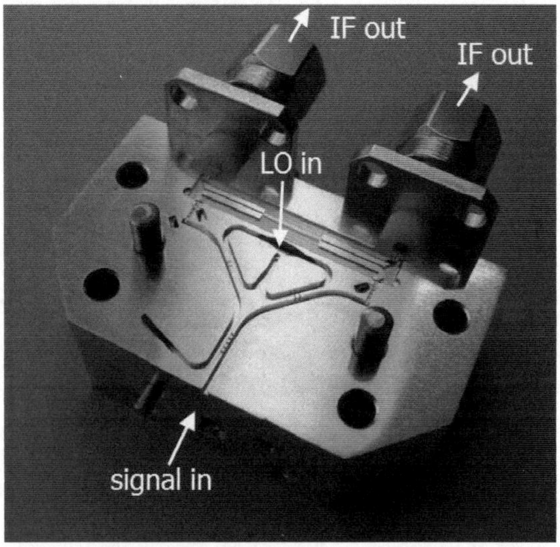

Fig. 1: One half of the E-plane splitblock of the fully integrated 230 GHz 2SB mixer.

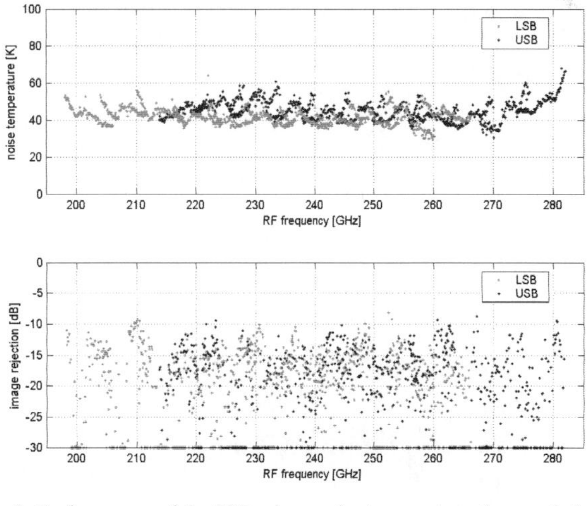

Fig. 2: Performances of the 2SB mixer: noise temperatures (upper plot) and image rejections (lower plot). Lsb measurements are plotted in green, usb measurements are shown in blue.

The measured performances of this mixer are shown in Fig. 2. The achieved noise temperatures lie between 35 K and 55 K and the image rejections lie well under 10 dB.

Due to its mechanical design this prototype 2SB mixer can easily be stacked in horizontal and vertical directions and is therefore suitable to serve as the base mixer unit for forming a multi-pixel mixer array.

MULTI-PIXEL DESIGN

For the overall design of the 7-pixels prototype the distance of the pixels in horizontal as well as vertical direction has been fixed in view of the sky coverage of the pixels to 25 mm. So each component had to fulfil the requirement of not exceeding a footprint of 25 mm x 25 mm per pixel in the horizontal and the vertical directions, respectively. The dimension in the z-direction was not subjected to any constraint. The signal had to be input from the front side of the prototype and to be output on its rear side, in order to allow a stacking of the 7-pixels prototypes to build a multi-beam receiver. Therefore, the LO signal is fed in from one side.

MIXERS AND LO DISTRIBUTION

In a first approach the 7-pixels mixer array has been designed as one unit consisting of three parts as shown inFig. 3. The 2SB SIS mixer base unit has been replicated 7 times horizontally and integrated into one E-plane splitblock which is represented by the two lower parts of the unit shown inFig. 3.

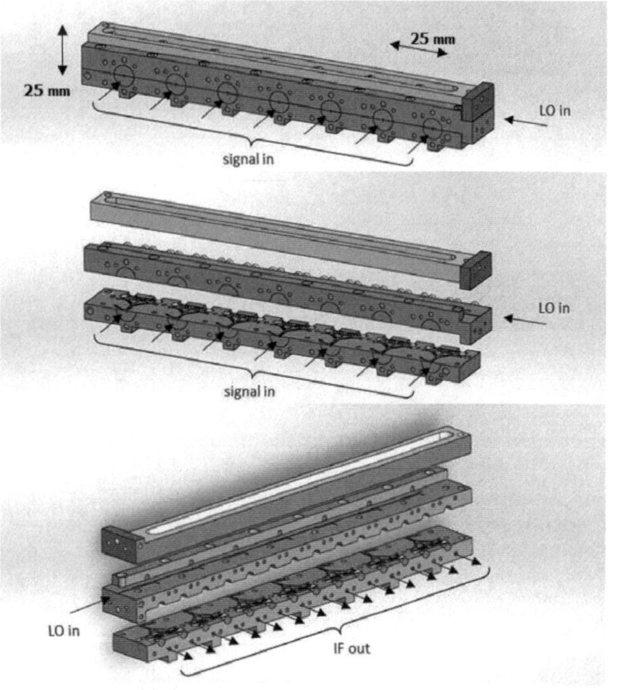

Fig. 3: Initial design of the mixer array and the LO distribution.

The LO signal distribution has been added in the plane between the upper and the middle parts. The LO injection is effected through the flange on the side of the unit into the continuous waveguide formed by the upper and the middle parts. Seven couplers then distribute the signal into seven branches. The couplers have been designed with different coupling factors taking into account the losses between the different pixels such that every branch receives the same LO power percentage. Each branch descends through the middle part and injects the LO signal from the top into each mixer, respectively.

The mixers formed between the middle and the lower parts receive the RF signal by their flanges on the front side of the unit. As seen above, the LO signal enters from the top. And the IF signals leave the unit by its rear side.

FEASIBILITY STUDY

All couplers in the LO distribution path as well as the RF and LO couplers in the mixers are designed as branchguide couplers with very small dimensions and requiring a high precision of a few tenths of microns. The slots of these couplers are usually defined by spark erosion. Although this is a regularly used technique for the fabrication of a single-mixer unit, the fabrication of this 7-pixels mixer unit presents a big challenge not only because of the large number of slots to be realized, but the requirements for the positioning of the individual pixels over a total length of about 17 cm, so that the half slots of two opposing parts fit within a few microns to each other.

In order to evaluate the fabrication feasibility of such a unit, a simplified workpiece has been defined (seeFig. 4). This workpiece consists of two halves of an E-plane splitblock and features seven simplified coupler structures spaced respectively 25 mm apart from each other. The couplers are represented each by two slots of 0.1 mm width with a distance of 0.9 mm. No real waveguides, but only two cavities on each side of the slots have been realized. After the fabrication of the pieces one of the cavities of each coupler structure has been milled away, in order to allow a view on the matching of the coupler slots when both parts are mounted together.

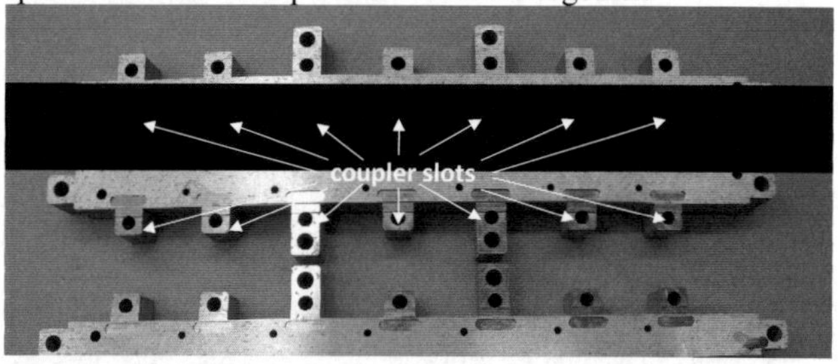

Fig. 4: Simplified workpiece for the fabrication feasibility study.

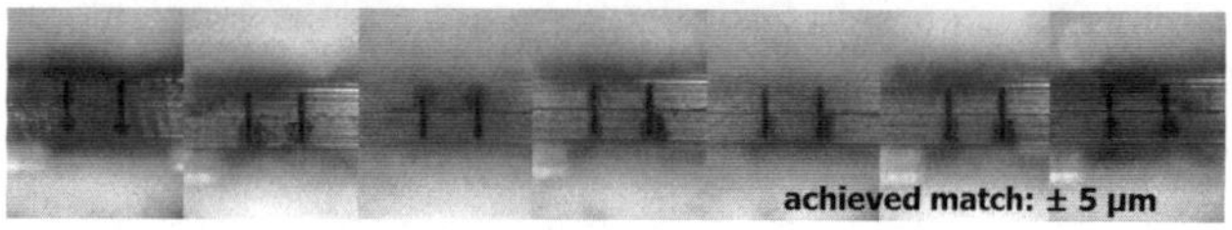

Fig. 5: Cross-sectional view of the seven coupler structures.

Fig. 5shows a cross-sectional view of the seven coupler structures. The achieved match between the coupler halves lies approximately within ± 5 µm, which corresponds to the usually achieved precision for the fabrication of a single mixer unit. Based on these results we decided to keep the LO coupler design.

The overall design of the 7 pixels unit however, we judged not suitable for a mechanical fabrication. Especially, the middle part of our design featuring coupler slots on both of its sides seemed to risky. Therefore we decided to modify our design.

ITERATED DESIGN

The iterated design is shown inFig. 6. It now consists of an LO signal distribution coupler and seven individual mixer units. The overall height of the unit is the same as for the initial design, so that the distance of the pixels did not change. Apart from relaxing the requirements for the mechanical fabrication, this solution offers the possibility to test each mixer individually prior to integration which facilitates the fabrication of the unit.

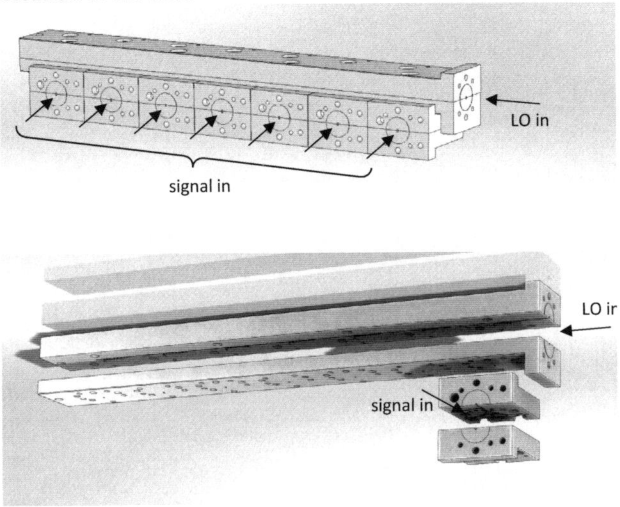

Fig. 6: Iterated design of the mixer array and the LO distribution.

7-PIXELS PROTOTYPE RECEIVER

A schematic view of the such defined 7-pixels prototype receiver is shown in Fig. 7. The signal is input on the front side into each horn. The horns are attached to the mixer blocks which in turn are mounted on an LO coupler feeding each mixer with the same amount of LO power. Finally, the mixers are connected at their outputs to the cold IF amplifiers.

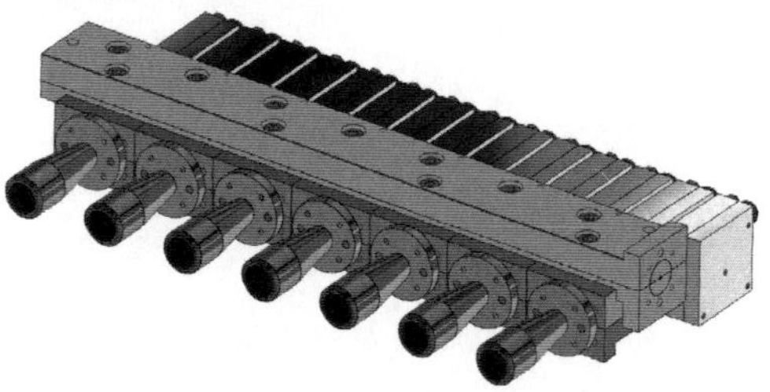

Fig. 7: 7-pixels prototype receiver.

7-PIXELS OPTICS

An optical system has been developed in order to allow testing the 7-pixelsprototype at the IRAM 30m telescope. The aim of this multi-pixels receiver optics is to transform the required angular beam spacing on the sky (~23 arc second, corresponding to a spacing of almost two half power beam widths) to the physical spacing between the feeds (25mm, given by the size of a pixel footprint). This transformation is achieved by using a pair of focusing mirrors, which form a Gaussian telescope and ensure a frequency independent illumination of the telescope. Additionally, each pixel has its own individual optics(which will be cooled to 4K). The role of this individual optics array is to maximize the coupling between each feed and the telescope by re-imaging the sub-reflector onto each of the conical corrugated feed horns. This individual optics (seeFig. 8) is fully reflective. It consists ofdouble-faced multi-mirrors: the beams are first reflected on a row of focusing mirrors, which fold them with an 80 degrees reflection angle onto a row of flat mirrors. The flat mirrors then reflect the beams towards the feeds. Such an array of individual optics is scalable to any number of pixels, wherein the side of focusing mirrors ofeach double-faced multi-mirrors serves for one row of pixels and its other side of flat mirrors serves for an adjacent row of pixels.

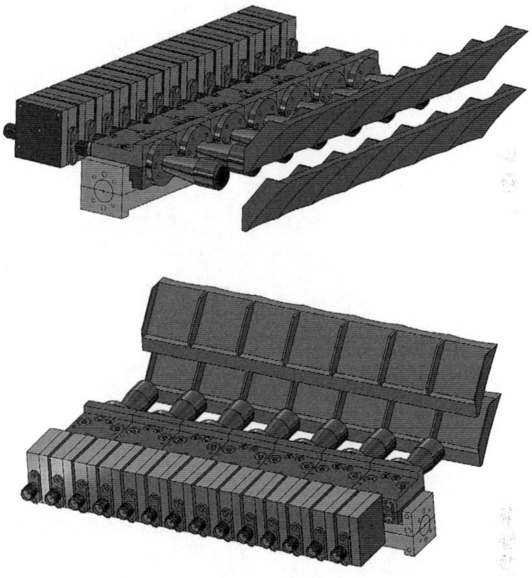

Fig. 8: Views of the 7-pixels array with its individual optics, composed of two multi-mirrors.

CONCLUSIONS

A 7-pixels prototype for an 1 mm array receiver has been developed. The array with the mixers and the LO distribution has been designed and optimized after a feasibility study with respect to its mechanical fabrication. The iterated design is currently fabricated and will be delivered in May 2016. When all subcomponents have been received, they will be integrated in a dedicated test cryostat and tests will then follow during summer 2016.

ACKNOWLEDGMENT

The research leading to these results has received funding from the European Commission Seventh Framework Programme (FP/2007-2013) under grant agreement No 283393 (RadioNet3).

REFERENCES

[1] K.F. Schuster, C. Boucher, W. Brunswig, M. Carter, J.-Y. Chenu, B. Fouilleux, A. Greve, D. John, B. Lazareff, S. Navarro, A. Perrigouard, J.-L. Pollet, A. Sievers, C. Thum, and H. Wiesemeyer, "A 230 GHz heterodyne receiver array for the 30 m telescope," Astronomy & Astrophysics, vol. 423, pp. 1171-1177, 2004

[2] M. Carter, B. Lazareff, D. Maier, J.-Y. Chenu, A.-L. Fontana, Y. Bortolotti, C. Boucher, A. Navarrini, S. Blanchet, A. Greve, D. John, C. Kramer, F. Morel, S. Navarro, J. Peñalver, F.F. Schuster, and C. Thum, "The EMIR multi-band mm-wave receiver for the IRAM 30-m telescope," Astronomy & Astrophysics, vol. 538, pp. 13, 2012

[3] J.Y. Chenu, A. Navarrini, Y. Bortolotti, G. Butin, A.L. Fontana, S. Mahieu, D. Maier, F. Mattiocco, P. Serres, M. Berton, O. Garnier, Q. Moutote, M. Parioleau, B. Pissard, and J. Reverdy, "The Front-End of the NOEMA Interferometer," IEEE Trans. on Terahertz Science and technology, vol. 6, pp. 223-237, 2016

[4] D. Maier, J. Reverdy, L. Coutanson, D. Billon-Pierron, C. Boucher, and A. Barbier, "Fully integrated sideband-separating Mixers for the NOEMA receivers," in *Proc. Int. Symp. On Space Terahertz Technology, 2014*, pp. 80-84

Ultra Low Noise 600/1200 GHz and 874 GHz GaAs Schottky Receivers for SWI and ISMAR

P. J. Sobis[1*], V. Drakinskiy[2], N. Wadefalk[3], P-Å. Nilsson[2], D. Nyberg[1], A. Hammar[1], T. Bryllert[4], H. Zhao[2], A. Emrich[1], J. Grahn[2], and J. Stake[2,4]

[1]*Omnisys Instruments AB, Göteborg, Sweden*
[2]*GigaHertz Centre, Department of Microtechnology and Nanoscience (MC2), Chalmers University of Technology, Göteborg, Sweden*
[3]*Low Noise Factory AB, Göteborg, Sweden*
[4]*Wasa Millimeter Wave AB, Göteborg, Sweden*
* Contact: ps@omnisys.se

Omnisys Instruments is responsible for the 600/1200 GHz broadband front-end receivers and back-end spectrometer hardware for the Submillimeter Wave Instrument (SWI) part of the Jupiter Icy moons Explorer (JUICE) mission, and for the development of the dual-polarization 874 GHz spectrometer channels for the airborne icecloud imager instrument ISMAR. We will present our development of these highly integrated heterodyne receivers which are based on membrane integrated GaAs Schottky diode mixer and multiplier circuit technology, and InP HEMT MMIC LNA technology from Chalmers University of Technology. Preliminary results at room temperature on the 1200 GHz breadboard prototypes show on a typical DSB receiver noise below 3000 K in the 1030 GHz-1220 GHz frequency range with only 1-3 mW of LO power. For the 874 GHz receiver flight modules a record low double sideband noise of 2500 K was obtained with only 2.3 mW of LO pump power. Both the 1200 GHz and 874 GHz subharmonically pumped Schottky mixer designs have been based on the broadband SWI 600 GHz channel mixer design, which had a repeatable receiver noise performance below 1200K with less than 2 mW of pump power at room temperature. All together these results are setting new standards for critical receiver hardware operating at room temperature used in instrumentation for atmospheric research and remote sensing applications.

This work was carried out in the GigaHertz Centre financed by the Swedish Governmental Agency of Innovation Systems (VINNOVA), by the Swedish National Space Board (SNSB), and by the European Space Agency under contract 4000112396/14/NL/BW "Terahertz Receiver Technology for Future Science Missions", and contract AO/1-7484/13MH "875 GHz Receiver Front-End for an Airborne Icecloud Imager Demonstrator".

1200GHz and 600GHz Schottky receivers for JUICE-SWI

A. Maestrini *(1), L. Gatilova (1)(2), J. Treuttel (1); F. Yang (3), Y. Jin (2), A. Cavanna (2), D. Moro Melgar (1), F. Tamazouzt (1), T. Vacelet (1), A. Féret (1), F. Dauplay (1), J-M. Krieg (1), C. Goldstein (4)

(1) LERMA, Observatoire de Paris, PSL Research University, CNRS, UMR 8112, Sorbonne Universités, UPMC Univ. Paris 06, F-75014 Paris, France
(2) CNRS - Laboratoire de Photonique et Nanostructures, Route de Nozay, F-91460 Marcoussis, France,
(3) Southwest University, State Key Lab, Nanjing, China,
(4) Centre National d'Etudes Spatiales, 18 avenue Edouard Belin, F-31401 Toulouse cedex 9, France,
* Contact: alain.maestrini@obspm.fr

Abstract— ***The Sub-millimeter Wave Instrument (SWI) for ESA Jupiter Icy Moons Explorer (JUICE) will be the first planetary instrument to feature 1200 GHz and 600 GHz heterodyne receivers. These receivers will investigate the temperature structure, composition and dynamics of Jupiter's stratosphere and troposphere, and the exospheres and surfaces of the icy moons.***

This paper will present the ongoing work at LERMA in partnership with LPN to design, fabricate and test, at room temperature and at cryogenic temperature, a 520-630GHz and a 1080-1280GHz Schottky receivers. Both receivers rely on 132-160GHz sources provided by Radiometer Physics GmbH that deliver respectively about 30-60mW and 60-120mW. LERMA, in partnership with LPN-CNRS, has designed and fabricated the high frequency parts of the LOs as well as the sub-harmonic mixers, using a Schottky process based entirely on e-beam lithography.

LERMA 520-630GHz Schottky receiver is the most sensitive to date with a DSB receiver equivalent noise temperature in the range of 1070K-1500K at room temperature, and a DSB receiver equivalent noise temperature in the range of 620K to 885K at 134K ambient temperature. LERMA 1080-1280GHz receiver has a DSB equivalent noise temperature of ~3500-7500K across the band at room temperature.

INTRODUCTION

The Sub-millimeter Wave Instrument (SWI) onboard the European Space Agency Jupiter Icy Moons Explorer (JUICE) [1] will investigate the temperature structure, composition and dynamics of Jupiter's stratosphere and troposphere, and the exospheres and surfaces of the icy moons. SWI is a heterodyne receiver with a spectral resolution of 10^7 and a frequency accuracy of 10^{-8}. SWI will receive RF signals in two frequency bands, 530-625GHz and 1080-1275GHz respectively, with a dual-axis steerable 30cm equivalent aperture off-axis antenna. The receiver front-end is based on sub-harmonic Schottky mixers pumped by frequency multiplier-based Local Oscillators (LOs). To achieve the science goals of SWI, the sensitivity of the receivers is specified at 120K-150K ambient temperature: for the 530-625GHz and 1080-1275GHz receivers the DSB equivalent noise temperature should be respectively lower than 2000K (goal is 1500K) and 4000K (goal is 3000K). This paper will present briefly the design of both receivers with an emphasis LERMA 1080-1280GHz bias-able sub-harmonic mixer.

RECEIVER DESIGN & TECHNOLOGY

Fig. 1 gives a schematic of SWI frontend. The LOs are made with a low-phase-noise VCO-based K-band synthesizer referenced to a 100MHz ultra-stable signal followed by an E-band tripler, one or several E-band amplifiers, and two or three cascaded frequency doublers. Each sub-harmonic mixer had the same Intermediate Frequency (IF) at 3.5-8.5GHz with an external LNA. The RF signal is coupled to the mixer with an external smooth-wall feed-horn designed by H. Gibson for Radiometer Physics GmbH[1].

The last stages frequency multipliers (at 300GHz and 600GHz output frequency) and the mixers are based on Schottky devices fabricated by LERMA-Observatoire de Paris in partnership with Laboratoire de Photonique et de Nanostructures – CNRS[2]. The Schottky process is entirely based on e-beam lithography and allows for diodes of 0.2µm² on 2-5µm-thick GaAs membranes, with front-side beam-leads (metal membranes) [2]. The design and the fabrication of LERMA single-chip 300GHz frequency doubler and of LERMA 600GHz sub-

[1] Radiometer Physics GmbH., http://www.radiometer-physics.de

[2] LPN and Institut d'Electronique Fondamentale have merged on June 1st, 2016, and are now Centre de Nanosciences et de Nanotechnologies (C2N).

harmonic mixer have been detailed in [3]. LERMA dual-chip 300GHz doubler features a 90° hybrid 3dB coupler at the input, two identical 4-anode balanced doubler devices and a compact Y-junction at the output. This doubler is packaged in a mechanical block of only 10mm long. LERMA 600GHz doubler features two anodes in a classic configuration. The micro-electronic devices of these multipliers are fabricated on the same wafer, with a GaAs membrane thickness of 5μm and a doping level of 1E17 cm^{-3}.

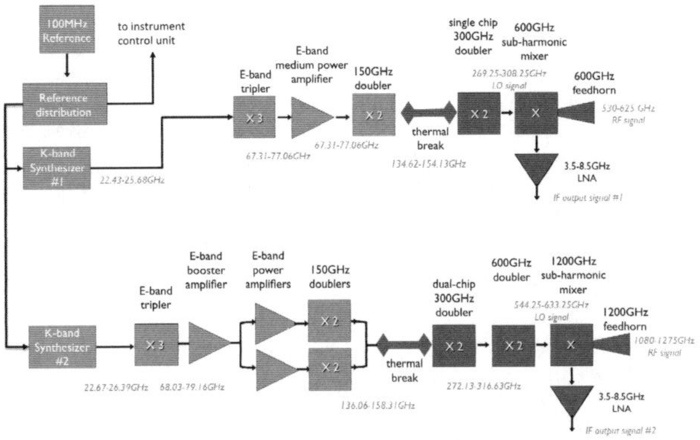

Fig. 1: schematic of SWI front-end. LERMA proposed solution for the mixers uses external feed-horns and external LNAs.

LERMA 1200GHz sub-harmonic mixer is made of a micro-electronic circuit fabricated on a 2μm thick GaAs membrane suspended in a micro-mechanical waveguide structure thanks to beam-leads (see Fig. 2). The micro-electronic circuit uses a novel configuration: it features a bias-able anti-parallel pair of Schottky diodes of 0.5fF each with a capacitor located on a T-shape mesa, which is RF grounded to the waveguide blocks thanks to lateral beam-leads attached to it (see Fig. 3).

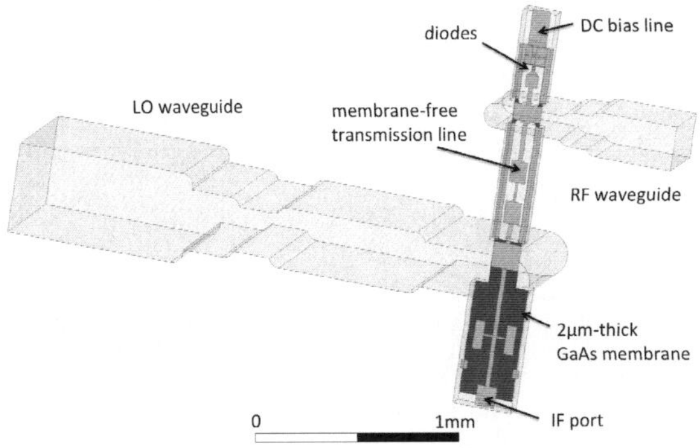

Fig. 2: 3D view of LERMA proposed 1200GHz bias-able mixer

This topology was found to provide a wideband LO and RF matching even with low levels of LO power: the mixer was designed to cover the full 1080-1280GHz band with only 1mW of available LO power. With 0.5mW of LO power the mixer can be pumped. This configuration has the additional advantage to separate the DC and the IF ports - like with the designs proposed in [4] - but with an in-line design.

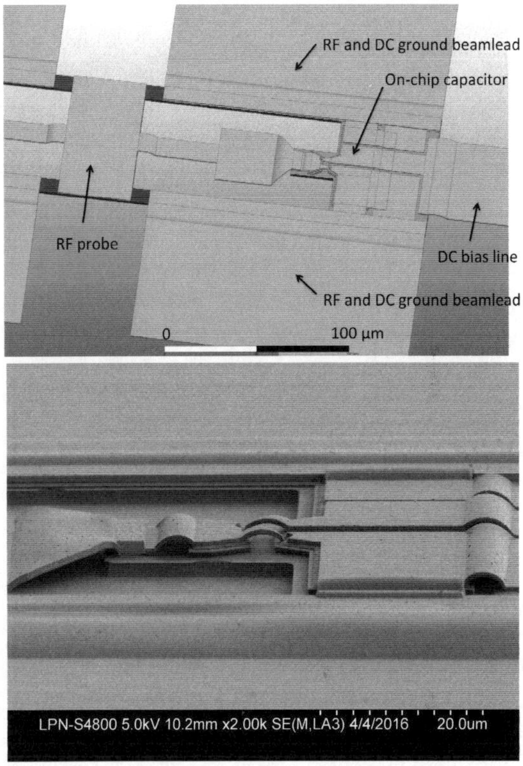

Fig. 3: 3D view (top) and SEM picture (bottom) of LERMA proposed 1200GHz bias-able mixer – detail of the diodes and bias-circuit

Preliminary measurements at room temperature in air (4.5cm air path with 60% humidity) give a DSB equivalent noise temperature of ~3500-7500K across the full 1080-1280GHz band. The 540-640GHz LO source could provide 0.5-2.0mW across the entire band at room temperature. The mixer was found to need very little LO power; it could be actually pumped with only 0.5mW at some frequency points, and we had to decrease the power provided by the LO source for most of the frequency points in order to get the best noise temperature.

Conclusions & Perspective

LERMA 600GHz Schottky receiver is at the state of the art and could provide unprecedented sensitivity for planetary missions like JUICE. LERMA 1200GHz receiver shows good to very good noise temperature at room temperature across the full RF band of JUCE-SWI. Cooling the last stages of the receiver at 120K is expected to improve the noise performance within the specs (4000K).

ACKNOWLEDGMENT

This research was founded by Centre National d'Etudes Spatiales under the R&T program R-S14/OT-0002-069-Mélangeurs Schottky THz Atmosphérique et Astronomie, by Centre National de la Recherche Scientifique, and by a PhD. grant financed entirely by LABEX ESEP (Exploration Spatiale des Environnements Planétaires).

REFERENCES

[1] JUICE Jupiter Icy Moons Explorer SWI Submillimetre Wave Instrument (SWI), Science Concept Document, JUI-MPS-SWI-PL-002, 13/05/2013.

[2] L. Gatilova, J. Treuttel, F. Yang, T. Vacelet, C. Goldstein, A. Maestrini, and Y. Jin , "THz Schottky Diode MMICs for Astronomy and the Physics of the Atmosphere", proceedings of the 25th International Symposium on Space Terahertz Technology 2014, ISSTT 2014, Moscow, Russia, 27-30 April 2014

[3] J. Treuttel, L. Gatilova, A. Maestrini, D. Moro-Melgar, F. Yang, F. Tamazouzt, T. Vacelet, Y. Jin, J. Mateos, A. Feret, C. Chaumont, C. Goldstein. "A 520-620 GHz Schottky receiver front-end for Planetary Science and Remote Sensing with 1070K-1500K DSB noise temperature at Room Temperature". *IEEE Trans.Terahertz Science and Tech.,* vol. PP, no 99, p.1-8 , 2015.

[4] E. Schlecht, J.V Siles, C. Lee, R. Lin, B. Thomas, G. Chattopadhyay, I. Mehdi, "Schottky Diode Based 1.2 THz Receivers Operating at Room-Temperature and Below for Planetary Atmospheric Sounding", *IEEE Transactions on Terahertz Science and Technology*, Vol. 4, No 6, pp. 661 - 669, 2014.

874-GHz Heterodyne Cusesat Receiverfor Cloud Ice Measurement –Flight Model Data

E.W. Bryerton*, J.L. Hesler, S.A. Retzloff, and T.W. Crowe

Virginia Diodes, Inc., Charlottesville, VA USA
* Contact:bryerton@vadiodes.com

This paper describes the design and flight model measurements of a 874-GHz Schottky receiver for ice cloud characterization that uses high-efficiency amplifiers and varactor multipliers in the LO chain to achieve low DC power with uncompromised sensitivity in a compact configuration optimal for CubeSat platforms. The total required DC power for the LO, including the 24.278 GHz dielectric resonator oscillator (DRO), is less than 3.75W. For the best flight model, the measured mixer noise temperature is 4000K DSB with 11.5dB conversion loss. Performance as a function of ambient temperature from 0-40C is also presented.

Fig. 1. Photograph of the 883-GHz IceCubeMLA 10 cm x 10 cm science plate

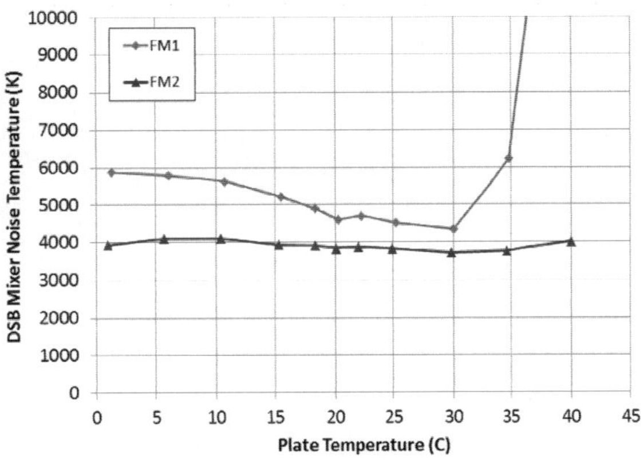

Fig. 2. Measured double sideband (DSB) mixer noise temperature as a function of science plate temperature for both the first (blue trace) and second (red trace) flight model.

W4 Session: THz Mixers & Detectors (II)

Ultra-low Noise TES bolometer Arrays for SAFARI Instrument on SPICA

P. Khosropanah[1*], T. Suzuki[2], R.A. Hijmering[1], M.L. Ridder[1], J.-R.Gao[1,3], H. Akamatsu[1], L. Gottardi[1], J. van der Kuur[1], B.D. Jackson[1]

[1]*SRON Netherlands Institute for Space Research, Sorbonnelaan 2, 3584CA Utrecht, The Netherlands*
[2]*Institute of Space and Astronautical Science (ISAS)/JAXA, 3-1-1 Yoshinodai, Chuo-ku, Sagamihara, Kanagawa 252-5210, Japan*
[3]*Kavli Institute of NanoScience, Delft University of Technology, Delft, the Netherlands*
* Contact: P.Khosropanah@sron.nl

SPICA (SPace Infrared telescope for Cosmology and Astrophysics) is a future space mission for mid- and far-infrared (IR) astronomy. By having a large (2.5 m) and cooled (< 8 K) telescope combined with ultra sensitive IR detectors, SPICA provides an opportunity to make natural background-limited observations over the wavelength range from 17 to 230 μm. One of the instruments aboard SPICA is SAFARI (SpicA FAR-infrared Instrument), which is a grating spectrometer covering the full 34-230 μm wavelength range. SAFARI detectors are transition edge sensor (TES) bolometers for three wavelength bands: S-band for 34-60 μm, M-band for 60-110 μm, and L-band for 110-230 μm. Each band requires a large number of pixels (~ 600-2000 pixels) and an extremely high sensitivity (electrical Noise Equivalent Power, NEPel ~ 2×10^{-19} W/√Hz at frequencies below ~ 100 Hz). SRON is developing ultra-low noise TESs based on a superconducting Ti/Au bilayer on a suspended SiN island with SiN legs. The pixel size is ~ 800×800 μm^2. Three types of TESs were fabricated on SiN islands with different sizes and with and without optical absorbers. These TESs have thin (0.20 μm), narrow (0.5-0.7 μm), and long (340-460 μm) SiN legs, and show T_c of ~ 93 mK and R_n of ~ 158 mΩ. They were characterized under AC bias using a Frequency Division Multiplexing readout (1-3 MHz) system. The TESs without absorber show NEPs as low as 1.1×10^{-19} W/√Hz with response time of below 1 ms,. For the TESs with absorber, we confirmed a higher NEPel (~ 5×10^{-19} W/√Hz) than that of TESs without the absorber, due to the stray light.

Readout of a 160 Pixel FDM System for SAFARI TES Arrays

R. A. Hijmering[1], R. den Hartog[1], M. Ridder[1], A.J. van der Linden[1], J. van der Kuur[1], J. R. Gao[1,2], B. Jackson[3], M. Bruijn[1]

[1]*SRON Netherlands institute for Space Research, Utrecht, The Netherlands*
[2]*Kavli Institute of NanoScience, Delft University of Technology, Delft, the Netherlands*
[3]*SRON Netherlands institute for Space Research, Groningen, The Netherlands*
*Contact: R.Hijmering@sron.nl

SAFARI is one of the focal-plane instruments for the Japanese/European far-IR SPICA mission proposed for the ESA M5 selection. It is based on three arrays with in total 3550 TES-based bolometers with noise-equivalent powers (NEP) in the range of 2 E-19 W/√Hz. The arrays are operated in three wavelength bands: S-band for 30-60 μm, M-band for 60-110 μm and L-band for 110-210 μm, and have background-limited sensitivity and high efficiency. At SRON we are developing Frequency Domain Multiplexing (FDM) for read out of large AC biased TES arrays for both the SAFARI instrument, for the far-IR SPICA mission, and XIFU instrument, for the X-ray Athena mission. In this paper we focus on the development of a FDM demonstration model for the SAFARI instrument. In FDM the TES bolometers are AC biased and readout using in 24 channels. Each channel contains 160 pixels of which the resonance frequencies are defined by in house developed cryogenic lithographic LC filters. To overcome the dynamic range limitations of the SQUID pre-amplifier, a special technique, baseband feedback (BBFB), is applied.FDM is based on the amplitude modulation of a carrier signal, which also provides the AC voltage bias, with the signal detected by the TES. BBFB attempts to cancel the error signal in the sum-point (located at the input coil of the SQUID), by feeding back a remodulated signal to the sum-point, and therefore improving the dynamic range of the SQUID pre-amplifier.

After reporting on the successful low-noise read-out of 38 bolometer TES pixels with an NEP level of 1 - 2E-18 W/√Hz we reported on a detailed study on the effects of electrical crosstalk using out first iteration of a prototype of a full 160 pixel FDM experiment. Using the obtained knowledge a second generation prototype of a full 176 pixel FDM experiment is developed in which the crosstalk elements of carrier leakage, mutual inductance and common impedance are minimized. The cold part of the experiment consists of a detector chip with 176 pixels with a design NEP of 7E-19 W/√Hz and two matching LC filter chips, each of which contains 88 carefully placed high-Q resonators, with in total 176 different resonance frequencies, and a single-stage SQUID. The warm electronics consist of a low-noise amplifier (LNA) and a digital board on which the generation of the bias carriers, the demodulation of the signal and remodulation of the feedback signal are performed. The effect of carrier leakage has been reduced by a factor two and the effects of mutual inductance have been removed. The common impedance has been reduced by design to 4nH, of which 3nH is from the input coil of the SQUID. It has successfully been further reduced to below 1nH by implementing screening of the input coil. The pixels have been connected in stages, a quarter, half and the full array, to be able to detect and solve any issues coming up. In this paper we will report on the results obtain with this 176 pixel FDM experiment.

The SpaceKIDs project: Development of Kinetic Inductance Detector Arrays for Space Applications

P. S. Barry[*1], J. Baselmans[2], A. Baryshev[4], A. Bideaud[1], B. Blázquez[3], J. Bueno[2], M. Calvo[5], S.Doyle[1], L. Ferrari[4], A. Gómez[6], J. Goupy[5], M. Griffin[1], M. Grim[2], P. Hargrave[1], T. Klapwijk[3], N. Llombart[3], J. Martin–Pintado[6], A. Monfardini[5], A. Neto[3], I. Walker[1], K. Wood[7] and S. Yates[4]

[1]Cardiff University, Queen's Buildings, The Parade, Cardiff, CF24 3AA, UK
[2]SRON Netherlands Institute for Space Research, Utrecht, The Netherlands
[3]Delft University of Technology, Delft, The Netherlands
[4]SRON, Groningen, The Netherlands
[5]Institut Neel, CNRS and Université de Grenoble, France
[6]Centro de Astrobiología (INTA-CSIC), Ctra. Torrejon-Ajalvir km.4, 28850 Torrejon de Ardoz, Spain
[7]QMC Instruments LTD, Cardiff, UK
* Contact: peter.barry@astro.cf.ac.uk

Kinetic Inductance Detectors (KIDs) offer the unique combination of excellent sensitivity to THz radiation along with minimal cryogenic complexity. The goal of the SpaceKIDs project is to work on the developments needed to enable this technology for both low-background (astrophysical) and high-background (Earth-observing) applications. Two laboratory demonstrator systems have been built to evaluate array characteristics and performance in an environment representative of the two applications. In this talk I will present an overview of the SpaceKIDs programme, describe in detail the performance of the demonstrator systems, and highlight some of the major results, which are set to have a significant impact on the design and characterisation of the next generation KID arrays.

Terahertz Superconducting Imaging Array (TeSIA)

S. Shi[1,2], J. Li[1,2], Z. Lin[1,2], J. Yang[1,2], Z. Lou[1,2], W. Zhang[1,2], Z. Wang[1,2], D. Liu[1,2], S. Li[1,2], W. Duan[1,2], W. Miao[1,2], Q. Yao[1,2]

[1]*Purple Mountain Observatory, Nanjing, Jiangsu, China.*
[2]*Key Lab of Radio Astronomy, Nanjing, JiangSu, China.*

Dome A, the highest point of the cold and dry Antarctic ice sheet, is a unique site for ground-basedTHz/FIR observations. The astronomical facilities China is planning to build there include a 5-m THz telescope namedDATE5. An instrument proposed for the DATE5 telescope is the THz superconducting imaging array (TeSIA)operating at the 350-μm window, with a pixel number of 32×32 and a background-limited sensitivity (NEP) of 10^{-16}W/Hz$^{0.5}$. For the development of TeSIA, a small-scale (8×8) array at longer wavelength (850μm or 345GHz) is firstlydeveloped. Microwave kinetic inductance detectors (MKIDs) based on TiN superconducting films of a critical transitiontemperature of 4.5K are chosen as the detectors of the system operating at 0.3K. In this talk, we will firstly introducethe design and performance of the system, and then present some results of video-rate imaging and testingobservations on a submillimeter-wave telescope.

Frequency Division Multiplexing with Superconducting Tunnel Junctions as Rectifiers and Frequency Mixers

G. de Lange*, I. Rodrigues, A. Detrain, M.D. Audley, and W.J. Vreeling

SRON Netherlands Institute for Space Research, Landleven 12, 9747AD Groningen, The Netherlands
*Contact: g.de.lange@sron.nl

The number of pixels in transition edge sensor (TES) imaging arrays is currently limitedby the complexity of the read-out scheme. The large wire count from the 300Kbiasand read-out electronics to the cryogenic detectors causes engineering problems associatedwith EMI and thermal design. For the read-out of TES detector arrays several read-out schemes exist. In this paper we focus on the frequency-division multiplexing (FDM) read-out [2]. Within FDM N channels of M TES pixels are operatedunder AC-bias and each pixel within a channel is operated at a unique frequency. Thefrequency selection of a pixel is achieved with a high-Q (superconducting) LC filter,located close to the pixel. With an AC-biasing frequency comb of M frequencies, Mpixels can be biased simultaneously. A DC-SQUID is used for the simultaneous readoutof M pixels. DC-SQUIDs need DC currents and DC flux-offsets and therefore aredifficult tomultiplex (separatewires are needed for each DC-current). Furthermore theSQUID amplifier has a limited dynamic range. The multiplexing factor M of currentstate-of-the-art SQUID devices is limited to a few tens to hundred, where feedback isapplied to the SQUID to improve the dynamic range. For large scale arrays with Kpixels, N=K/M parallel channels with each a separate SQUID amplifier chain haveto be used.In order to significantly reduce the wiring between the 300K electronics and thecryogenic level, we propose to apply an extra layer of frequency multiplexing forthe N channels, by frequency up- and down-conversion in the warm electronics andthe use of superconducting tunnel junction (SIS) frequency converters and RF-to-DCconverters at cryogenic temperatures.

In the paper we describe initial experimental results of the components that are part of a cryogenic multiplexing scheme. We have designed and characterized a six-channel cryogenic RF-to-DC multiplexing scheme consisting of commercial discrete components, in combination with a planar superconducting channelizing filter and SIS junctions. With the current multiplexer we can control six DC-bias currents with a single pair of wires, and one coaxial cable running from 4K to 300 K. The individual DC-current amplitudes are controlled by adjusting the RF power of individual frequencies in a frequency comb of 6 frequencies. The frequency span of the comb is 4-6 GHz.

As an initial demonstration we have connected the RF-to-DC converter output to the feedback coil of a DC SQUID. With this set-up we were able to control the feedback flux in the SQUID by adjusting the RF power and we could actually also demonstrate an RF controlled flux-locked loop of a DC-SQUID.

A 230 GHz Finline SIS Receiver with Wide IF Bandwidth

J. D. Garrett[1*], F. Boussaha[2], C. Chaumont[3], B. -K. Tan[1], and G. Yassin[1]
1 Department of Physics (Astrophysics), University of Oxford, Keble Road, Oxford OX1 3RH, U.K.
2 Observatoire de Paris, Pôle Instrumental/LERMA – 77 Avenue Denfert-Rochereau, 75014 Paris, France
3 Observatoire de Paris, Pôle Instrumental/GEPI – 77 Avenue Denfert-Rochereau, 75014 Paris, France
* Contact: john.garrett@astro.ox.ac.uk, phone +44 01865 273 303

Abstract— **We have developed anSIS receiver with a wide intermediate-frequency (IF) bandwidth.This is important for reducing image integration time and simultaneously measuring multiple spectral lines. The receiver is a finline mixer-based design, which allows for ultra-wide radio-frequency (RF) bandwidth and has lower mechanical requirements compared to radial stub designs. Simulations of this receiver showed quantum limited noise in the RF frequency range of 140 to 260 GHzand from DC to 10GHz in the IF spectrum.We measured the noise temperature by comparing the receiver's response to hot and cold loads. The best noise temperature was 37.9 K at 231.0 GHz, and all of the results were below 100 K from 213 to 257 GHz (the bandwidth of our local-oscillator). We measured the IF bandwidth using a spectrum analyser, and found good results from around 3-10 GHz. The lower frequency was restricted by our IF amplifier's bandwidth but the higher frequency limit was lower than we expected from simulations. We believe that this discrepancywas due to the inductance of the bondwires that we used to connect the mixer chip to the IF board. We are currently investigating techniques to reduce and compensate for this inductance.**

Introduction

Observations of millimetre-wave spectral lines are crucial for understanding the chemical and physical properties of star forming regions.However, large detailed surveys of the Milky Way and neighbouring galaxies areslow with the current technology. This is especially true for large interferometers, which have inherently small fields of view.

In order to decrease the time required to form an image, different techniques can be applied. Firstly, the sensitivity of the receivers can be improved. For a given signal-to-noise ratio (SNR), the integration time ($\Delta\tau$) is proportional to the noise temperature (T_N) squared.While improving the sensitivity could greatly reduce the integration time, modern superconductor-insulator-superconductor (SIS) receiversare approaching the quantum limit of sensitivity. It is now very difficult to lower the noise temperature by any significant fraction, hence other techniques must be considered. An alternative way to increase the receiver sensitivity is to increase the number of detectors (mixers) in the focal plane of the dish. If the mixers are operating independently, the imaging time is inversely proportional to the number of mixers. This involves creating a focal-plane array, which has its own challenges (e.g., [1]). Athird technique is to increase the intermediate-frequency bandwidth (IFBW) of the receivers. For a continuum source and a given SNR, integration time is inversely proportional to the IFBW. A large IFBW also has the added benefit that multiple spectral lines can be observed simultaneously, again helping to reduce imaging time. For most modern receivers, the IFBW is approximately 8 GHz and there are many efforts to raise this even further (e.g., [2-5]).

The IFBW of the SIS devices is typically limited due to (a) the intrinsic capacitance of the junction, and (b) RLC resonances forming within the planar circuit of the receivers [3]. Both of these issuescan result in the output impedance of the receiver dropping to zero for extended regions of the IF spectrum, which makes it impossible to match the receiver to the IF circuitry. Issue (a) can be addressed by making the junctions as small as possible(to reduce the intrinsic capacitance), by using multiple distributed junctions [5], and/or by tuning out the junction's capacitance with IF tuning boards. Likewise, issue (b) can be addressed by carefully designing the planar circuit to shift the resonance outside the desired IF spectrum. This mostly involves reducing the surface area of the wiring layer of the planar circuit to reduce the capacitance seen by the IF circuit.We shall address these issues in detail for the receiver presented in this paper.

Receiver Design

The receiver we created is a single-ended, dual sideband, single-junction SIS mixer (Fig. 1). We designed it to operate from 150 to 250 GHz with a large instantaneous IFBW from 0 to 15GHz. Full details of the design, simulation and fabrication can be found in a previous publication [4].

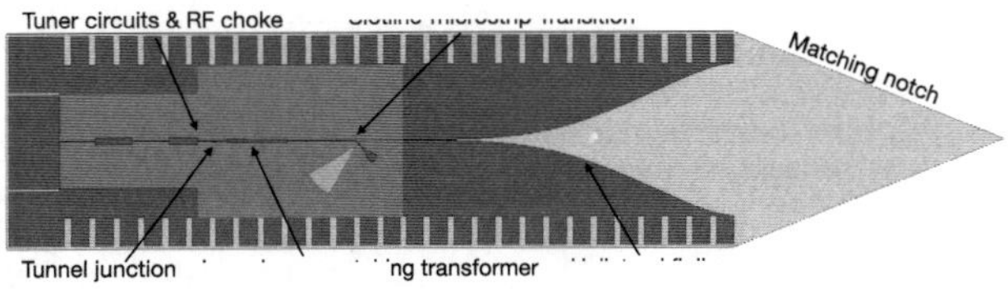

Fig. 1 Layout of the 230 GHz SIS receivers. This device is mounted in the E-plane of the waveguide.

Each component of the receiver was simulated using Ansys'sHigh-Frequency Structural Simulator (HFSS) and then imported intoCalTech's SuperMix package [6,7]. The simulation results show quantum limited performance for an RF bandwidth from 140 to 260 GHz, and an IF bandwidth from DC to 25 GHz.

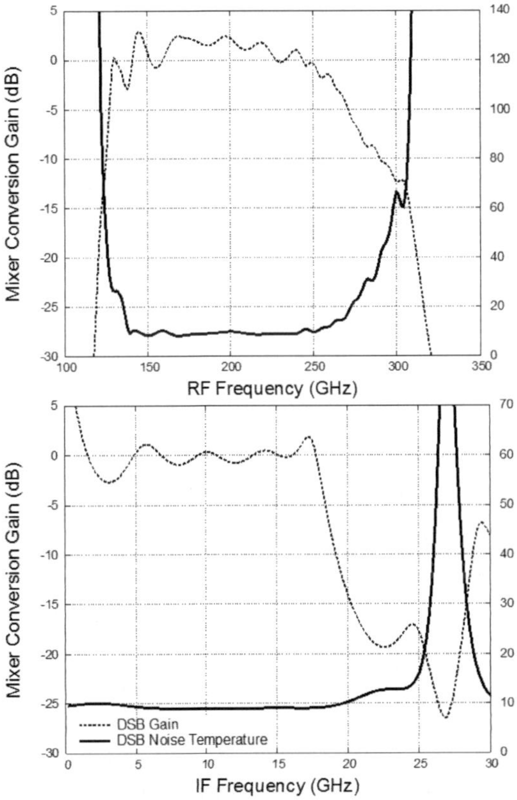

Fig. 2 RF and IF simulation results of the 230 GHz receiver using Ansys's HFSS and CalTech's SuperMix package.

FABRICATION

The mixer chip was fabricated by the Paris Observatory's clean room facility, on a 2 inch quartz wafer, using photolithography techniques [4]. The materials and layer thicknesses are listed in Table 1, and a subsection of a 2" wafer is shown in Figure 3.The junctions are circular, 1.5 μm^2Nb/AlO_x/Nb SIS tunnel junctions. Based on the materials and dimensions, this gives an estimated normal resistance of 14 Ω, a capacitance of 120 fF, andanωRC product of approximately 2.4.

TABLE I
DEVICE LAYERS.

Layer	Thickness	Material
Wiring	400 nm	Niobium
Dielectric	490 nm	Silicon Oxide
Ground	250 nm	Niobium
Substrate	100 um	Quartz

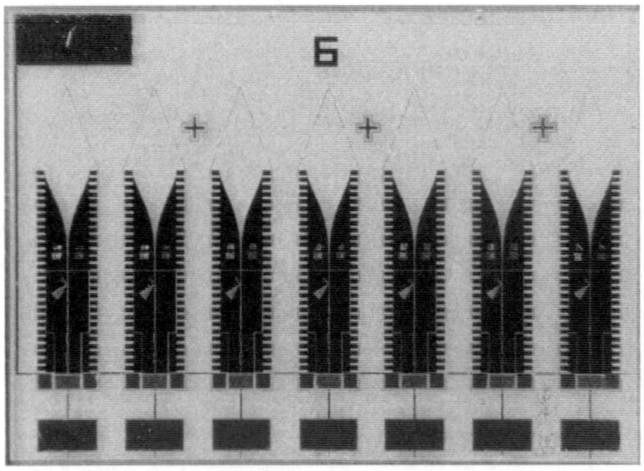

Fig. 3A subsection of a 2" wafer showing 7 completed receivers.

RF RESULTS

The noise temperature and gain weremeasuredusing hot and cold loads(room-temperature Eccosorb and liquid nitrogen, respectively) and measuring the IF current with a detector diode (Fig. 4). Our existing LO (frequency range: 213-257 GHz) was used to pump the the mixer. In the IF chain, we used a 4-6 GHz bandpass filter. We have not correctedthis data to account for the beam splitter loss, the vacuum window loss, or any other optical losses.The best noise temperature was 37.9 K at 231.0 GHz, and all of results were below 100 K from 213 to 257 GHz. Using the methods described in [8] and [9], weestimated theIF noise to be 11-14 K and theRF noise to be16-19 K.

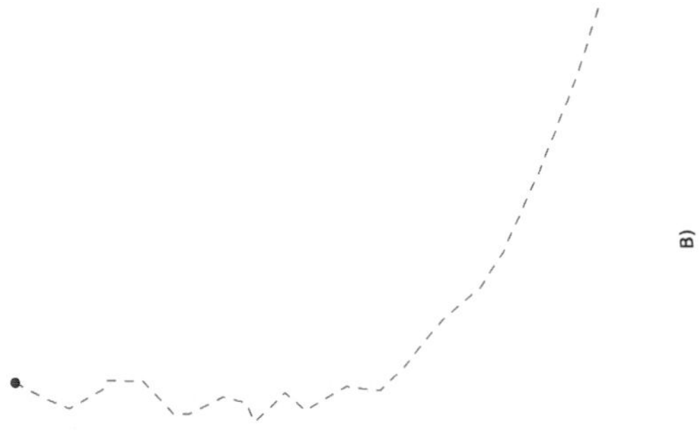

Fig. 4 Noise temperature and gain results for the 230 GHz receivers. This was measured using hot and cold loads with the current measured by a detector diode.

IF RESULTS AND DISCUSSION

The IF spectrum was measured usinga spectrum analyser (Fig.5). Ringing was present below 4 GHz, but this was due to reflections from the low-noise amplifier (LNA, bandwidth: 4-12 GHz).We could have used an isolator to prevent this, but they are typically very narrowband.The peak around 6 GHz was due to a standing wave in the IF chain and wasn't present in every measurement. On the high frequency side, the resultsare much higher than what was simulated by SuperMix (i.e., compared to Fig. 2). We tested 2 other devices as well and they both provided similar results.

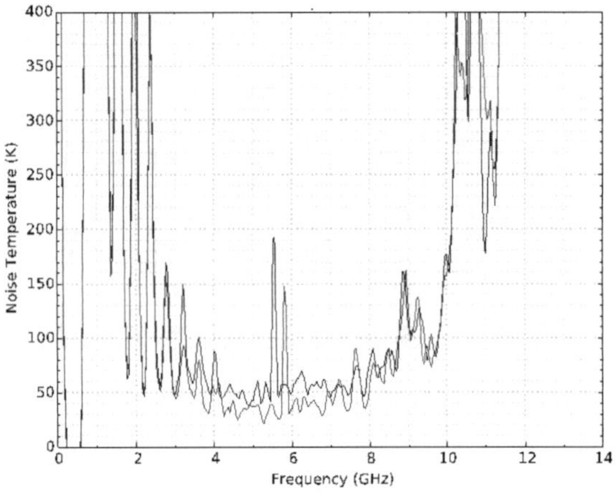

Fig. 5 IF spectrum at 220 GHz (blue line) and 230 GHz (green line).

To help understand the IFBW results, we measured the IF output impedance by using the technique described in [10] (Fig. 6). This technique uses a vector network analyser (VNA) to measure the reflection from the receiver, which can then be used to determine the output impedance. The system was calibrated by biasing the junction to create artificial open, short and matched load calibrations. In Fig. 6, the impedance appears to drop to zero around 9 GHz, but this is an artefact of the calibration method not working as well at these frequencies (i.e., strong reflections from the IF side of the junction cause a loss of sensitivity). This suggests that poor matching is occurring above 9 GHz, which is consistent with the spectrum analyser measurements.

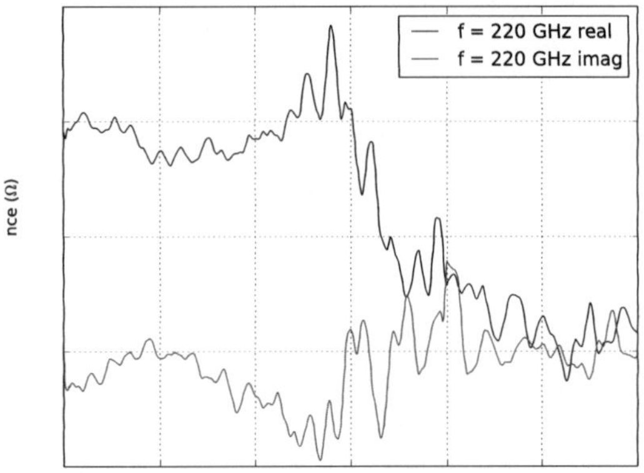

Fig. 6The receiver's IF output impedance for f_{LO} = 220 GHz. This was measured using the technique described in [10]. The results were filtered using a median and moving-average filter (order=3).

We believe that this mismatch is likely due to the inductance of the bondwires. To connect the mixer output to the IF tuning boards (microstrip circuit on aPCB board [11]), we used 25.4 μm diameter aluminium bondwires. A general rule of thumb for bondwires is 1 nHof inductance for every 1 mm. The bondwires for our devices were typically less than 0.5 mm, and usually two wires were used. If we include a series inductance of 2 nH in our SuperMix simulations, the IFBW shrinks dramatically as seen in Fig. 7. To further confirm these results, we tested the receiver with an extremely long bondwire (around 1 mm with a large arc). The IFBW for this device (Fig. 8) was much lower than the previous results (shown in Fig. 5). These results demonstrate that the bondwire inductance must be taken into consideration and that it is limiting the IFBW.

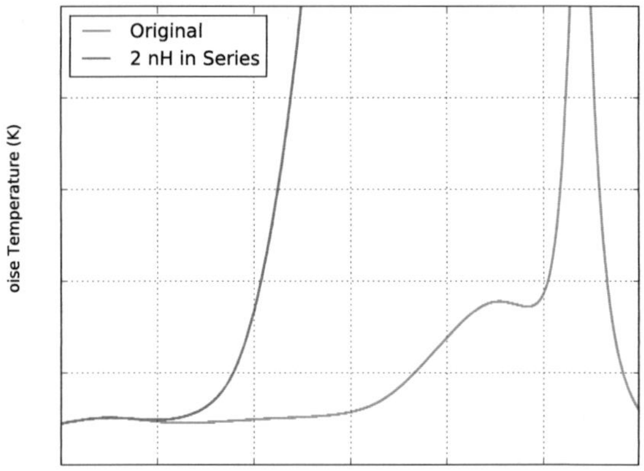

Fig. 7Simulated noise temperature shift due to the bondwires' inductance (assuming 2 nH).

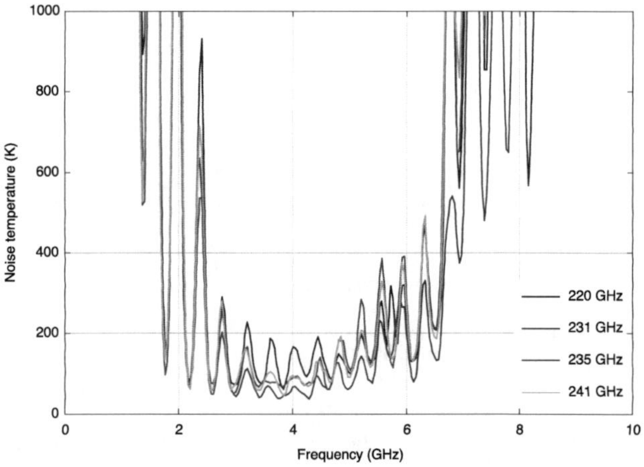

Fig. 8Experimentally measured noise temperature shift due to a long bondwire (1mm long with a large arc). Similar device to Fig. 5.

NEXT STEPS

To extend the IFBW of this device, we first attempt reducing and compensating for the bondwire inductance. This can be done by using larger bondwires (e.g., ribbon bonding versus wedge bonding), by using multiple bondwires (although space on the receiver is limited), and by compensating for the inductance with a shunt capacitance on the IF tuning board. To measure the entire IFBW of this device, we will also test the receiver with broadband LNAs, which cover the whole spectrum of interest (0-20 GHz). Finally, in the very near future, these devices will be used to populate a new 1x4 pixel array [1].

Conclusion

We have designed and tested a new finline-based SIS mixer. Considerable effort was made in the design to extend the IFBW over the existing state of the art. Simulations showed an RF bandwidth from 140-260 GHz and an IFBW from DC to 25 GHz.Experimentalresultsshowed high sensitivities withnoise temperatures down to 37.9 K, and a wideIFBW extending up to around 10 GHz. The IFBW waslower than simulated results due to the bondwires' inductance, which we have confirmed through simulations and experiments using longer bondwires. We are currently investigating solutions to reduce the effects of this inductance which will extend the IFBW.

References

[1] J. Leech *et al.*, "An SIS Mixer Based Focal-Plane Array at 230 GHz," in *Proceedings of the Twenty-Sixth International Symposium on Space Terahertz Technology*, Cambridge, MA, 2015.

[2] Y. Zhou *et al.*, "An SIS Unilateral Finline Mixer with an Ultra-Wide IF Bandwidth," in *Proceedings of the Twenty-First International Symposium on Space Terahertz Technology,* Oxford and Didcot, 2010, pp. 204-211.

[3] Y. Zhou *et al.*, "Performance of a 230 GHz Finline SIS Mixer with a Wide IF Bandwidth," in *Proceedings of the Twenty-Third International Symposium on Space Terahertz Technology,* Tokyo, 2012, pp. 111-116.

[4] J. Garrett *et al.*, "A 220 GHz Finline Mixer with Ultra-Wide Instantaneous Bandwidth," in *Proceedings of the Twenty-Sixth International Symposium on Space Terahertz Technology,* Cambridge, MA, 2015.

[5] C.-Y. Tong *et al.*, "A Distributed Lumped-Element SIS Mixer with Very Wide Instantaneous Bandwidth," *IEEE Trans. Appl. Supercond.,* vol. 15, pp. 490-494, 2005.

[6] J. Ward*et al.*, "SuperMix: A Flexible Software Library for High-Frequency Circuit Simulation, Including SIS Mixers and Superconducting Elements," in*Proceedings of the Tenth International Symposium on Space Terahertz Technology*, Charlottesville, VA, 1999, pp. 269-281.

[7] F. Rice *et al.*, "Fast Harmonic Balance of SIS Mixers with Multiple Junctions and Superconducting Circuits," in *Proceedings of the Tenth International Symposium on Space Terahertz Technology*, Charlottesville, VA, 1999, pp. 282-297.

[8] D. Woody, R. Miller, and M. Wengler, "85—115-GHz Receivers for Radio Astronomy," *IEEE Trans. Microw. Theory Tech.*, vol. 33, pp. 90-95, Feb. 1985.

[9] C.-Y. Tong, A. Hedden, and R. Blundell, "An Empirical Probe to the Operations of SIS Receivers—Revisiting the Technique of Intersecting Lines," in *Proceedings of the Nineteenth International Symposium on Space Terahertz Technology*, Groningen, 2008, pp. 28-30.

[10] P. Serres *et al.*, "The IF Output Impedance of SIS Mixers," *IEEE Trans. THz Sci. Technol.*, vol. 5, pp. 27-36, Jan. 2015.

[11] B.K. Tan, G. Yassin, and P. Grimes, "A Simple Method to Widen the IF Bandwidth of an SIS Mixer," in *Proceedings of the Twenty-fourth International Symposium on Space Terahertz Technology*, 2013.

T1 Session: THz Projects & Instruments (II)

Beyond *Herschel*: Key Scientific Requirements for Future Far Infrared Facilities

M. J. Griffin

School of Physics and Astronomy, Cardiff University, The Parade, CF24 3AA, UK
*Contact: matt.griffin@astro.cf.ac.uk

The THz spectral region is critical to the characterisation of the obscured Universe, in our own galaxy and also in nearby and distant galaxies. The *Herschel* Space Observatory, which operated between 2009 and 2013 has made many observational advances in this respect, and its results are setting the scene for ground-based facilities such as ALMA and for future THz space observatories.

I will focus on two particular *Herschel* science themes and their implications: characterisation of the ubiquitous filamentary structure which pervades the interstellar medium (ISM) and is clearly strongly associated with the formation and evolution of pre-stellar cores and protostars, and the spectroscopy of high-redshift galaxies revealing the mechanisms by which galaxies in the early Universe formed their stellar populations. Both of these have significant implications for our understanding of star formation on local and global scales and over cosmic time, and also raise questions that can only be addressed observationally by future facilities.

The picture of the star-formation process that emerges from *Herschel* is one in which ISM filaments form via the dissipation of kinetic energy in large-scale flows, followed by fragmentation of the densest filaments into prestellar cores through gravitational instability if the filament mass per unit length exceeds a critical value. But this scenario requires further observational study to establish the roles of turbulence and magnetic fields in the formation, growth and maintenance of filamentary structure, the condensation of dense cores and the formation of protostars within them. As well as kinematic observations with ALMA and other telescopes, this will requires characterisation of the magnetic fields through sensitive THz polarimetry on a variety of angular scales, covering the relevant ranges of mass and density from clouds to filaments and cores, and covering a significant sample of galactic regions to establish statistically significant results and to understand the influence of the local cloud environment on star formation. Both ground-based and spaceborne facilities will be needed. To date, no FIR space observatory has had the necessary polarimetric capability, and the development and implementation of a future space polarimeter is therefore important to pursue.

Herschel also allowed us to probe the interstellar medium by means of THz spectroscopy in our own and nearby galaxies, but, because its sensitivity was fundamentally limited by thermal emission from its passively-cooled 80-K telescope, it lacked (except in a few tantalising cases) the sensitivity to study the detailed physics and chemistry of the ISM in high-redshift (distant) galaxies. Future FIR satellite missions will be able to operate with actively cooled apertures and to take advantage of a new generation of superconducting detectors with sensitivity orders of magnitude better than those used on *Herschel*. Such missions are needed to investigate in spectral and spatial detail the material and processes involved in galaxy formation and evolution. SPICA is the next step, with a huge increase over *Herschel* in spectroscopic observing speed. Ultimately, a FIR space interferometer will allow us to study the high-redshift Universe with the same capabilities that we currently have for the local Universe.

Millimetron SpaceObservatory as a Scientific Instrument with Excellent Astronomical Capabilities

A. V. Smirnov*, Thijs de Graauw, N.S. Kardashev, T.A. Kosmovich, S. F. Likhachev, S.V. Pilipenko, D.E. Radchenko

Astro Space Center, Lebedev Physical Institute of the Russian Academy of Sciences, Moscow, Russia
* Contact: asmirn@asc.rssi.ru

Millimetron Space Observatory (MSO) is the next generation space instrument based on 10-meter cryogenically cooled deployable telescope which will be placed at Lagrangian L2 orbit of the Earth-Sun system. Cooling down of the entrance optics to temperatures below 10K with passive and active cooling systems will considerably reduce of magnitude the thermal background emission and thus will provide an unparalleled achievement in the sensitivity. FIR camera, imaging spectrometers and heterodyne instrument will provide high-resolution imaging and spectroscopy and allow investigating the coldest objects in the Universe, the chemical composition and physical properties of gas in the different objects ranging from protoplanetary discs to galaxies at different epochs and many others. Using MSO as an element of Space-Earth Very Large Baseline Interferometer will provide an unprecedented sub-microarcsecond angular resolution which is necessary to study the most compact objects in the Universe, such as the surroundings of black holes, pulsars and gamma-ray bursts. The MSO is proposed as a Russian-led mission with an extensive international consortium.We will show thatpioneeringinstrument -MSO, with present technology, is feasible as a new scientific instrument with excellent astronomical capabilities. We will present an overview and progress in the development of the payload module.

ICEMuSIC – A new Instrument Concept for Mm-wave Observations of Ice Clouds, and Temperature and Humidity Sounding from Space

Peter Hargrave[1*], Simon Doyle[1], Jochem Baselmans[2], Clare Lee[3], Chawn Harlow[3], Anthony Baran[3], Stefan Buehler[4]

[1]*School of Physics and Astronomy, Cardiff University, UK*
[2]*SRON, Netherlands*
[3]*UK Met Office, Fitzroy Road, Exeter, UK*
[4]*University of Hamburg, Hamburg, Germany*
*Contact: hargravePC@cardiff.ac.uk. +44 2920 876682

We present a new mm-wave satellite instrument concept for the study of the role of ice clouds in the global climate system, and for atmospheric temperature and humidity sounding. Ice cloud parameters in particular are poorly understood, and represent the biggest uncertainty in current global climate models.

We employ broadband antennae coupled to filterbank spectrometers, enabled by recent developments in kinetic inductance detector technology. Preliminary atmospheric radiative transfer modelling predicts potentially game-changing performance for the retrieval of atmospheric temperature and humidity, with significantly improved accuracy and vertical resolution.

We present the current instrument concept, the predicted performance, and outline future developments for the satellite instrument.

Terahertz Intensity Interferometry for Very High Angular Resolution Observations

H.Matsuo[1*], H. Ezawa[1], H. Kiuchi[1], M. Honma[1], Y. Murata[2]

[1]*National Astronomical Observatory of Japan, Mitaka, Tokyo 181-8588, Japan*
[2]*Institute of Space and Astronautical Science, Sagmihara, Kanagawa 252-5210, Japan*
* Contact: h.matsuo@nao.ac.jp

High angular resolution imaging in far-infrared and terahertz frequenciesare foreseen either using large single dish telescopes orinterferometers. In this presentation I will focus on a realization ofrelatively long baseline interferometer in terahertz frequencies thatenable milli-arcsecond angular resolution.

Intensity interferometry was first demonstrated by Hanbury-Brown and Twiss.However, low correlation efficiency and missing phase informationrestricted their application to aperture synthesis imaging.In terahertz frequencies, a large number of bunched photons enables delaytime measurements, and high efficiency aperture synthesis imaging can berealized.We have already demonstrated the delay time measurement by an intensityinterferometer experiment in microwave frequency (Ezawa et al. ISSTT2015 proceedings).

In this presentation, an application of the intensity interferometry to space terahertz interferometry will be discussed in detail.Under a low background condition with cryogenic space telescopes, photoncounting detectors become advantageous for high dynamic range intensitymeasurements, so the interferometer technology can be named as a photon counting terahertz interferometry (PCTI).A photon rate of 100 M photon/s from a typically bright terahertz source would enable the delay time accuracy better than 0.1 ps in an integration time of 100 s. The required NEP for a photon counting is just less than 10^{-17} W/Hz$^{0.5}$ when a time resolution is 1 ns.This sensitivity can be realized by using a low leakage current superconducting tunnel junction detectors.

Combination of PCTI and a double Fourier interferometer is proposed toachieve a good u-v coverage for aperture synthesis imaging in terahertzfrequencies.Shorter baselines are covered by the double Fourier interferometer andlonger baselines by PCTI.With maximum baseline length of 20 km at 3 THz, angular resolution of 1milli-arcsecond can be obtained.In terahertz frequency region there are more than 100,000 catalogedpoint sources brighter than 1 Jy, which will be targets for the spaceterahertz interferometer.

Since PCTI can be applied to the very long baseline intensity interferometer (VLBII), satellite control will be similar to space VLBI with heterodyne receivers. Moreover, photon counting detectors realize much higher sensitivityand wider bandwidth to enable imaging thermal emission sources, such asstars and exo-planets.

NOEMA: a Powerful mmArray in the Northern Hemisphere

F. Gueth[1*], on behalf of the whole IRAM staff

IRAM, 300 rue de la Piscoine, 38400 St Martin d'Hères, France
* Contact: gueth@iram.fr

NOEMA (Northern Extended Millimeter Array) is –after ALMA– the most ambitious project in the field of mm astronomy world-wide. It will decuple the capabilities of the IRAM Plateau de Bure interferometer by: adding 6 new 15-m antennas to the array, to reach a total of 12 dishes; equipping all antennas with state-of-the-art wide-band new receivers; installing an FPGA-based new-generation, extremely flexible correlator system; and extending the baselines from 800 m to 1.6 km.

NOEMA will have a major science impact in many astronomical fields, including studies of solar system, star formation, interstellar medium, or high-z galaxies. It will nicely complement ALMA by providing full-sky coverage with comparable sensitivities in the critically important mm domain. NOEMA will also provide unique features, in particular for spectral surveys and astrochemical studies.

The construction of NOEMA is now well advanced, with the two first antennas (the 7th and 8th in the array) completed, and the two next ones under construction. The deployment of the new receivers on all antennas has started, and the installation of the new correlator is scheduled for early 2017.

This talk will summarize the scientific and technical aspects of the NOEMA array, give an up-to-date status of its construction timeline, and present the long-term perspectives for future instrumental developments. A special emphasis will be given to the receiver systems, which include 4 bands covering the 70-370 GHz atmospheric windows, all of them with 2 polarizations and 2SB (2x8 GHz) mixers (i.e. 32 GHz per band).

T2 Session: Quantum Cascade Lasers

Integrating THz Quantum Cascade Lasers to Flexible Dielectric-Metallic Waveguides: Moving beyond Free Space Optics

H.E. Beere[1*], R.Wallis[1], R. Degl'Innocenti[1], D.S. Jessop[1],Y. Ren[1], O. Mitrofanov[2],C.M. Bledt[3], J.E,Melzer[3],J.A. Harrington[3] and D.A. Ritchie[1]

[1]*Cavendish Laboratory, University of Cambridge, J J Thomson Avenue, Cambridge, UK*
[2]*Department of Electronic and Electrical Engineering, University College London, London, UK*
[3]*School of Engineering, Rutgers University, 607 Taylor Road, Piscata, USA*
* Contact: heb1000@cam.ac.uk

The terahertz (THz) quantum cascade laser (QCL) holds great potential in a whole host of applications including laboratory spectroscopy, imaging, sensing and particular astrophysics. QCLs are both compact and robust, exhibiting high power emission (1W in pulsed and 100mW in *cw*mode) with a frequency and bandwidth which can be tailored by suitable device design.However, a number of technological issues still hinder the full exploitation of this class of THz source; like temperature of operation, use of free-space optics, and non-Gaussian beam profiles.

In this work, we present a novel approach that simultaneously targets the latter two issues by developing a THz fibre based approach. The choice of THz waveguide (WG) is critical, since traditional approaches exhibit high material losses in this frequency range (>5dB/m). This work utilises a hollow flexible polystyrene-lined silver waveguide (PS-MWG) as it demonstrates broadband transmission with low dispersion, relatively low losses (<1dB/m), and the ability to support fundamental low order Gaussian-like modes. A key challenge is to efficiently couple the QCL output beam, from the cryostat, to the THz waveguide, external to the cryostat. We adopted an integrated metallic waveguide (MWG) *in-cryo* approach as it demonstrates robust mechanical stability and ease of alignment. Additionally, the diameter can be tailored to allow the spatial beam distribution to be readily matched to that of the *ex-cryo* PS-MWG.

Far-field measurements from the MWG coupled to a metal-metal (MM) THz QCL (*in-cryo* approach) showed a reasonable degree of beam divergence (half angle ~20°) and evidence of the formation of the TE_{11} mode from the Gaussian-like power profile. This is significantly better than the highly diffracted far-field pattern observed from a typicalMM QCL (half angle >30° with strong interference patterns). The coupling efficiency was estimated to be ~70%, with negligible reduction in QCL performance observed. The PS-MWG is a 1mm diameter fused silica capillary with a ~600nm layer of silver deposited on the inner walls. Onto this, a 10μm thick layer of polystyrene (PS) is deposited which determines the transmission band of the structure, centred around 3THz. Simulations show a significant spatial overlap between the TE_{11} mode of the MWG and the HE_{11} mode of the PS-MWG at a separation of 3 mm, for the case of a 1mm diameter for each WG. The resulting measured profile confirmed the excitation of a low divergence (~7°) HE_{11} Gaussian-like beam at a distance >500mm from the facet of the QCL. A high coupling efficiency (>90%) was achieved between the internal MWG and external PS-MWG without the use of any additional optical components, compared to just ~15% between a bare MM QCL and PS-MWG directly. Additionally, a beam cleaning effect was observed, whereby a sub-optimal, multi-lobed beam exiting the cryostat reverts to the single-lobed HE_{11} mode after transmission through the PS-MWG. Comparison with another commonly used collimating technique, a high resistivity silicon lens attached to the laser facet, was also investigated.This was shown to excite a higher order, non-Gaussian mode in the PS-MWG, with a lower coupling efficiency (~70%).

Frequency Instabilities of Terahertz Quantum-Cascade Lasers Induced by Optical Feedback

H.-W. Hübers[1,2*], H. Richter[1], R. Eichholz[1], M. Wienold[1,2], K. Biermann[3], L. Schrottke[3], and H. T. Grahn[3]

[1]*German Aerospace Center (DLR), Institute of Optical Sensor Systems, Rutherfordstr. 2, 12489 Berlin, Germany*
[2]*Humboldt-Universität zu Berlin, Department of Physics, Newtonstr. 15, 12489 Berlin, Germany*
[3]*Paul-Drude-Institut für Festkörperelektronik, Hausvogteiplatz 5–7, 10117 Berlin, Germany*
[*]Contact:heinz-wilhelm.huebers@dlr.de

Heterodyne spectroscopy is a unique tool for applications in remote sensing, in particular in astronomy and planetary research. Its major strength is the high spectral resolving power up to $\nu/\Delta\nu = 10^7$. This allows to resolve the line shape of many atomic and molecular emission or absorption lines. A key component which determines to a large extent the frequency resolution of such a spectrometer is the local oscillator (LO). Since 2014, a THz quantum-cascade laser is operated as the LO in the German Receiver for Astronomy at Terahertz Frequencies (GREAT) on board of SOFIA, the Stratospheric Observatory for Infrared Astronomy. This LO is based on a 4.7 THz QCL in a compact mechanical cryocooler [1,2].

For laser diodes, it is well-known that optical feedback into the laser cavity will shift the lasing frequency as well as change its output power. In the case of a QCL, such an effect may deteriorate its performance when used as an LO. The influence of optical feedback on the emission frequency of QCLs is investigated. The emission frequency is measured with high spectral resolution (< 1 MHz) and a time resolution down to 1 ms by mixing the output from two QCLs with almost the same frequency in a Schottky diode and analyzing the difference frequency. QCLs operating at 3.4 and 4.7 THz with different designs of the active medium are investigated, and the data are analyzed using the theory originally developed for laser diodes. A shift of the QCL frequency as well as mode hopping is observed, both depending on the operation parameters of the QCLs. A quantitative analysis shows that a fraction of only 3×10^{-5} of the output power of the QCL is sufficient to change its emission frequency by 70 MHz. The induced variation occurs on a time scale of less than 1 ms. The results emphasize the importance of reducing optical feedback for high-resolution spectroscopy with THz QCLs.

References

1. L. Schrottke, M. Wienold, R. Sharma, X. Lü, K. Biermann, R. Hey, A. Tahraoui, H. Richter, H.-W. Hübers, and H. T. Grahn, "Quantum-cascade lasers as local oscillators for heterodyne spectrometers in the spectral range around 4.745 THz", Semicond. Sci. Technol. **28**, 035011, 5 pages (2013).
2. H. Richter, M. Wienold, L. Schrottke, K. Biermann, H. T. Grahn, and H.-W. Hübers, "4.7-THz local oscillator for the GREAT heterodyne spectrometer on SOFIA," IEEE Trans. Terahertz Sci. Technol. **5**, 539–545 (2015).

This work was supported in part by the German Federal Ministry of Economics and Technology (grant no. 50 OK 1104) and by the Deutsche Forschungsgemeinschaft. R. E. acknowledges support by the Helmholtz Research School on Security Technologies.

Double Metal Quantum Cascade Laser with 2D Patch Array Antenna on a BCB Substrate with Gaussian Beam Shape for Local Oscillator Applications at 1.9THz

M. Justen[1*], C. Bonzon[2], K. Otani[2], U. Graf, M. Beck[2] and J. Faist[2]
[1] I.Physikalisches Institut, Universität zu Köln, 50937 Köln, Germany
[2] Institute for Quantum Electronics, ETH-Zürich, CH-8093 Zürich, Switzerland
* Contact: justen@ph1.uni-koeln.de

Double metal waveguide QCLs are attractive for applications because of their low power dissipation and good c.w. operation properties, especially at lower operation frequencies. Due to the sub wavelength facet dimensions, the farfield of these devices is generally very poor. An on-chip 2D patch antenna array dramatically improves the farfield and allows high power coupling to free space.

2x2 parallel fed and 3x3 serial fed patch antenna arrays on a benzocyclobutene (BCB)polymer layer are combined with a70µmwide, dry etched, double metal waveguide quantum cascade laser, operating at about 1.9THz. The BCB surrounds the QCL ridge and is planarized to fit precisely its height.

The low dielectric constant (ε_r=2.45) and the high thickness (h = 13.6µm) of the BCB substrate allow a very wide operation bandwidth of the antennas (1.7…2.1THz). On the other hand, under these conditions special care has to be taken in the design of the microstrip power- and phase- distribution network of the array to avoid losses due to surface waves. We achieve a total antenna efficiency of 70…75% where half of the losses can be attributed to the losses in the BCB substrate.

The patch antenna arrays emit anarrow,highly symmetric beam perpendicular to the antenna plane. The beam has a (power) FWHM angle of 48°(2x2) and 29°(3x3). We calculate a very high beam Gaussisity of ≈90%. The measured beam shape and Gaussicity agree very well with the simulations. The high beam quality is important for the use of the QCLs as strong local oscillator source in multi pixel heterodyne receiver setups at THz frequencies.

Our QCL development is supported by the German Federal Ministry of Economics and Technology via the German Space Agency (DLR), grant 50 OK 1103 and by Collaborative Research Council 956, sub-project area D, funded by the Deutsche Forschungsgemeinschaft (DFG).

Simultaneous frequency locking and monitoring using a bi-directional THz quantum cascade laser

B. Mirzaei,[1,*] N. van Marrewijk,[1] D. Hayton,[2] J. R. Gao,[1,2] T. Y. Kao,[3] Q. Hu,[3] and J. L. Reno[4]
[1]*Kavli Institute of Nanoscience, Delft University of Technology, Lorentzweg 1, 2628 CJ, Delft, The Netherlands*
[2]*SRON Netherlands Institute for Space Research, Groningen, the Netherlands*
[3]*Department of Electrical Engineering and Computer Science, Massachusetts Institute of Technology(MIT), Cambridge, Massachusetts 02139, USA*
[4]*Center for Integrated Nanotechnologies, Sandia National Laboratories, Albuquerque, NM 87185-0601, USA*
*Contact: b.mirzaei@tudelft.nl, phone +31616897318

Abstract— **We have performed frequency locking of a dual, forward reverse emitting 3rd order distributed feedback quantum cascade laser (QCL) at 3.5 THz. By using both directions of THz emission in combination with two gas cells and two power detectors, we can for the first time perform frequency stabilization, while monitor the frequency locking quality independently. We also characterize how the use of a less sensitive pyroelectric detector can influence the quality of frequency locking, illustrating experimentally that the sensitivity of the detectors is crucial. Using both directions of THz radiation has a particular advantage for the application of a QCL as a local oscillator, where radiation from one side can be used for frequency/phase stabilization, leaving the other side to be fully utilized as a local oscillator to pump a mixer.**

INTRODUCTION

Terahertz (THz) quantum cascade lasers (QCLs) have been demonstrated as local oscillators for high resolution spectroscopy both in the lab [1] and more recently, in a real astronomic instrument [2]. In general, since the QCL is not inherently frequency stable, a system of frequency or phase locking [3], [4] is required. So far, the radiation emitted from only one direction of the QCL has been used for both pumping a mixer and stabilizing the frequency of the source [5]. In this way, to achieve frequency locking, part of the beam power is unavailable which can be half of the total power available from the laser, making it very difficult to pump a mixer.

It is known that both a standard Fabry-Perot QCL and a distributed feedback (DFB) QCL can emit radiation from both forward and backward directions [6]. To take full advantage of the total radiating power available from a QCL, it is very beneficial to make use of the radiation from both directions. For example, one direction acts as a local oscillator source whilst the other is used for frequency or phase locking. Although it seems obvious that one should take advantage of both beams, in practice no one has ever reported the use of a QCL in this configuration as local oscillator at the high end of THz frequencies (e.g. 4.7 THz), where the available power is still relatively low.

This conference paper is based on our recent publication in J Infrared Milli Terahz Waves [7]. We develop a measurement setup that allows the detection of the radiation simultaneously from both directions. We start with the basic characterization of the radiation beam patterns and emitted power of a 3.5 THz, 3rd-order distributed feedback (DFB) QCL [8]. We demonstrate a practical application of the dual emitting QCL by applying 2 gas cell based frequency discriminators, one for each emission. Specifically, one side is used to realize frequency locking whilst the other side is used to monitor frequency stability. We find that the sensitivity of the detectors is crucial for both frequency locking and frequency monitoring. Finally, we describe briefly an experiment to make use of one side of radiation to carry out the frequency locking and other side of the radiation to pump a superconducting NbN hot electron bolometer (HEB) mixer [9].

QCL AND MEASUREMENT SETUP

We use a 3rd-order DFB THz QCL based on a four-well resonant phonon depopulation design [10] developed at MIT (Fig.1. a). It emits a single mode at 3.490 THz, as measured by a Fourier Transform Spectrometer (FTS) with a resolution of 0.6 GHz. The device comprises 27 lateral corrugated grating periods over a ~1 mm long active region, which is 10 μm thick and 50 μm wide. We make a symmetric sample holder (Fig. 1. b) such to make the same conditions for the out coming beams from both directions. The only difference is that the bonding wire appears only on one end of the laser.

The setup for the key measurement of this paper is illustrated in Fig. 2. The QCL is mounted in a pulse tube cryocooler that reaches ~4 K without load and typically ~12 K with the ~3 W electrical power dissipated by the QCL. The QCL is positioned in such a way where one end of the laser with the bonding pad and wire points to the backward direction. To allow both forward and backward radiation to exit the cryostat, two windows are installed. The front window (corresponding to the forward direction) is a 3 mm thick high density polyethylene (HDPE) with a transmission of 71% measured at the laser frequency, while the rear window is a 1 mm thick ultra-high-molecular-weight polyethylene (UHMW-PE) with a transmission of 89% obtained at the same frequency. The QCL is placed in the centre of the cryocooler with roughly an equal distance of ~ 80 mm to the windows.

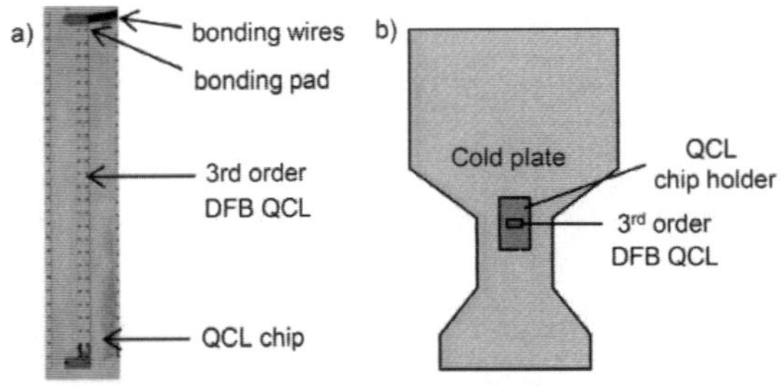

Fig. 1. (a) Photo of the 3rd-order DFB QCL used for the experiment. One end with the bonding wire/pad is positioned towards the backward direction in the setup shown in Fig. 2. (b) Sketch of the QCL sample holder. The QCL (red) is mounted on a Cu chip holder (dark grey). The chip holder is attached to a cold plate (light grey) connected to a cryocooler.

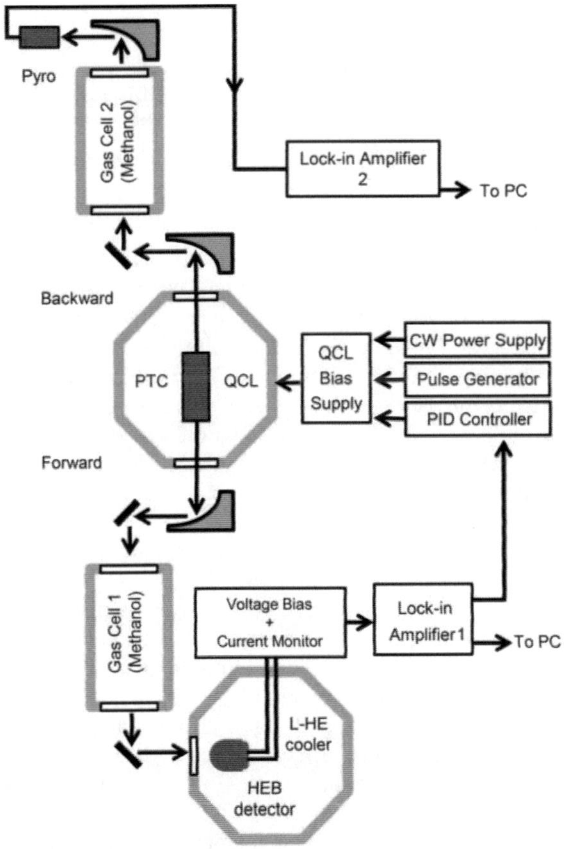

Fig. 2. Schematic of the measurement setup. The QCL is operated in a pulse tube cryocooler (PTC). The combination of a gas cell and a HEB detector is applied to generate an error signal to a PID controller for frequency locking (forward), and a 2nd gas cell with a pyroelectric detector to monitor the quality of frequency locking (backward).

Both forward and backward radiation are collimated and reflected through each of two gas cells with lengths of 41 cm and 27 cm, respectively. Due to the abundance of absorption lines in the THz, methanol is used as the reference gas in both gas cells.

The forward radiation beam is then reflected into a Si lens/antenna coupled superconducting NbN HEB [1], [9], which is operated as a bolometric power detector. It produces an error signal that is fed, via a lock-in amplifier, to a Proportional Integral Derivative (PID) controller. The PID controller makes a correction signal that is added into the QCL bias voltage to hold the error signal at zero, and therefore to stabilize the frequency. The feedback bandwidth, limited by the lock-in amplifier time constant, is ~10 Hz although the PID bandwidth is much higher (~1 kHz). As indicated by the measured frequency noise power spectral density [11], a bandwidth of ~10 Hz is in practice sufficient to stabilize the average laser frequency and to remove low-frequency jitters.

The backward radiation beam after passing through the gas cell 2 is focused onto a room temperature pyroelectric detector that is used for monitoring the quality of the frequency locking. We read out the signals from both detectors via two separate lock-in amplifiers connected to a PC. Since we have the same gas and roughly the same pressures in the gas cells, we expect to see a similar changing behaviour from the signals detected by both detectors. The two detectors however have very different sensitivities. The HEB has a noise equivalent power (NEP) of 10-12 ~10-13 [12], whereas the pyroelectric detector has an NEP of ≥10-9 [13]. Also, the former works at 4 K, while the latter operates at room temperature.

EXPERIMENTAL RESULTS AND DISCUSSION

We start with the measurements of the far-field QCL beam patterns in both directions by using a small aperture pyroelectric detector scanned within a plane normal to the direction along the laser structure indicated in Fig. 2b (z-axis). Fig. 3 shows the measured beam patterns of the radiation from both directions.

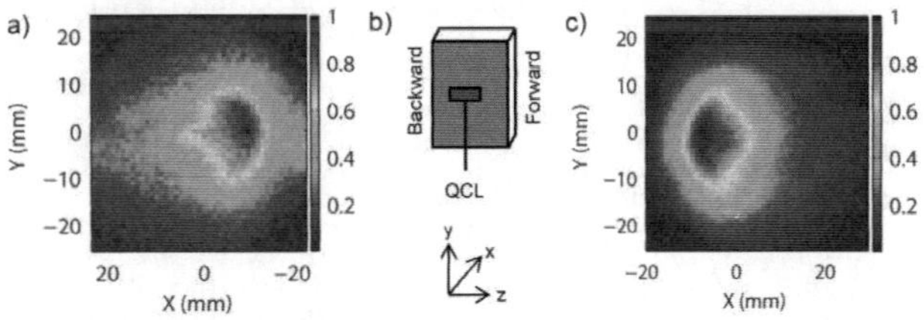

Fig 3. (a) Measured beam pattern (normalized) from the backward radiation. The observation plane (x,y) is about 90 mm to the QCL. (b) Orientation of the QCL. The arrows indicate the positive x, y and z directions. (c) Measured beam pattern (normalized) from the forward radiation. The observation plane (x,y) is also about 90 mm to the QCL.

We find that the backward direction emits less power and has only 56% power from the forward direction, obtained after correcting the effect due to two different transmissions of the windows. The difference by nearly a factor of two in power is attributed to the bonding pad/wire on the laser in the backward direction. The power result is consistent with the beam pattern measurement, where the S/N ratio is worse in the backward direction.

Prior to frequency locking, we measure methanol absorption lines by sweeping the QCL bias voltage from 13.5 V to 14.5 V, which tunes the frequency electrically, as confirmed by a separate FTS measurement. Both gas cells are filled with methanol at a pressure of ~1.7 mbar. Results are plotted in Fig. 4, where the signals from both detectors are recorded with two lock-in amplifiers simultaneously.

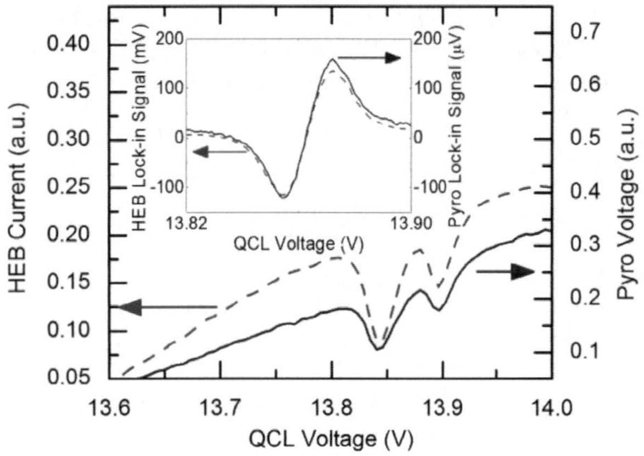

Fig. 4. Absorption lines of methanol at 1.7 mbar. The lines are measured with an HEB (red dashed) and a pyro (blue), respectively. The inset shows the derivative of an absorption line around 13.8 V measured with the HEB (red dashed) and the pyro (blue) by a lock-in amplifier when QCL is modulated with a small AC signal.

As expected, the absorption lines appeared at exactly the same bias voltages. The derivative of the absorption line at ~13.84 V was also measured by applying a small 70 Hz, 10 mVp-p AC modulation [14] (inset of Fig. 4). They change linearly with the QCL bias voltage over a range close to the absorption line center.

In this way, we can make use of an absorption line for frequency stabilization of the QCL because its frequency is known to be fundamentally stable [11], [14]. Any fluctuations in the frequency of the QCL below the bandwidth of ~10 Hz will cause proportional changes in the derivative output. In practice, we set the QCL bias voltage so as to have its frequency close to the center of a specific absorption line and then feed the derivative signal as the error input to the PID controller. The controller produces a feedback to the QCL bias voltage to keep its frequency aligned to the center of the absorption line where the derivative is equal to zero.

Now we focus on the key experiment of this paper using the setup in Fig. 2 by applying this method to gas cell 1 by feeding the HEB's derivative signal to the PID controller to stabilize the frequency, while utilizing output from the gas cell 2 to monitor the quality of frequency locking. A time series of the error signals measured simultaneously from both lock-in amplifiers is plotted in Fig. 4.

In the time interval from 0 to 9 seconds (mode 1), the QCL was free running and the error signals recorded in both detectors are relatively large, which is primarily due to the ~1 Hz frequency of the pulse tube cooler. Afterwards (mode 2, 10-24 sec), the PID is turned on reducing the error signal from the HEB by a factor of 20. In the same time interval, the fluctuations of the pyroelectric signal is not as strongly suppressed as the one for the HEB. To understand this, we actually block the radiation to the pyroelectric detector and record its error signal, while the frequency locking is maintained by the HEB (referred as the mode 3 in Fig. 5). We find that the intrinsic noise level of the pyroelectric detector dominates in both cases; no matter whether there is a radiation signal to the pyroelectric detector or not. Thus, we realize that the error signal from the pyroelectric detector does not directly correspond to the frequency locking quality, but rather to the noise floor of the detector.

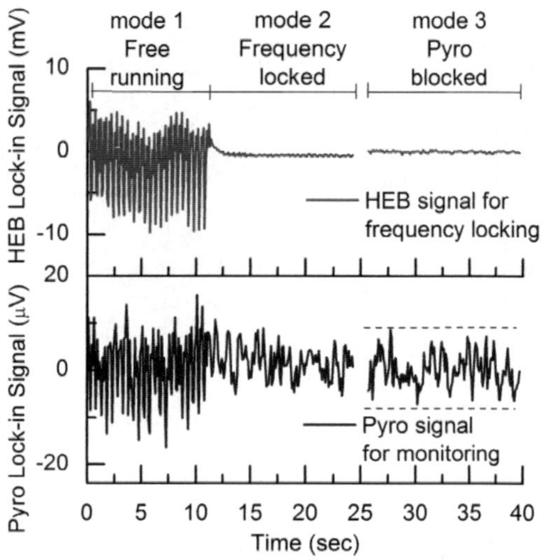

Fig. 5. The lock-in amplifier signal from the HEB (top, red) and the pyroelectric detector (bottom, blue), reflecting the frequency stability of the QCL. Frequency locking is engaged to the forward radiation after 12 sec using the HEB signal (control), while the pyroelectric detector monitors the frequency of the backward radiation. After 30 sec the radiation to the pyroelectric detector is blocked. The dashed line represents the pyroelectric detector noise limit.

Since the derivative signal is linear versus the QCL voltage around the locking frequency, we can estimate the frequency fluctuation, knowing the QCL voltage tuning coefficient of about 0.6 GHz/V determined from a separate FTS experiment. We find a free running QCL linewidth of around 800 KHz. After turning the frequency locking on, this linewidth is reduced to about 40 kHz. This analysis is based on the observation from the HEB. In contrast, if we make use of the error signal from the pyroelectric detector, we would record a linewidth of 300 kHz, which contradicts obviously with the first result.

To verify the importance of the noise level of the detector in such a frequency locking experiment, we modify the experiment slightly and take the error signal from the pyroelectric detector for the frequency locking and the HEB's signal for the monitoring. The results, plotted in the same manner as in Fig. 5, are shown in Fig. 6.

Looking at the mode 2, the error signal from the pyroelectric detector has been reduced considerably relative to the free running case from 800 KHz to 100 KHz. However, compared with the results by using the HEB for the frequency locking in Fig. 5, the residue on the locked signal is large. We attribute these fluctuations to the intrinsic noise of the pyroelectric detector. In this case, the PID controller cannot distinguish the changes between the QCL frequency and the noise from the detector. Consequently the feedback signal to the bias of the QCL cannot be appropriately applied. The lack of suppression in the frequency fluctuations can be monitored by the HEB. Since the (intrinsic) noise floor of the HEB is at least three or four orders of magnitude lower than that of the pyroelectric detector [12], the error signal in this case reflects more accurately the quality of the frequency locking. Because of the higher sensitivity of the HEB, these fluctuations are exclusively due to the frequency fluctuations of the QCL and they show only a mild reduction in the linewidth of the QCL from 800 KHz to 300KHz.

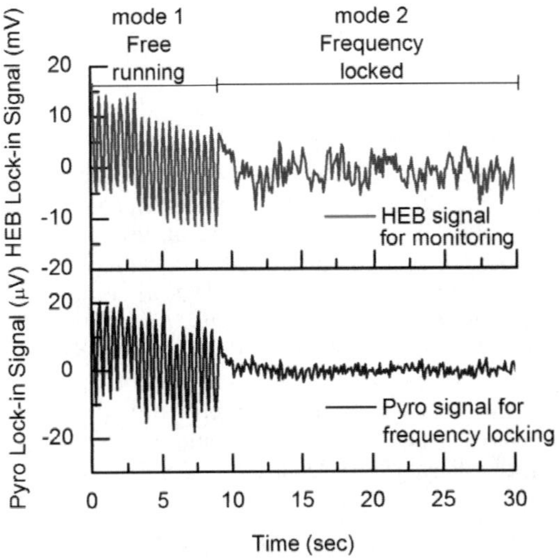

Fig. 6. The lock-in amplifier signal from the HEB (top, red) and the pyroelectric detector (bottom, blue) reflecting the frequency stability of the QCL. Frequency locking is engaged to the backward radiation after 9 sec using the pyroelectric detector signal (control), while the HEB monitors the frequency of the forward radiation.

It is worthwhile to stress that our experiment represents the first one to make use of the bi-directional radiation from a single THz QCL for a frequency locking experiment, where the laser can be locked, while the quality of the locking can be evaluated in the same time. It is also the first to experimentally demonstrate the importance of the detector sensitivity in a frequency lock loop.

A key demonstration of the advantage in using a dual emitting QCL is to show that a superconducting NbN HEB mixer can be appropriately pumped using one side of the laser whilst the other side is used for frequency locking. We perform such an experiment by using a standard NbN HEB mixer, which has a NbN area of 2 μm×0.2 μm, corresponding to a power requirement of 200 nW at the detector itself [1]. We apply a setup simplified with respect to the one in Fig. 2 by removing the gas cell 1 in the forward direction. We then lock the frequency of the QCL using the backward beam. At the same time, we apply the forward beam to pump the superconducting mixer. We find that it can pump the HEB to its nearly optimum operating points. With further optimization of the optics to match the beam to the HEB we expect that the forward beam can provide sufficient power to pump the HEB to its optimum operating points, while the frequency locking is realized with the backward beam.

In this way we can in essence make use of 100% available power from a frequency locked QCL. This approach is certainly beneficial for the case where a QCL is applied as a local oscillator for a superconducting mixer. This approach will be even more attractive for the cases where a QCL is applied as a local oscillator for a semiconductor Schottky mixer and an array of mixers, both of which require high power.

CONCLUSIONS

By making use of the radiation from the forward and backward directions of a 3rd-order DFB QCL at 3.5 THz, we demonstrate for the first time that we can introduce the frequency locking, while can monitor the quality of the locking simultaneously. Furthermore, by applying two power detectors with a different noise level, we show that the frequency locking quality, namely the linewidth derived from the error signal, depends strongly on the noise level of the detector used. In the case of applying a high noise power detector for the locking, the PID controller not only corrects the frequency fluctuations of the laser, but also compensates the noise from the detector by adjusting the QCL frequency, which can lead to a much wider locked linewidth than what is indicated by the (locking) detector.

REFERENCES

J. L. Kloosterman, D. J. Hayton, Y. Ren, T. Y. Kao, J. N. Hovenier, J. R. Gao, T. M. Klapwijk, Q. Hu, C. K. Walker, and J. L. Reno, "Hot electron bolometer heterodyne receiver with a 4.7-THz quantum cascade laser as a local oscillator," *Appl. Phys. Lett.*, vol. 102, 011123, 2013.

H. Richter, M. Wienold, L. Schrottke, K. Biermann, H. T. Grahn and H. W. Hubers, "4.7-THz Local oscillator for the GREAT heterodyne spectrometer on SOFIA," *IEEE Trans. Terahertz Sci. Technol.*, vol. 5, pp. 539-545, 2015.

A. L. Betz, R. T. Boreiko, B. S. Williams, S. Kumar, Q. Hu, and J. L. Reno, "Frequency and phase-lock control of a 3 THz quantum cascade laser," *Opt. Lett.*, vol. 30, no. 14, pp. 1837–1839, 2005.

M. Zhu and J. L. Hall, "Stabilization of optical phase/frequency of a laser system-application to a commercial dye laser with an external stabilizer," *J. Opt. Soc. Am. B.*, vol. 10, pp. 802-816, 1993.

Y. Ren, D.J. Hayton, J.N. Hovenier, M. Cui, J.R. Gao, T.M. Klapwijk, S.C. Shi, T-Y. Kao, Q. Hu, and J.L. Reno, "Frequency and amplitude stabilized terahertz quantum cascade laser as local oscillator," *Appl. Phys. Lett.*, vol. 101, 101111, 2012.

B.S. Williams, "Terahertz Quantum-Cascade Lasers," *Nature Photon.*, vol. 1, pp. 517-525, 2007.

N. van Marrewijk, B. Mirzaei, D. Hayton, J. R. Gao, T. Y. Kao, Q. Hu, and J. L. Reno, "Frequency Locking and Monitoring Based on Bi-directional Terahertz Radiation of a 3rd-OrderDistributed Feedback Quantum Cascade Laser," *J. Inf. Mil. THz Waves.*, vol. 36, pp. 1210–1220, 2015.

M. I. Amanti, M. Fischer, G. Scalari, M. Beck, and J. Faist, "Low-divergence single-mode terahertz quantum cascade laser," *Nature Photon.*, vol. 3, pp. 586-590, 2009.

J. R. Gao, M. Hajenius, Z. Q. Yang, J. J. A. Baselmans, P. Khosropanah, R. Barends, and T. M. Klapwijk, "Terahertz superconducting hot electron bolometer heterodyne receivers," *IEEE Trans. Appl. Supercond.*, vol. 17, pp. 252-258, 2007.

Q. Qin, B. S. Williams, S. Kumar, J. L. Reno, and Q. Hu, "Tuning a Terahertz Wire Laser," *Nature Photon.*, vol. 3, pp. 732-737. Nov. 2009.

Y. Ren, J.N. Hovenier, M. Cui, D.J. Hayton, J.R. Gao, T.M. Klapwijk, S.C. Shi, T-Y. Kao, Q. Hu, and J. L. Reno, "Frequency locking of single-mode 3.5-THz quantum cascade lasers using a gas cell," *Appl. Phys. Lett.*, vol. 100, 041111, 2012.

Y. Ren, W. Miao, Q. J. Yao, W. Zhang, and S. C. Shi, "Terahertz direct detection characteristics of a superconducting NbN bolometer," *Chin. Phys. Lett.*, vol. 28, 010702, 2011.

R. N. Schouten, "A new amplifier design for fast low-noise far-infrared detectors using a pyroelectric element," *Meas Sci. Technol.*, vol. 9, pp. 686-691, 1998.

H. Richter, S. G. Pavlov, A. D. Semenov, L. Mahler, A. Tredicucci, H. E. Beere, D. A. Ritchie, and H.-W. Hübers, "Submegahertz frequency stabilization of a terahertz quantum cascade laser to a molecular absorption line," *Appl. Phys. Lett.*, vol. 96, 071112, 2010.

Spectral Modulation of Terahertz Quantum Cascade Lasers with Radio Frequency Injection Locking

H. Li*and J. C. Cao

Key Laboratory of Terahertz Solid State Technology, Shanghai Institute of Microsystem and Information Technology, Chinese Academy of Sciences, 865 Changning road, Shanghai 200050, China
*Contact: hua.li@mail.sim.ac.cn

The authors demonstrate the electrical beat note analysis and radio frequency (RF) injection locking of a continuous wave (cw) terahertz quantum cascade laser (QCL) emitting around 3 THz (~100 μm). In free running, the beat note frequency of the QCL shows a shift of ~180 MHz with increasing the drive current. The beat note, modulation response, injection pulling, and terahertz emission spectral characteristics in different current regimes are investigated. The results show that in the current regime close to the laser threshold we obtain narrower beat note, broader modulation bandwidth, and stronger response to the RF modulation at the cavity round trip frequency. As a result, in the beat note spectra in the low current regime we observe the strong pulling effect under the RF injection. In the meantime, we find that the terahertz emission lines can be strongly modulated by RF injection.

The authors are grateful to Dr. Stefano Barbieri for useful discussions. This work was supported by the "Hundred-Talent" Program of Chinese Academy of Sciences, the National Natural Science Foundation of China (61575214), the Shanghai Municipal Commission of Science and Technology (15560722000).

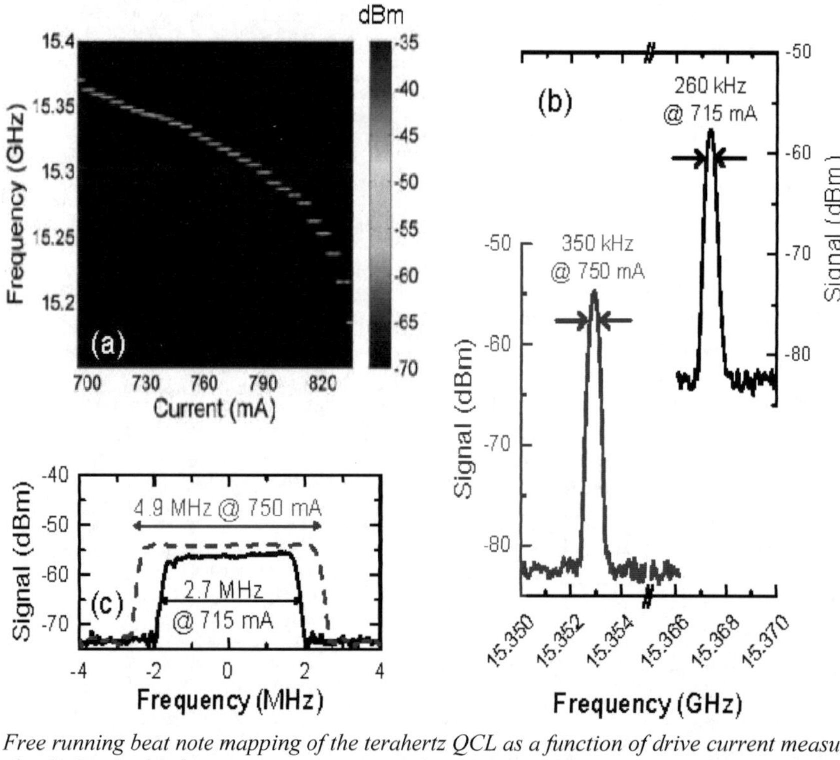

Fig. 1: (a) Free running beat note mapping of the terahertz QCL as a function of drive current measured at 10 K in cw mode. (b) Single shot beat note spectra recorded at 715 mA (black curve) and 750 mA (red curve) in free running. The arrows show the 3-dB linewidths of the free running beat note spectra. The resolution bandwidth is set as 300 kHz. (c) Beat note spectra measured with the "Max-Hold" function of the spectrum analyzer at 715 (black curve) and 750 mA (red curve) in a time duration of 3 minutes. Note that the centre frequencies are subtracted for clarity.

T3 session: THz Sources & Optics

Design Considerations for Amplifier-Multiplier Chain (AMC) for Low Noise Local Oscillator

Cheuk-Yu Edward Tong[1*], Paul K. Grimes[1], and Chih-Chiang Han[2]
[1]*Harvard-Smithsonian Center for Astrophysics, Cambridge, MS 02138, USA*
[2]*ASIAA, Taipei, Taiwan, ROC*
*Contact: etong@cfa.harvard.edu

Abstract— **It is well known that broadband Amplitude Modulation (AM) noise from an Amplifier-Multiplier Chain (AMC) degrades the noise performance of a low-noise THz receiver. We have examined the role played by the source oscillator, frequency multipliers and power amplifiers in the generation of AM noise. An AMC has been designed to provide 0.2 mW of output power, tunable over 210 – 270 GHz. This module is used as the Local Oscillator for the 240 GHz SIS receiver for the Submillimeter Array. The receiver demonstrates no noise degradation with this AMC LO, when compared to operation with a Gunn oscillator based LO. Based on our experiences, we have established design rules for building AMCs for use as low-noise LO.**

INTRODUCTION

With the advent of commercial Amplifier-Multiplier Chains (AMC) operating in the millimeter and THz frequency range, the use of an off-the-shelf AMC as the Local Oscillator (LO) in low noise THz receivers becomes very attractive. In addition to the presence of low level spurious tones in the output of an AMC, these modules also present the challenge of relatively high output Amplitude Modulation (AM) noise. This AM noise is broadband in nature, covering a bandwidth, which can be many GHz wide. Thus when an AMC is used as an LO for a THz receiver, the AM noise would appear as added noise at the receiver output, increasing its noise temperature.

Starting with a discussion of the effects of AM noise, we will look at the role of 3 key components of an AMC, namely the oscillator, the diode frequency multiplier and the power amplifier, in the introduction and propagation of AM noise. We will present noise measurements on a 210-270 GHz custom built AMC that we have designed for the Submillimeter Array.

AM NOISE

THz LO modules often produce broadband noise in addition to the pure CW signal (carrier) required to pump THz mixers. At high frequency offsets from the carrier (MHz to GHz), such added noise behaves as an amplitude modulation of the carrier – AM noise. When a noisy LO operates a single-ended mixer, this broadband AM noise is down-converted into the IF, degrading the noise performance of the mixer.

Referring to Fig. 1, consider the case when the LO signal has significant AM noise within the RF bandwidth of the signal. Let T_{AM} be the effective noise temperature of this broadband noise, corresponding to a noise spectral density of $k_B T_{AM}$ W/Hz, where k_B is the Boltzmann constant. For a 1% LO coupler (or beam splitter), the mixer noise, which is referred to the optical input port, is increased by $T_{AM}/100$.

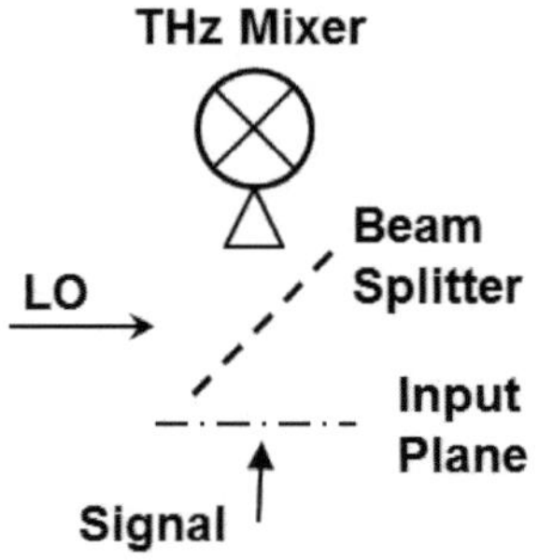

Fig. 1 Typical configuration of a THz low noise mixer in which the LO is injected optically via a beam splitter. Noise temperature of the mixer is referred to the input plane of the signal beam.

In the case of high AM noise, the noise level is conveniently described by the Excess Noise Ratio (ENR), which is usually used to measure the noise power output from a noise source. ENR is defined by the following equation:

$$\mathrm{ENR} = 10\log_{10}\left(\frac{T_{\mathrm{noise}}}{T_0} - 1\right) \qquad (1)$$

where T_0 is taken to be 290 K per IEEE standard (Haus, 1963). Thus, a noise source, with T_{noise} = 580 K, would have an ENR of 0 dB, corresponding to a noise spectral density of -171 dBm/Hz. If a source with an ENR of 0 dB is used to pump a low noise mixer through a 1% LO coupler, it would add 6 K to the receiver noise temperature.

Fig. 2 compares the measured noise temperatures of a 240 GHz receiver obtained using a Gunn oscillator-based LO and those obtained using an off-the-shelf commercial AMC as the LO. The AMC-LO introduces 10 - 25 K of additional noise. This corresponds to an ENR of 4 – 9 dB.

It may be argued that the use of a balanced mixer can eliminate the problem of AM noise from AMCs. However, besides the issue of added complexity, broadband balanced mixers typically provide only ~20 dB of LO isolation, which is comparable to the attenuation of the LO beam splitter or coupler. The main advantage of the balanced mixer is the lower LO requirement, which in turn means a potentially lower AM noise content. Clearly, low AM noise is still a necessary requirement for the LO module of a low noise receiver system even with a balanced mixer.

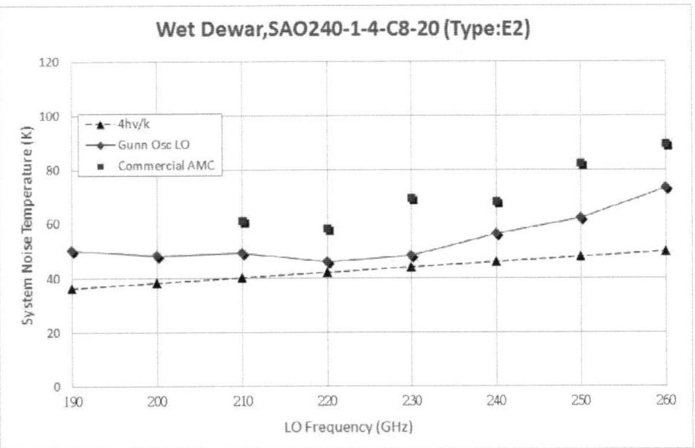

Fig. 2 Measured double-side-band noise temperatures of a 240 GHz receiver, operated by either a Gunn oscillator-based LO or a commercial AMC for a 4 – 8 GHz IF. The receiver noise is close to $4h\nu/k$ when the Gunn oscillator LO is in use.

Sources of AM Noise in an AMC

In order to achieve low noise operation for the 240 GHz receiver for the Submillimeter Array (Tong 2016), we have designed and constructed a YIG oscillator based AMC. This AMC is specified to provide more than 0.2 mW over the frequency range of 210 – 270 GHz. The schematic of the AMC is given in Fig. 3. Starting from the YIG oscillator operating in the frequency range of 11.6 to 15 GHz, the chain performs a frequency multiplication x18 (=x2x2x3). A single power amplifier, operating between 22 and 30 GHz, is used.

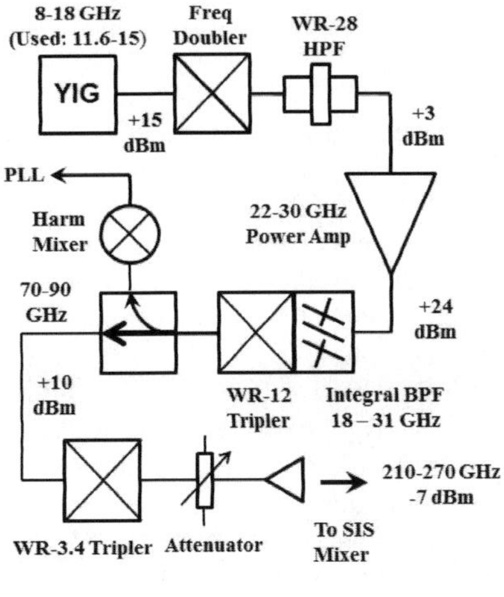

Fig. 3 Schematic of the 210-270 GHz AMC module designed for low noise operation of the 240 GHz receiver for the Submillimeter Array. The phase lock loop (PLL), essential for the interferometer, operates from the 70 – 90 GHz stage.

We have performed measurements of ENR at different stages of this AMC in an effort to trace how AM noise is generated and propagated within the module. The 3 main sources of AM noise are the YIG oscillator, the frequency multipliers and the power amplifier.

A. YIG Oscillator

Although commercial AMCs generally are not supplied with a primary oscillator, their noise output is important for the low noise operation of the AMC, as any noise from the oscillator will propagate through the AMC. YIG oscillators are chosen as the primary source in our system because they provide low phase noise needed for the interferometric operation of the Submillimeter Array.

Most solid-state oscillators incorporate integral output buffer amplifiers to boost output power and to reduce or eliminate load pulling. This amplifier is the source of AM noise at large frequency offsets from the carrier.

In order to measure the ENR emerging from the YIG oscillator, we used a narrow-band YIG-tuned filter followed by a microwave receiver to examine the noise spectral density at different offsets from the carrier. Care was taken to attenuate the output of oscillator to below the input compression level of the YIG-tuned filter. The measured ENR is calibrated using a noise source with a known ENR. The result of the measurement is given in Fig. 4 for a YIG frequency of 12 GHz. The observed ENR is ~24 dB.

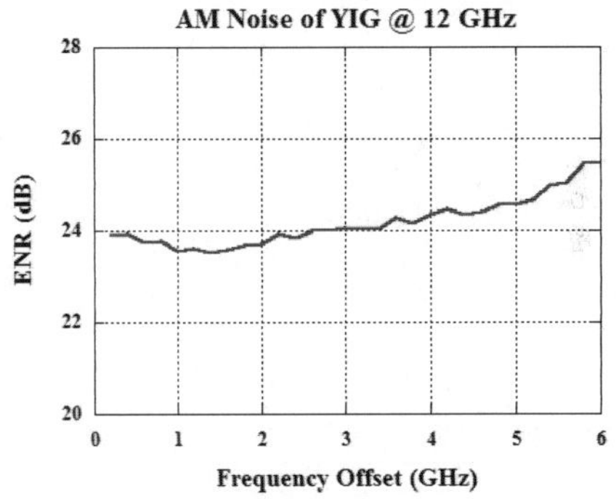

Fig. 4 Measured ENR as a function of frequency offset of the noise side-band. The YIG oscillator operates at 12 GHz and the measured noise is in the upper-side-band with frequency above 12 GHz.

The output noise temperature of an amplifier, T_{out}, is given by the following equation:

$$T_{out} = G_{amp}\left(T_{amp} + T_{inp}\right) \quad (2)$$

where G_{amp} and T_{amp} are the gain and noise temperature of the amplifier respectively, and T_{inp} is the input noise temperature to the amplifier. Assuming some typical numbers for T_{amp} and T_{inp}, we infer that the YIG oscillator used has an output buffer amplifier with a gain of ~20 dB.

Given that the oscillator output power is around +15 dBm, an ENR of 24 dB corresponds to a noise spectral ratio of -165 dBc/Hz, orders of magnitude lower than the specified phase noise level of the oscillator. If we integrate the AM noise over the entire bandwidth of the oscillator, the total amount of AM noise power is only around -50 dBm. In order to limit the impact of this broadband noise, as well as to remove any out-of-band spurious signal, any broadband oscillator should be band-limited.

B. Frequency Multiplier

As far as the propagation of AM noise is concerned, a diode frequency multiplier behaves like a frequency mixer or a frequency up-converter. This effect is illustrated in Fig. 5. An input noise sideband to the multiplier, at a frequency offset of ΔF to the carrier, appears at the output at the same frequency offset to the carrier.

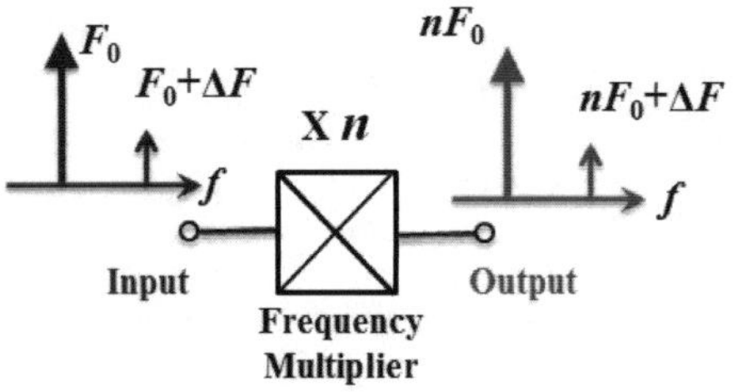

Fig. 5 Frequency up-conversion property of a frequency multiplier. A small signal side-band at a frequency offset of ΔF from the carrier at the input of the multiplier appears with the same offset at the output.

The implication of this feature is that a bandpass filter, of width B, placed at the input of a frequency multiplier yields an output AM noise bandwidth of B as well. As a result, the impact of AM noise on a THz mixer is greatest when the carrier is close to either edge of the bandpass filter. In such case, it is likely that more added noise may be observed at the IF output of the mixer to an IF of B. It has been proposed that a YIG-tuned filter, with very narrow bandwidth, can be used to restrict the AM noise bandwidth. Unfortunately, because of its relatively low power handling capacity (typically up to +10 dBm), a YIG-tuned filter cannot be placed after the main power amplifier of an AMC. Besides the maximum frequency of operation of YIG-tuned filter is 40 – 50 GHz.

We also find that the carrier-to-AM noise ratio does not seem to change much after frequency multiplication. This means that a passive frequency multiplier add only small amount of AM noise. Since the frequency multiplier presents a conversion loss to the carrier, the output ENR of the multiplier is approximately the same as the input ENR minus the conversion loss (in dB).

C. Power Amplifier

Equation (2) shows that the power amplifier is the main source of output AM noise in an AMC: not only does it amplify the input AM noise, the power amplifier also has relatively high noise temperature. It is well known that by operating a power amplifier into saturation, one can reduce somewhat the output AM noise. This is because a power amplifier has reduced small signal side band gain when it is driven beyond its compression point. We have performed measurements of the gain and output ENR of the power amplifier in our AMC. In order to measure the small signal properties of the power amplifier, a back-to-back pair of waveguide coax adapters, sized so that the carrier was 2.5 GHz below the waveguide cut-off, was used as a high pass filter to block the carrier. The measurement results are presented in Fig. 6. Just below the 1-dB compression point of the amplifier, at an output power of +23 dBm, the small signal sideband gain drops below the gain of the carrier. The ENR also starts to drop at around similar level. At the 1-dB point, the ENR has dropped by about 3 – 4 dB. When the amplifier is driven harder, at 2 dB beyond the 1-dB point (at +25 dBm output power), ENR registers a further 2 dB drop, while the small signal gain drops by close to 5 dB.

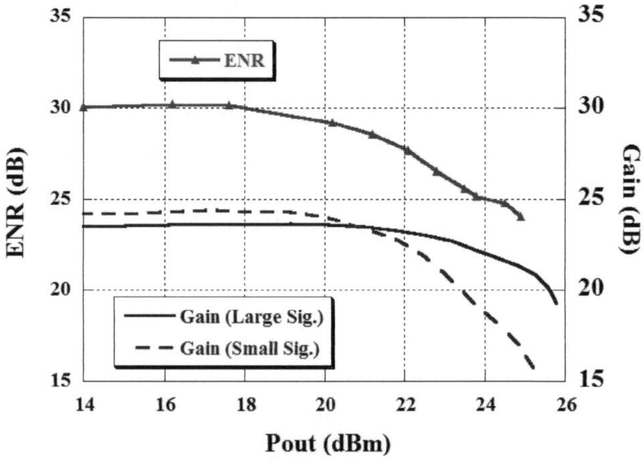

Fig. 6 Gain and ENR of the 22-30 GHz power amplifier operating at 24 GHz. The small signal gain and ENR are measured at a frequency of 29 GHz, with the 24 GHz carrier blocked by a WR-22 waveguide filter.

One may be tempted to think that a series of low gain power amplifiers, all working well into saturation would drastically limit the amount of AM noise. Nevertheless, the role of the power amplifier in an AMC is to boost the power level between 2 successive frequency multipliers. As such, the required gain is more or less fixed, otherwise one risks under-pumping the frequency multiplier following the power amplifier, leading to undesirable effects related to the generation of spurious signals. The better way is to use more efficient multipliers, which requires less power amplification in the AMC. For maximum power efficiency, we propose that the power amplifier should be operated at an output level 1 – 2 dB beyond its 1-dB output compression point.

In the case of our 210-270 LO module, the 22 – 30 GHz power amplifier is required to yield around +24 dBm to drive the next stage WR-12 tripler, with an input power of +3 dBm from the frequency doubler preceding it. Thus, a 21 – 22 dB gain at 2-dB beyond the 1-dB compression point is appropriate, corresponding to an output ENR of 24 – 25 dB, as shown in Fig. 6. Although this value appears high, the power amplifier is followed by two triplers, each with a conversion loss of about -13 dB. Finally, there is a WR-3.4 variable attenuator used to set the output power level. This attenuator further reduces the output ENR of the entire module. While we are not able to measure the output ENR directly, we believe that it is around 0 dB close to the carrier.

As discussed earlier, a 0 dB ENR would introduce a noise degradation of 6 K, which is not acceptable for an SIS receiver with 40 – 50 K noise temperature. To mitigate this problem, we have added a 4 GHz wide bandpass filter at the output of the YIG oscillator, the tuning range of which is from 11.6 to

15 GHz (a 3.4 GHz bandwidth). This filter allows us to further reduce the ENR at high frequency offsets from the carrier, such that the LO module can operate an SIS mixer with a minimum IF of 4 GHz.

PERFORMANCE OF 240 GHZ AMC

The 210 – 270 GHz AMC is designed to operate the 240 GHz SIS receiver for the Submillimeter Array. Radiation from the AMC was optically coupled to the SIS receiver through a 1% beam splitter. The module produced more than enough power (>0.2 mW) to pump the SIS mixer which has a 3-junction array (Tong 2013). As a comparison, we have also employed an LO unit based on a Gunn oscillator. To cover the entire frequency band, two Gunn oscillators were used: a 105 – 120 GHz Gunn oscillator pumping a doubler, and an 80 – 90 GHz Gunn oscillator pumping a tripler.

The double-side-band receiver noise temperature of the receiver was measured as a function of LO frequency for the AMC and the Gunn-based LOs, with the IF spanning 4 to 12 GHz. The LOs were tuned from 210 to 270 GHz. In Fig. 7, we compared the noise temperatures obtained with the two LO modules at 3 different LO frequencies as a function of IF. Clearly, the noise performance of the AMC is comparable to that of the Gunn-based LO.

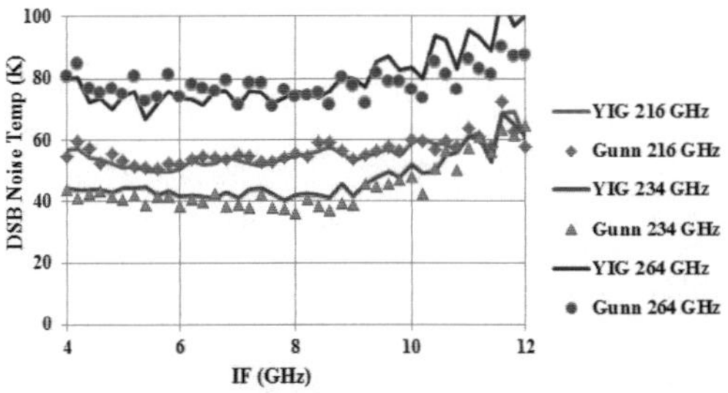

Fig. 7 Comparison of noise temperatures of SMA-240 receiver as a function of IF for 3 different LO frequencies. The receiver is either operated by the YIG-based AMC or by a Gunn oscillator based LO.

In Fig. 8, we plot, as a function of LO frequency, the difference in measured noise temperatures between the two types of LO, averaging over the IF band of 4 – 12 GHz. It can be seen that the AMC produces higher noise temperature only at the high end of the band, by ~5 K. This is a significant improvement compared to the results shown in Fig. 2. The higher noise at 270 GHz can be explained by the lower efficiency of the AMC at the high frequency end. In addition, the gain of different components of the AMC is higher at the low end of the band, which means that when the AMC operates at the very high end, noise from the lower side band is amplified more strongly by the AMC.

DESIGN RULES FOR LOW NOISE AMC

To achieve low noise performance, a number of design iterations were taken. Based on our experience, we have drawn up a number of design suggestions, which seek to minimize the output AM noise of the AMC:

1. The gain of any power amplification should be limited, and the power amplification should take place at as low frequency as possible, so that the ENR of the amplifier will be reduced by subsequent frequency multipliers.
2. For maximum power efficiency as well as optimal ENR, the power amplifier should be operated at an output level which is 1 – 2 dB below that of its 1-dB output compression point.
3. The use of higher efficiency frequency multipliers reduces the required gain of power amplification, which in turn reduces the ENR of the AMC.
4. The highest ENR occurs at frequency offsets close to the carrier because of the way frequency multipliers up-convert the noise side-bands. It is, therefore, important to have tight filtering at every stage in the AMC, particularly at the output of the YIG oscillator, which drives the AMC.

Back-to-back waveguide to coax adapter pairs can be very useful as a high rejection high pass filter, providing 90 dB rejections to unwanted harmonics as well as noise sideband only a few GHz below the waveguide cutoff.

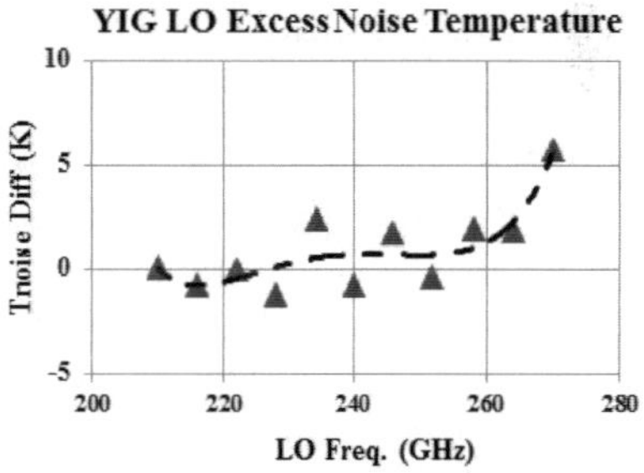

Fig. 8 Excess noise temperature introduced by AMC compared to Gunn based LO. Noise temperature is averaged over 4 – 12 IF.

CONCLUSIONS

We have studied the propagation of AM noise in Amplifier-Multiplier Chains intended for use as an LO for low noise THz receivers. The study allows us to establish basic design rules which can minimize the output AM noise of an AMC. Using these rules, we have designed a 210 – 270 GHz AMC module driven by an 11.6 – 15 GHz YIG oscillator. This module is now being used to operate the 240 GHz SIS receiver for the Submillimeter Array, demonstrating excellent noise performance, comparable to the case when the LO is replaced by a Gunn oscillator based module.

ACKNOWLEDGMENT

The authors would like to thank SAO Receiver Lab staff: Jack Barrett, Robert Kimberk and Steve Leiker for their superb technical assistance, without which this work would not have been possible.

REFERENCES

H.A. Haus et al, "Description of the noise performance of amplifiers and receiving systems," *Proc. IEEE,* vol. 52, pp. 436-442, March 1963.

C.E. Tong, P. Grimes and P.S. Leiker, "A wideband 240 GHz receiver for the Submillimter Array," in *Proc. SPIE Astronomical Telescopes & Instrumentation,* Paper 9914-13, Edinburgh, U.K. June 2016.

C.-Y.E. Tong, P. Grimes, R. Blundell, M.-J. Wang, and T. Noguchi, "Wideband SIS receivers using series distributed SIS junction array," *IEEE Trans. THz Sci. & Tech.,* vol. 3, pp. 428-432, July 2013.

A 600GHz tripler with >5mW and 6% efficiency

Hugh. J. Gibson[1*], Oleg Cojocari[2], Ion Oprea[2]
*1 Gibson Microwave Design EURL *, Antony 92160, France*
2 ACST GmbH, 63457 Hanau, Germany
* Contact: hugh.gibson@radiometer-physics.de

***Abstract*— We present test measurements for a 600GHz tripler with an output power in excess of 5mW and efficiency close to 6%. Simulations were presented at the ISSTT 2015 "A 600GHz high power tripler for space applications". The design uses a new method of diamond-film attachment, which allows very effective heat-sinking of the Schottky varactor diodes to create a high power tripler which could be used to drive a super-THz multiplier, or a sub-harmonic mixer for space use. Thermal management was a primary concern as we believe this is the major limiting factor for the high efficiency of the device at this frequency and power. The RF and thermal aspects of the design will be reviewed and compared with the measured results. As far as we are aware, this is the highest power single multiplier device (with no power combination) yet made at this frequency. We believe output powers of nearly 8mW from a single device may be possible, given sufficient input power.**

INTRODUCTION

Schottky varactor multipliers have a very long history yet they remain the only way to produce power at the THz range that is reliable, monochromatic, compact, efficient and not reliant upon cooling. Producing useable amounts of power at THz frequencies requires very high power driver stages in the sub-THz range. This is often just as challenging as the final, highest frequency stage and is the subject of this paper.

Our novel design allows relatively high powers to be reached in the frequency ranges from 200GHz – 700GHz. This frequency range and power remains out of reach of high power GaAs or GaN MMIC amplifier technology, or indeed of conventional multipliers, due to the very high input power required. Using such a high powered driver multiplier, one can easily provide in excess of 1mW at frequencies over 1THz with a suitable final multiplier. Power combination using multiple devices remains possible, further boosting output power. The same technology can be used to create a wider band design at the expense of some efficiency.

The main activity of this work was to analyse the limits of high power, high frequency varactor sources. We wanted to test assumptions made during the simulations that multipliers are essentially thermal limited (given adequate breakdown voltage) and to verify the role of velocity saturation in combination with overheating in our simulations. Support was from ESA AO/1-6649.

TRIPLER RECAP

The tripler design has already been discussed in ISSTT 2015 "A 600GHz High Power Tripler for Space Applications". The design is quite standard balanced, 4-anode design, with DC bias. The three unusual aspects of this design are using a thin diamond membrane to dissipate heat away from the anodes, a thin low-dielectric support membrane and large (but thin) heat-dissipative mesas with a direct heat-path to the diamond. Each of these modifications is less effective in isolation, but in combination allows very effective thermal management.

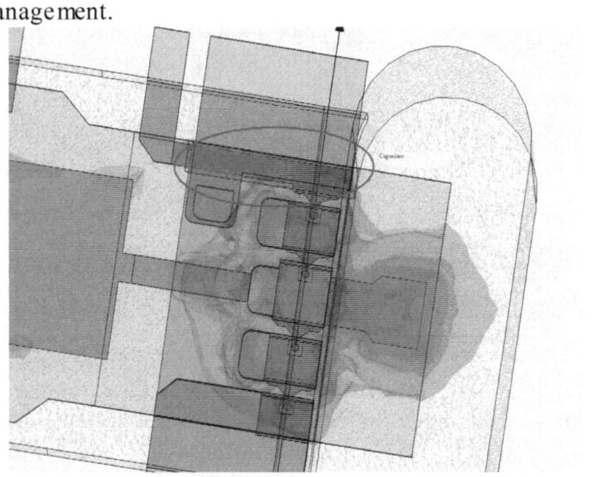

Fig. 1 The tripler shown with gold straps to the CVD diamond. The area circled is a single-layer capacitor for DC bias. Contours are 600GHz E-field

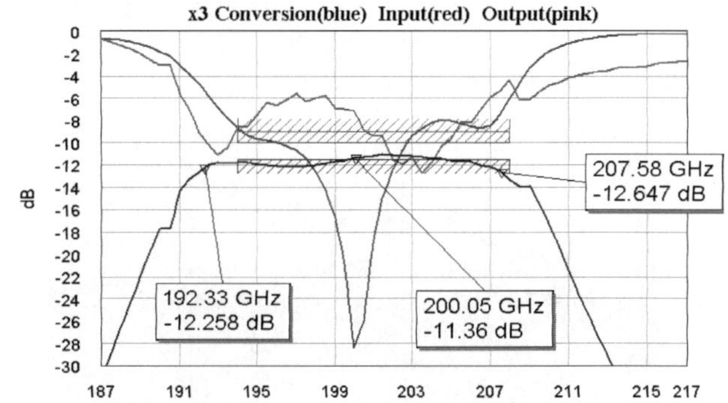

Fig. 2 Simulated conversion efficiency and input and output match (dB scale) for an input power of 100mW and 10V bias

MEASUREMENT RESULTS

Manufacture of the THz-MIC was done by ACST GmbH, using their modified FD-processTechnology details will be covered in other papers by Ion Oprea (ACST).The main goal of this project was to demonstrate that very high output power was possible at these high frequencies using standard varactor technology if one had a good design and one could maintain low anode temperatures.

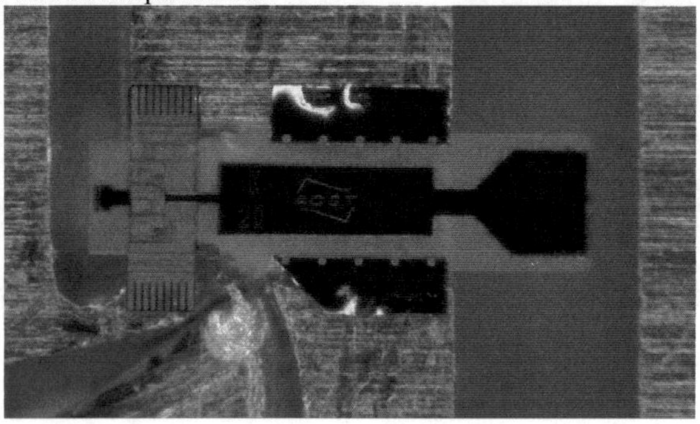

Fig. 3 Device mounted in the split-block, ready for testing

Testing took place at RPG Radiometer Physics, Nov. 2015 using a specially fabricated driver chain, producing close to 100mW at 200GHz [1].

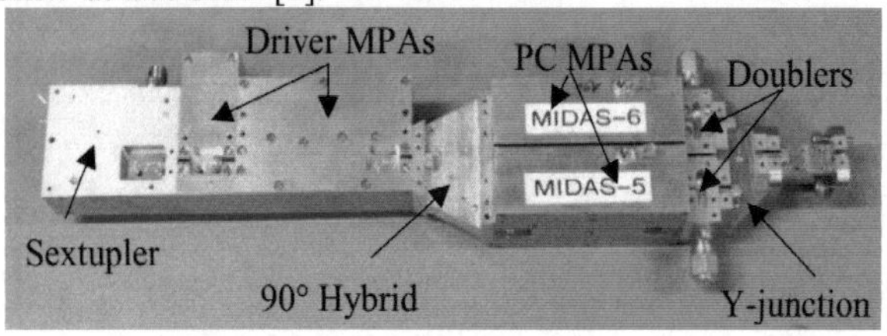

Fig. 4 Power combined amplifier/doublers for the high power 200GHz test source (courtesy of Bertrand Thomas, RPG GmbH)

At 100mW input power, the anode temperatures were expected to rise in temperature by about 60°C from room-temperature. Maximum anode temperature was expected to be 390K. While this may still seem high, it is greatly reduced compared to normal anode temperatures and emphasizes the large thermal problem faced. After using a diamond heatspreader, the thermal conductivity of gold is the new thermal limitation.

Fig. 6 Double-directional coupler and harmonic mixers, providing full input match characterization (similar to a VNA)

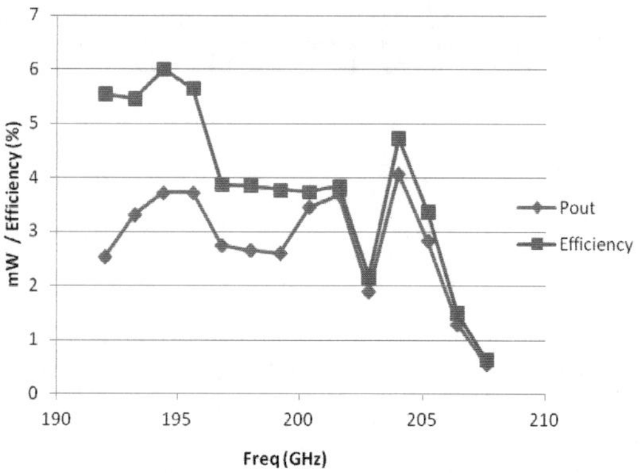

Fig. 7 Plot of measured Output power (blue, mW) and Efficiency (red, %). Frequency scale is for the tripler input.

Output power was measured using an Erickson PM4 power meter with no corrections applied, either for waveguide losses or power-meter calibration. Maximum power was seen at around 610GHz, with 4.96mW measured. After power-meter corrections this corresponds to well over 5mW.

Fig. 7 For output power measurements, the PM4 meter is directly connected to the tripler using a taper-transition to overmoded WR10.

COMPARISON BETWEEN MEASURED AND SIMULATED

The multiplier at first glance provided quite acceptable output power, but detailed comparison between simulation (Fig 2) and measurement (Fig7) shows a major discrepancy in the efficiency which seems to be optimal only at the lowest frequencies. This was especially strange, as the value of Cj0 of the fabricated THz-MIC was smaller than ideal, so we expected a high frequency shift (see later section "block manufacture error").

Detailed measurements were taken at the highest power point (610GHz), as the principal objective of the project was to confirm simulation and measurement at high powers. A plot was taken of output power, against bias point. This same plot shows the value of DC current.

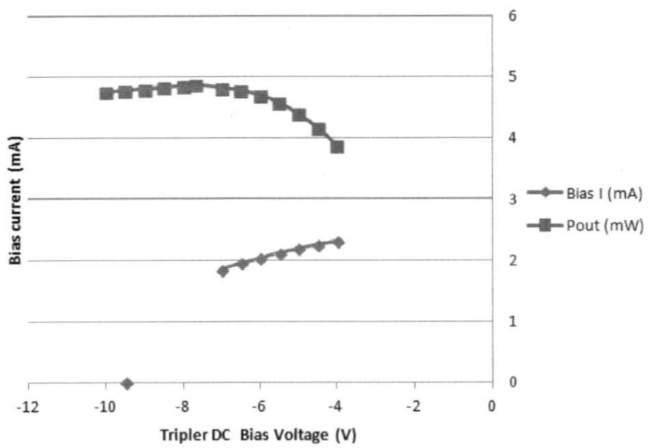

Fig. 8 Measured output power (at 610GHz) vs. DC bias, showing output power (red) and rectified DC bias current (blue)

Maximum efficiency occurs when voltage peaks are just staring to forward bias the varactors and a small rectified DC current is seen (as expected see Fig 8). What was interesting to us was that the efficiency remained quite respectable, even at lower bias voltages. Indeed, we tested the device with even lower voltages (and higher currents) and performance was less impaired than expected. This might be a useful indication that we can use even fewer anodes and live with an abnormally high bias current (to avoid the voltage breakdown limitations). The tripler seems extremely robust. It was tested it with high currents (several mA) and also with DC bias over 12V with no incident. The device remained fully functional throughout and resisted all attempts to destroy it.

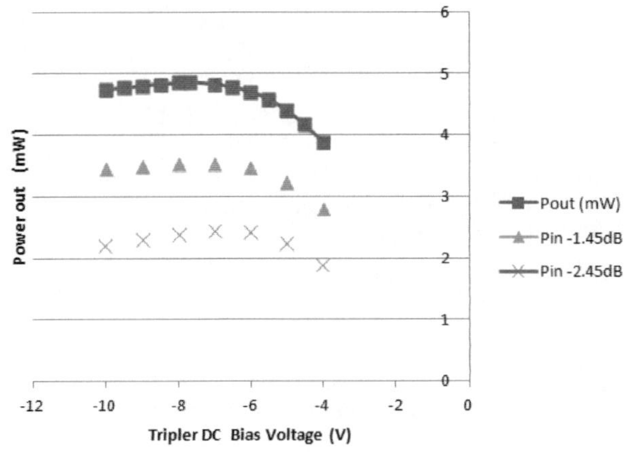

Fig. 9 Power output vs. Bias voltage plotted for different input powers

Another test performed was to reduce the input power and measure output power against DC bias voltage (Fig 9). The data shows the optimal DC bias point reducing with reduced input power (as expected). At full power, -8V bias is optimal, whereas at near ½ power, -6V is better. All these power curves are very similar to predicted and give us confidence in our thermal and RF models. However, the efficiency plots against frequency do not look correct, and indeed an error was subsequently found:-

MACHINING ERROR DISCOVERY AND SIMULATION

On further inspection of assembly photos, a fabrication error was discovered in the block, making the input waveguide backshort too long.

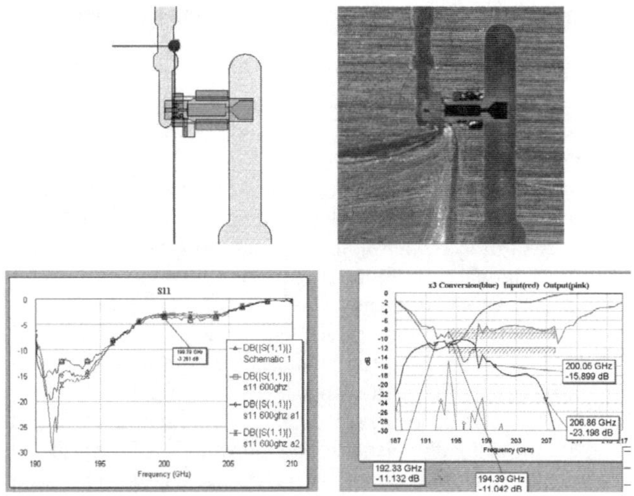

Fig. 10 machining error discovered, and updated simulation including error

The bottom left plot of Fig10 shows the measured S11 for the tripler, using the measurement setup from Fig 6. Below 197GHz, the input mach is good and efficiency is high (using the VNA measurement in Fig 6). After simulating this machining error (and also the lower Cj0 of the MMIC), we saw an almost exact duplication of this input match (red curve, bottom right of Fig 10).

We then plotted the complex impedance of the input match on a Smith chart (de-embedding the correct amount to set the reference phase to the same position as the simulation). The correlation between simulation (left) and measured (right) in Fig 11 was very close. The marker point is the same frequency.

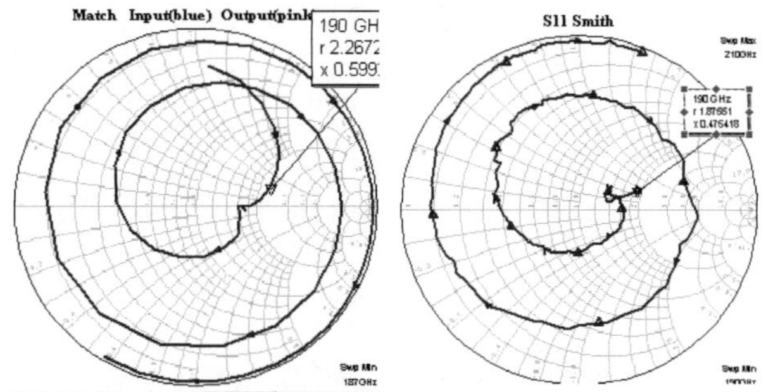

Fig. 11 S11 (input of tripler) simulated (left) and measured (right)

We will build a corrected block and measure again, as we expect still better performance with all parameters correct. Unfortunately, this was not possible in the time available before the ISSTT conference. As the main point of the project was to verify the simulation models and to test the output power at 600GHz, all goals were reached, even with a machining error, and also with an MMIC with less Cj0 than optimal.

FOLLOW-UP

We saw in the measurement data that biasing the device at quite low voltages and accepting quite high DC rectified current was less detrimental than we expected. A high power tripler, using just 2 anodes, well heat-sunk using the grounded beam-leads would seem a logical follow-up, as one can achieve this without requiring diamond heat-sinking for the central diodes. The ACST varactors have extremely high break-down voltages, exceeding -10V per anode. This is a key feature in allowing relatively few diodes to be used, which reduces losses even further and simplifies the structure considerably. The use of very thin mesas, with thickness of approximately 1 skin-depth (in GaAs) would also seem to be very helpful. The correct choice of doping concentration is also obviously very important. A design which is optimised for as large an anode diameter as possible is also beneficial, as this helps reduce the current density in the GaAs, and keep the critical series resistance value low

Observed efficiency reduction in similar devices, which have traditionally been attributed to purely velocity saturation effects may be a mixture of effects, including overheating. Carrier mobility and velocity saturation is strongly temperature dependent, and partly explains why Schottky multipliers and mixers perform substantially better when cooled. We therefore believe there is every advantage in keeping anodes as cool as possible. This aspect seems often overlooked in current designs.

CONCLUSIONS

A single (no power combining) 600GHz tripler has been constructed with over 5mW of output power. As far as the author's knowledge, this is the highest power device yet made at this frequency. Efficiency of near 6% is seen and is completely in agreement with the simulations of the device (when taking into account machining errors) and also in-line with other similar reported devices for narrow-band use from other groups (440GHz x3 with 8% efficiency [2]). We expect still higher output powers with a correctly machined block and when the tripler can be driven by a higher powered 200GHz source (when available). A maximum output power of nearly 8mW is expected before the device is thermally or breakdown-voltage limited.

ACKNOWLEDGMENT

The project was funded from ESA project AO/1-6649 under the guidance of Tapani Narhi and Petri Piironen. The 200GHz source was designed by Bertrand Thomas. All hardware (except the THz-MIC) was built by RPG Radiometer Physics GmbH. Special thanks for the skills of A. Walber, J-H Freerksema for testing assistance at RPG, to N. Bauer for skilful THz-MIC mounting and the fine machining skills of B. Elsner and others at the RPG workshop.

REFERENCES

1. B. Thomas H.Gibson M. Phillip A. Walber I. Opera O. Cojocari T. Narhi "*Compact Power-combined LO Sources for SWI Onboard JUICE*" ESA/ESTEC 36th Antenna Workshop October 2015
2. E.W. Bryerton J.L. Hessler S.A. Retzloff T.W. Crowe "*874-GHz Heterodyne Cuesat for Cloud Ice Measurement –Flight Model Data" ISSTT 2016 Nanjing China*

Broadband Direct-Machined Corrugated Horn for Millimeter Observation

Shigeyuki Sekiguchi*†, Masahiro Sugimoto†, Kenji Mitsui†, Shibo Shu*†, Yutaro Sekimoto*†, Norio Okada†, Kouichi Kubo†, Toshikazu Takahashi†, Tom Nitta‡
* Department of Astronomy, The University of Tokyo, Hongo, Tokyo, Japan
† National Astronomical Observatory of Japan, Mitaka, Tokyo, Japan
‡Division of Physics, Faculty of Pure and Applied Sciences, University of Tsukuba, Tsukuba, Ibaraki, Japan
§ Graduate School of Science and Engineering, Saitama University, Saitama, Japan

Abstract—**We developed a broadband 4 pixels corrugated horn array with simple corrugation design. Because of the simple geometry of corrugations, the horn array can be directly machined from a bulk of aluminum with an end-mill. The design achieved over an octave-band (80 - 180 GHz) with good beam shape and -20 dB cross-polarization have been realized at all frequencies. All horns of the array have got similar beam pattern and return loss with simulation. This design provides an octave bandwidth of the corrugated horn array at reasonable cost.**

I. Introduction

Corrugated horn has been widely used in the millimeter-wave astronomy as a cryogenic feed, thanks to its good symmetric beam, low sidelobe, and low cross polarization [1]. It has been deployed for frequency band receivers of Atacama Large Millimeter/submillimeter Array (ALMA) [2]–[4]. A broader bandwidth, easier methods of array fabrication, and manufacturing with low cost are required for the large focal plane feeds.

Bandwidths of corrugated horns used in millimeter-wave astronomy are usually less than $\sim$ 40 %. A ring-loaded structure expanding the internal corrugation has been used as one of the technique to extend the bandwidth [5]. An octave bandwidth ring-loaded corrugated horn array fabricated by Si platelet with micro-machining has been demonstrated [6].

To increase mapping speed of observation, an array of feeds is necessary [7]. Platelet or stacking thin sheets technique is one of the method to fabricate corrugated horn array easily [8]. Several research groups have developed the corrugated horn array by platelet or stacking of the silicon or aluminum [9]–[11]. However, it needs to bind all plates tightly so that the weight of the platelet corrugated horn is limited.

It is preferable to manufacture a corrugated horn array at reasonable cost, in order to cover a larger focal plane area with a wide field-of-view. Electroforming is a primarily corrugated horn fabrication process, which manufactures horn shape accurately [1]. However, it is hard to fabricate corrugated horn array with electroforming since it takes time and cost.

We have developed a millimeter wave corrugated horn array which overcomes these items.

II. Design

We have designed broadband conical corrugated horn with 80 - 180 GHz bandwidth. The cross-section of horn is shown in Fig.1. Horn aperture diameter is 12.4 mm and horn backside is a circular waveguide with the 2.4 mm diameter. The broadband bandwidth has been achieved with simple plane 24 corrugations which have thin grooves and tooth structure.

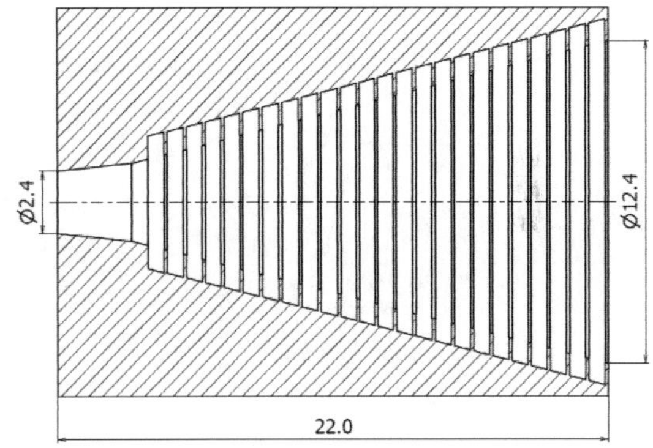

Fig. 1. Cross-section of a broadband corrugated horn. The unit of length is millimeter.

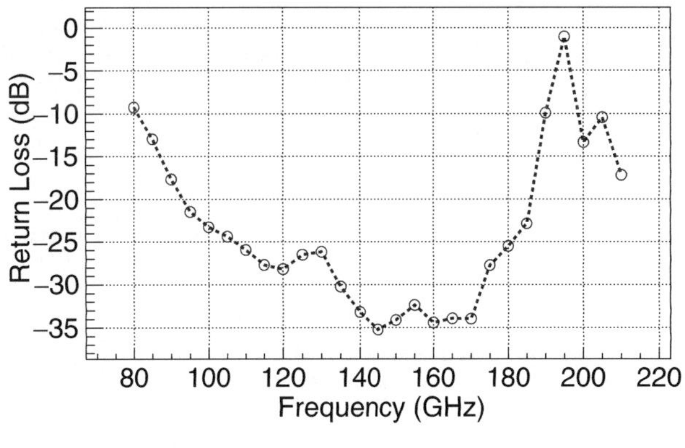

Fig. 2. Return Loss of a broadband horn.

Horn properties, return loss and beam pattern, are simulated using ANSYS HFSS software [12]. Return loss shown in Fig.2 is realized less than -15 dB level in most design frequency. Beam pattern, in Fig.3, is also simulated every 10 GHz in 80 - 180 GHz range. E-plane, H-plane of co-polar and diagonal plane of cross-polarization are superimposed. This simulation shows excellent performance in terms of broadband, symmet-

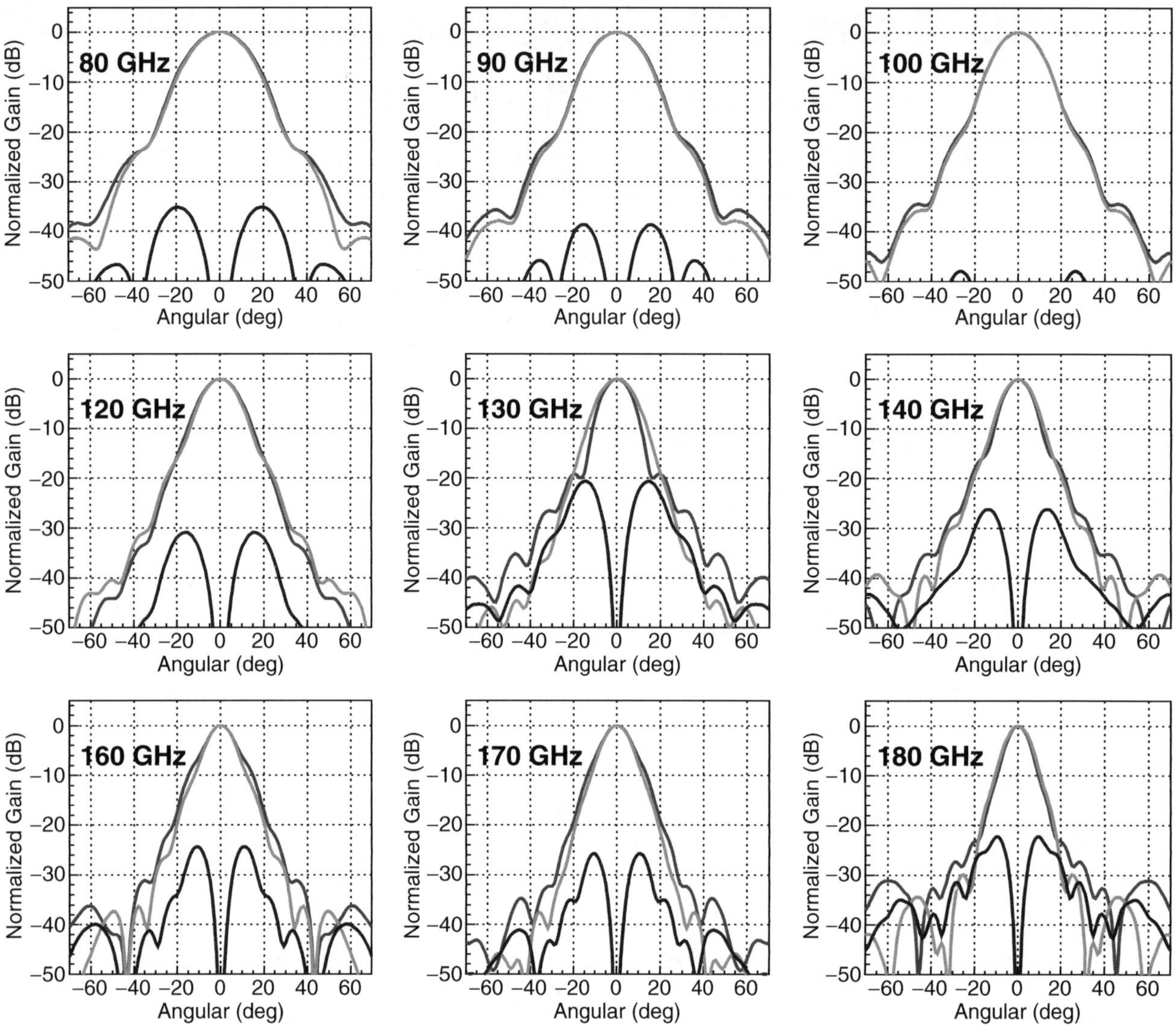

Fig. 3. Simulated beam pattern of an 80 - 180 GHz horn. Green, red and blue lines are E-plane and H-plane of Co-polarization and Cross-polar diagonal plane, respectively.

ric beam pattern, low cross polarization, and low side lobes.

III. Manufacturing

A corrugated horn array has been fabricated by direct machining in the mechanical engineering shop of the National Astronomical Observatory of Japan. The corrugations are shaved by single special end-mill from the metal bulk because the pattern is only simple plane. The material of the horn is aluminum (A6061), which is suitable for cryogenic use. Aluminum is easy to manufacture and its not only weight also thermal conductivity and thermal capacity help us to reduce the cooling time of focal plane. Fig.4 shows the fabricated 4 pixel 80 - 180 GHz corrugated horn array.

The choke structure for coupling the planar OMT has been machined on the circular waveguide side. Since the critical temperature of aluminum is around 1.2 K, it works as a superconducting magnetic shield when the corrugated horn array is coupled with superconducting detectors such as Microwave Kinetic Inductance Detectors (MKID). It takes

Fig. 4. 4 pixel 80 - 180 GHz horn array.

about one hour to cut the corrugations of one horn. Thus, it is possible to fabricate the horn array at reasonable cost.

IV. Prototype Detector

A prototype 80 - 160 GHz horn array detector coupled with MKID has been also designed. A large focal plane with this corrugated horn array has been designed for CMB B-mode observation [13]. The signal coming from horn is received by a planar ortho-mode transducer (OMT) antenna and detected by MKIDs. The OMT, MKIDs and other circuits fabricated on a silicon on insulator wafer are reported in a separate paper [14]. It detects 2 bands (80 - 115 GHz and 125 - 160 GHz) and dual polarization signals in single pixel horn. The size is 30 mm × 30 mm × 27 mm and total mass is around 50 g. For a prototype of large focal plane camera, we plan test observations using this detector with Nobeyama 45 m radio telescope.

Acknowledgment

This work was supported by MEXT/JSPS KAKENHI Grant Numbers 25247022, 15H05743, 15H05891.

References

[1] P. J. B. Clarricoats and A. D. Olver, Corrugated horns for microwave antennas. IEE Electromagnetic Waves Series 18, 1984.

[2] K. Kimura, H. Iwashita, S. Asayama, M. Sugimoto, G. Kikuchi, and H. Ogawa, "Antenna performance of a directly dug corrugated feedhorn for the 150-GHz band," *Int. J. Infrared Millim. Waves*, vol. 29, no. 8, pp. 713–723, 2008.

[3] M. Naruse, T. Ito, Y. Sekimoto, H. Toba, N. Satou, M. Sugimoto, W. Shan, Y. Iizuka, T. Kamba, M. Kamikura *et al.*, "Near-field beam pattern measurement of qualification model of alma band 8 (385–500 ghz) cartridge receiver," *Exp. Astron.*, vol. 24, pp. 89–107, 2009.

[4] A. Gonzalez and Y. Uzawa, "Tolerance Analysis of ALMA Band 10 Corrugated Horns and Optics," *IEEE Transactions on Antennas and Propagation*, no. 7, pp. 3137–3145, jul.

[5] F. Takeda and T. Hashimoto, "Broadbanding of corrugated conical horns by means of the ring-loaded corrugated waveguide structure," *IEEE Trans. Antennas Propag.*, vol. 24, no. 6, pp. 786–792, 1976.

[6] J. P. Nibarger, J. A. Beall, D. Becker, J. Britton, H.-M. Cho, A. Fox, G. C. Hilton, J. Hubmayr, D. Li, J. McMahon *et al.*, "An 84 pixel all-silicon corrugated feedhorn for CMB measurements," *J. Low. Temp. Phys.*, vol. 167, no. 3, pp. 522–527, 2012.

[7] S. Padin, "Mapping speed for an array of corrugated horns." *Applied optics*, vol. 49, no. 3, pp. 479–83, jan 2010.

[8] R. W. Haas, D. Brest, H. Mueggenburg, L. Lang, and D. Heimlich, "Fabrication and performance of MMW and SMMW platelet horn arrays," *Int. J. Infrared Millim. Waves*, vol. 14, no. 11, pp. 2289–2293, 1993.

[9] J. W. Britton, J. P. Nibarger, K. W. Yoon, J. A. Beall, D. Becker, H.-M. Cho, G. C. Hilton, J. Hubmayr, M. D. Niemack, and K. D. Irwin, "Corrugated silicon platelet feed horn array for CMB polarimetry at 150 GHz," in *Proc. SPIE*, vol. 7741, p. 77410T, 2010.

[10] F. Del Torto, M. Bersanelli, F. Cavaliere, A. De Rosa, O. D'Arcangelo, C. Franceschet, M. Gervasi, A. Mennella, E. Pagana, A. Simonetto *et al.*, "W-band prototype of platelet feed-horn array for CMB polarisation measurements," *J. Instrum.*, vol. 6, no. 06, p. P06009, 2011.

[11] L. Lucci, R. Nesti, G. Pelosi, and S. Selleri, "A Stackable Constant-Width Corrugated Horn Design for High-Performance and Low-Cost Feed Arrays at Millimeter Wavelengths," *IEEE Antennas Wirel. Propag. Lett.*, vol. 11, pp. 1162–1165, 2012.

[12] HFSS, http://www.ansys.com/Products/Electronics/ANSYS-HFSS.

[13] Y. Sekimoto, A. Dominjon, T. Noguchi, N. Okada, K. Mitsui, Y. Obuchi, M. Sekine, S. Sekiguchi, S. Shu, T. Nitta, and M. Naruse, "Design of MKID focal plane array for LiteBIRD," in *Proceedings of ESA antenna workshop*, 2015.

[14] S. Shu, S. Sekiguchi, M. Sekine, Y. Sekimoto, T. Nitta, and A. Dominjon, "Development of Octave-band Planar Ortho-Mode Transducer with MKID for LiteBIRD Satellite," in *Proceedings of The 27th International Symposium on Space Terahertz Technology*, 2016.

The Global phase grating

F. Defrance[1]*, M. Casaletti[2], J. Sarrazin[2], M.C. Wiedner[1], H. Gibson[3], G. Gay[1], R. Lefèvre[1], Y. Delorme[1]

[1]*Observatoire de Paris, Laboratory for Studies of Radiation and Matter in Astrophysics, 61 av de l'Observatoire, 75014 Paris, France*

[2]*Laboratoire d'Electronique et Electromagnétisme, Sorbonne Universités, UPMC Univ Paris 06, UR2, L2E, F-75005 Paris, France*

[3]*Gibson Microwave design, 9 rue Pasteur, 92160 Antony, FRANCE*

**fabien.defrance@obspm.fr*

Abstract—**In heterodyne array receivers, phase gratings are useful to divide the local oscillator (LO) signal into several beams to pump the mixers of an array. We have developed a new iterative algorithm to generate phase gratings without any geometrical constraints. Two prototypes (a reflective one and a transmissive one) were fabricated at 610 GHz and their splitting efficiency validates the design, the simulation, and the manufacturing process of these phase gratings.**

INTRODUCTION

Early THz heterodyne systems concentrated on high spectral resolution and large bandwidth (with many spectral channels), but they usually only had one spatial pixel (i.e. HIFI instrument on Herschel satellite [1]). Recently, arrays of heterodyne receivers have been developed to simultaneously measure spectra at several positions in the sky. In heterodyne receivers, each pixel has a mixer. So, observing with several pixels involves pumping all these mixers with an LO. The most efficient way of doing it is to split the LO beam into several beams to specifically illuminate each mixer.

Phase gratings are the perfect tool for achieving this goal and they are already used in some heterodyne receivers such as CHAMP or upGREAT [2]. However, the existing phase gratings for THz frequencies, the stepped and Fourier gratings [3,4], have a constrained geometry and are limited in the beam patterns they can efficiently produce. For the next generation of THz heterodyne receivers we need a phase grating able to efficiently produce any kind of beam pattern above 1 THz.

Stepped gratings, such as Dammann gratings, have discrete steps able to shift the phase of the signal. Fourier gratings [5,6] use a spatial phase modulation given by Fourier series expansion. They are smoother than stepped gratings and can usually reach higher efficiency. Because these two kinds of phase gratings have geometrical constraints, the efficiency and the far-field pattern that can be generated is limited. To overcome this limitation, the algorithm presented hereafter has been developed to be able to design phase gratings without any geometrical constraints, and reach a good efficiency for any far-field distribution. To validate this method, we have designed, simulated and built two Global phase grating prototypes (one in transmission and the other in reflection) able to split the LO beam into four beams.

DESIGN PROCESS

The design process of the phase grating is based on an iterative algorithm which calculate the shape of the grating. Then, this profile is simulated with an electromagnetic software in order to evaluate its efficiency and improve it if possible.

A. Phase profile design

The iterative algorithm is based on alternate projections using an inverse Fourier transform and a Fourier transform. It converges to an aperture distribution that matches both the desired radiated field pattern and the grating's physical constraints. This iterative algorithm has been used to generate the phase profile for two phase grating prototypes (a reflective one and a transmissive one) generating four beams of similar intensity at 610 GHz, making an angle of ±12.6° with respect to the specular reflection axis.

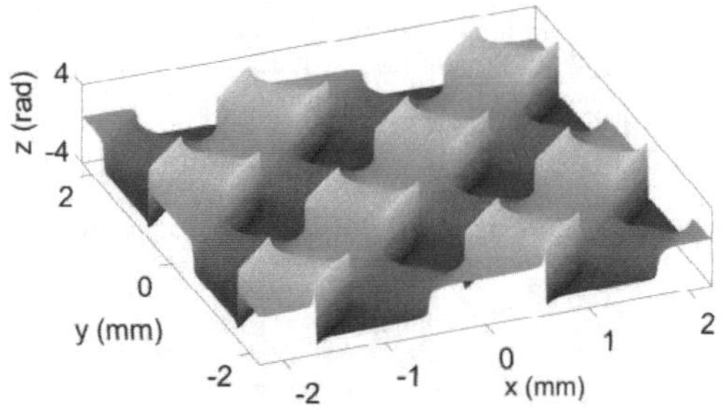

Fig. 1 Phase grating profile calculated by the iterative algorithm.

Both prototypes are based on the same phase profile (fig. 1), but are differently scaled. The transmissive grating is made of TPX® and is illuminated by an orthogonal incident beam, while the reflective phase grating is made of brass and is illuminated by an oblique incident beam making an angle of 25° with the normal of the grating.

B. Electromagnetic simulation

The transmissive and reflective grating prototypes have been simulated with a commercial electromagnetic simulation software Feko™, which is based on the Method of Moments (fig. 2). In both cases, only a part of the gratings has been simulated (because of limited computing power). So, the results of the simulation can only predict the diffracted beams' efficiency but not the beams' widths. Both simulations used the Multilevel Fast Multipole Method (MLFMM).

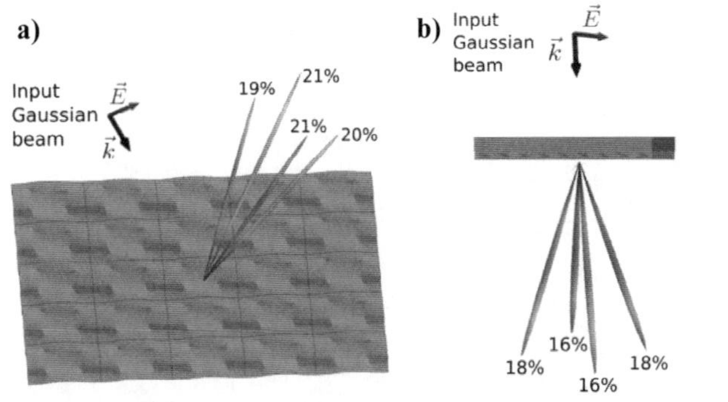

Fig. 2 Far-field beam patterns calculated by the electromagnetic simulations in reflection (a) and in transmission (b).

The efficiency predicted by the simulation is 81 % for the reflective grating and 68 % for the transmissive grating.

TEST OF THE GRATINGS

A. Mechanical test

The phase grating prototypes were milled by a 100 μm diameter end-mill with a surface accuracy of 6 μm (fig. 3). The size of the reflective grating is 44.8 mm x 49.4 mm and the transmissive grating is circular with a diameter of 44.8 mm. In both cases these dimensions were chosen to be large compared to the incoming Gaussian beam, whose beam waist is 10 mm.

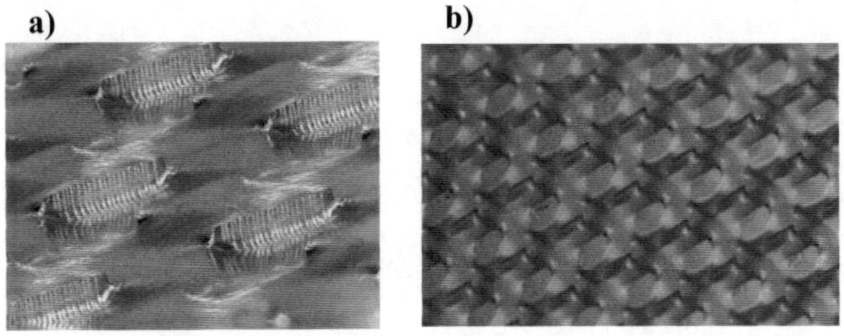

Fig. 3 Pictures of the manufactured (a) reflective and (b) transmissive gratings taken with a microscope.

B. Efficiency test

The prototypes were tested with a 610 GHz source and a Golay cell power meter to measure the intensity of the output beams. The source and the phase gratings were positioned on a rotating platform and the horizontal radiation pattern of the gratings was measured by rotating the platform. The distance between the Golay cell power meter and the LO was kept constant, so the measured beams had always the same radius and were comparable. The power measured by the power meter with and without the phase gratings was compared to the calculated relative intensity of each output beam, as well as to the total efficiency of the gratings (fig. 4). The measured efficiency of the prototypes is 78 ± 4 % for the reflective grating and 62 ± 4 % for the transmissive grating. These efficiency values are quite good and very close to the ones predicted by the electromagnetic simulations.

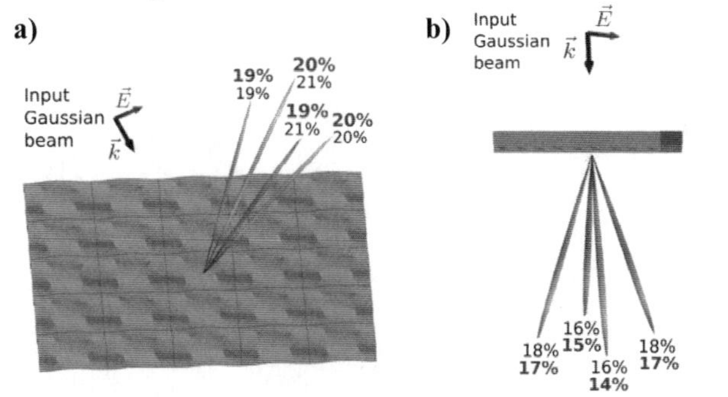

Fig. 4 Far-field beam intensities measured (in red) and compared to the simulation results, in reflection (a) and in transmission (b).

CONCLUSION

The two first Global phase grating prototypes are successful in dividing the LO beam into 4 similar intensity beams. Moreover, the experimental beam intensities are very close to the simulation predictions (within 2 %). These good results validate the design and fabrication processes of this new kind of grating. Therefore, Global gratings are ready to be used in the next generation of array receivers.

ACKNOWLEDGMENT

This work was funded by the National Centre for Space Studies (CNES) and the Laboratory for Studies of Radiation and Matter in Astrophysics (LERMA).

REFERENCES

[1] F. P. Helmich et al., "Herschel-hifi: The heterodyne instrument for the far-infrared," *EAS Publications series*, vol. 52, pp. 15–20, 2011.

[2] C. Risacher, R. Güsten, et al., "The upGREAT heterodyne array receivers for far infrared astronomy," *39th International Conference on Infrared, Millimeter, and Terahertz waves (IRMMW-THz)*, pp. 1–2, 2014.

[3] H. Dammann and E. Klotz, "Coherent optical generation and inspection of two-dimensional periodic structures," *Optica Acta*, vol. 24(4), pp. 505–515, 1977.

[4] J. N. Mait, "Design of Dammann gratings for two-dimensional, non-separable, non-centrosymmetric responses," *Opt. Lett.*, Vol. 14, pp. 196-198, 1989.

[5] U. U. Graf and S. Heyminck, "Fourier gratings as submillimeter beam splitters," *IEEE Trans. on Antennas and Propagation*, vol. 49(4), pp. 542–546, 2001.

[6] S. Heyminck, U. U. Graf, et al., "Array-receiver lo unit using collimating Fourier-gratings," *12th International Symposium on Space Terahertz Technology*, pp. 563–570, 2001.

[7] Victor Arrizón, Markus Testorf, Stefan Sinzinger, and Jürgen Jahns, "Iterative optimization of phase-only diffractive optical elements based on a lenslet array," *J. Opt. Soc. Am. A*, vol. 17, pp. 2157-2164, 2000.

[8] R. W. Gerchberg, W. O. Saxton, "A practical algorithm for the determination of the phase from image and diffraction plane pictures," *Optik*, vol. 35(2), pp. 237-246, 1972.

Modal Analysis of Far-Infrared Multimode Horns and Waveguides for Ultra-Low-Noise Detectors for Astronomy

Jiajun Chen, Stafford Withington, Christopher N. Thomas

Cavendish Laboratory, University of Cambridge, CB3 0HE, Cambridge, UK
Contact: j.chen@mrao.cam.ac.uk

We have developed the theory and the numerical procedures for modelling optical systems of ultra-low-noise detectors for far-infrared and submillimeter astronomy. A variable temperature blackbody load has been engineered to test the optical efficiencies of ultra-low-noise Transition Edge Sensors using these procedures. Multimoded horns and waveguides are used in the experiment. To improve on initial simulations and have a more comprehensive understanding of the experimental results, we have run a set of simulations to examine the loss due to mis-matching of waveguide modes and the attenuation due to the waveguide itself, and hence to understand how they would affect the throughput of the waveguide.

This method enables us to assess all performance information for multimoded optical systems easily and accurately, while most commercial simulators are not able to handle multi-mode simulations. It is also fast to run over wide frequency ranges.

Research on High Precision Antenna For DATE5

Zheng Lou[1,2,*], Yingxi Zuo[1,2], Yuan Qian[1,2], Hairen Wang[1,2], Ji Yang[1,2], Jingquan Cheng[1,2],and Shengcai Shi[1,2]

[1]*Purple Mountain Observatory, Chinese Academy of Sciences, Nanjing, 210008, China*
[2]*Key Laboratory for Radio Astronomy, Chinese Academy of Sciences, Nanjing, 210008, China*
*Contact: zhenglou@pmo.ac.cn, phone +86-25-83332213

Abstract—**Dome A 5m Terahertz Explorer(DATE5) is a proposed telescope to be deployed at Dome A, Antarctica to explore the excellent terahertz observation condition unique to the site. The telescope needs to realize and maintain an overall reflector surface accuracy of 10μm rms and a blind pointing accuracy of 2 arcseconds under the extreme site conditions and unmanned operating mode. Two candidate antenna designs have been proposed, one of which is based on an all-CFRP reflector and slant-axis mount and the other based onaluminum panels on carbon fiber backup structures and altazimuth mount. Both aluminum and CFRP sandwich prototype panels have been fabricated and tested in a climate chamber. The aluminum panel shows desired surface accuracy from room temperature down to -60°C. CFRP panels with different sandwich structures are fabricated and tested. All of them achieve a surface accuracy around 5μm rms at room temperature, but their surface accuracy differs significantly when they are cooled, indicating the importance of panel structural design. The aluminum and CFRP panels are equipped with different type of de-icing heaters, and their performances have been verified respectively. Structural and thermal analyses on the overall antenna shows that both of the two candidate designsmeet the specifications of DATE5. Moreover, demonstration experiments on near-field radio holography are performed and a repeatibility accuracy of 3μm rms is achieved.**

INTRODUCTION

Dome A 5m Terahertz Explorer (DATE5) is a proposed telescope to be deployed at Dome A, Antarctica to explore the excellent terahertz observation condition unique to the site [1]. One of the key challenges of the telescope is to realize and maintain the required 10micronsrms overall reflector surface accuracy under the extreme site conditions at Dome A[2].To realize this objective, some key technologies of the telescope need to be studiedincluding prototyping of reflector panels and corresponding environmental experiments, de-icing techniques, and surface profile measurement technology, etc. In the preliminary design phase of DATE5, two candidate designs for the main reflector were proposed. Thefirst design is based on aluminum panels on carbon fiber reinforced polymer (CFRP) backup structures.In the second design, panels are made ofsandwich structures with CFRP front and back skins and an aluminum honeycomb core, and the backup structures are made of CFRP truss. According to the surface error budget for DATE5, the manufacturing error of each reflector panel should not exceed 5 microns rms, and the gravitational and thermal deformations of the panels should be kept on the level of a few microns rms, which is very challenging considering the extreme site conditions at Dome A.

Prototyping of both aluminium and CFRP panels have been carried out to verify the corresponding manufacture technologies.The aluminumprototype panel (0.4m × 0.6m in size) was directly milling machined. Several CFRPprototype panels (1m× 0.6m in size) were fabricated on the same high-precision invar mould. The figuresof two types of panels wereboth measured on a three coordinate measurement machine (CMM).The measured surface errors are 3.2 and 5.1 microns rms at the room temperaturefor the aluminium and CFRP panels respectively. However, the thermal deformationof the two types of panelsdue to the large seasonal soak temperature variationis still not clear. To verify the performance of the panel under extreme cold environment, the surface figures of both prototype panelsweremeasured in a climate chamber whose interior temperature was varied from room temperature down to -60°C.

Reflector panels working at Dome A run high risks of icing[3], and active de-icing mechanisms has to be implemented on the panels. Aluminum and CFRP panels are equipped with different types of de-icing heaters, and their respective performances needs to beverified experimentally.

DATE5 will employnear-field radio holography to align the reflector panels. In order to eventuallyrealize anoverall surface accuracy of 10μm rms, the repeatability error of the holographic measurements should be less than 3μm rms.To evaluate the feasibility of near-field holography for DATE5, a dual-channel radio holography receiver operating at the 3-mm waveband were developed[4]. Demonstration experiments on a 1.45m test antenna were performed and the results were reported and analysed at the end of this paper.

PROTOTYPING FOR ALUMINUM PANEL

The prototype for the aluminum panel is a 400×600mm fan-shaped panel with back stiffening ribs. The panel is directly milling machined out of aluminum alloy (LC4) on a high speed machining centeras a single piece. After the fabrication process, the surface figure of the panel was first measured on a CMM. The measured surface is best fitted with a parabolic surface with a focal length slightly larger than 2000 mm and the residual error is 3.2 μm rms. The front and back view of the panel is shown in Fig. 1.

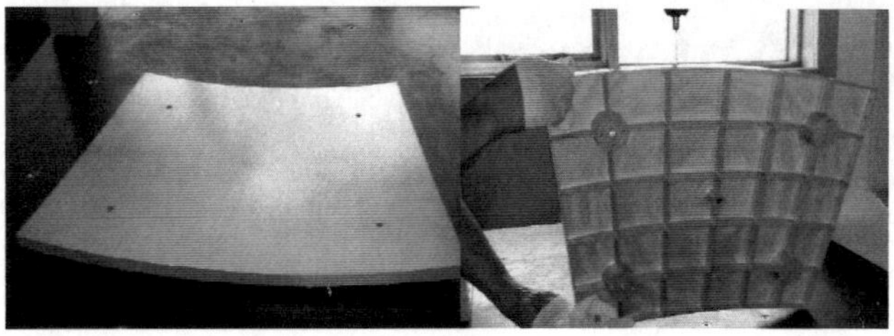

Fig. 1Prototype aluminium panels for DATE5

A. Surface Figure at Low Temperatures

Since the CMM cannot operate in a climate chamber under low temperatures, we measure the panel surface using photogrammetry. The measurement setup is shown in Fig.2.

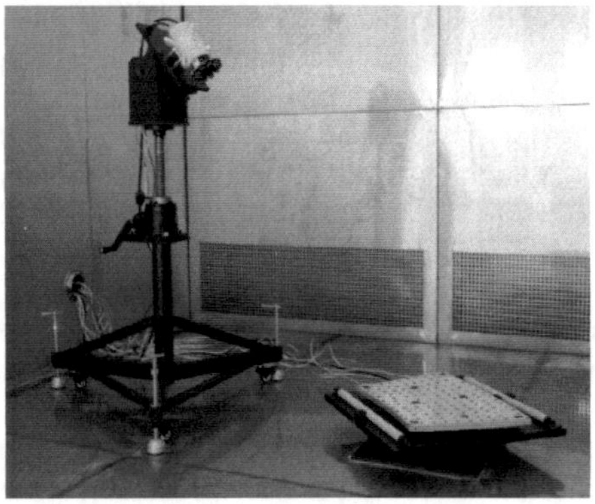

Fig. 2Experimental setup of the photogrammetry measurement in a climate chamber.

The photogrammetry measurement system demonstrated a repeatability error of 1.3, 1.4 and 1.0 microns rms in the x, y and z direction respectively at room temperature in the laboratory. However, when operating in the climate chamber at low temperatures the repeatability errors degrades to 3.6, 3.5 and 2.1 microns rms in the x, y and z direction respectively. The degradation is mainly due to the mechanical vibration and air turbulence in the climate chamber.

Panel surface figures are measured at several typical ambient temperatures from 25°C to -60°C.The measured surface rms errors before and after removing the defocusing component is listed in Table I for various ambient temperatures. It can be seen from TableI that at the lower temperatures the surface error increases only slightly after removing the defocusing component, indicating that the major contributors to the surface errors are the manufacture error and a defocusing error. No additional surface error caused by internal stress is observed under temperatures down to -60°C.

TABLE I RMS SURFACE ERROR UNDER VARIOUS AMBIENT TEMPERATURES FOR ALUMINUM PANEL (UNIT: MICRON).

Ambient Temperature	T=26°C	T=0°C	T=20°C	T=-40°C	T=-55°C
Fixed-focus Fitting (f=2000mm)	3.2	3.6	3.6	3.9	4.6
Best fitting	3.0	3.6	3.5	3.6	3.8

B. Panel De-icing

The anti-icing system implemented on the aluminum prototype panel composes of a 30 identical heater pads and a feeding network. The heaters are polyimide film heaters tailored into square pads fitting into the pockets of the panel back, as shown in Fig. 3. Such type of films is suitable to work at temperatures as low as -100°C.

Fig. 3Aluminum panel anti-icing system.

First, we test the performance of the panel anti-icing system at room temperature in the laboratory. An input heat flux of 60W/m^2 is applied to the panel and the back and side of the panel are thermally insulated to simulate the situation on the real telescope. The temperature distribution of the panel front surface obtained by an infrared camera is shown in Fig.4. The average temperature increase is measured to be 3.0°C and the peak-to-valley temperature difference across the panel is less than 0.5°C. Similar results are found at low temperatures in a climate chamber. In such cases, panel temperatures are measured by RTD sensors distributed across the panel.

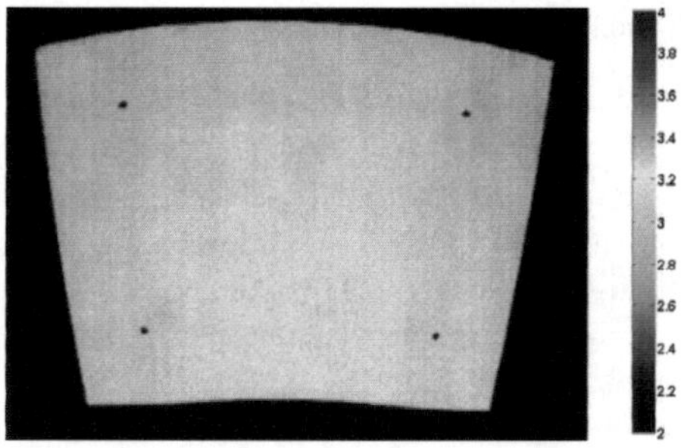

Fig.4Panel temperature increase over ambient.

PROTOTYPING FOR CFRP PANELS

The CFRP prototype panels are sandwich structures using thin CFRP front and back skins (2.5mm thick) and a honeycomb aluminum core (65mm thick), providing high specific stiffness[5]. Two CFRP prototype panels with different sandwich structures are fabricated, one with CFRP side plates on the edge of panel(Panel A) and the other without (Panel B), as shown in Fig.5.

Fig. 5The CFRP prototype panels. Left: with side seal, right: without side seal.

A. Surface Figure at Low Temperatures

The same photogrammetric measurement system as for aluminum panel is used. Panel surface figure are measured at several typical ambient temperatures from 40°C to -40°C with the results summarized in Table II.

TABLE IIRMS SURFACE ERROR UNDER VARIOUS AMBIENT TEMPERATURES FOR CFRP PANEL.

Ambient Temperature		T=40°C	T=15°C	T=-10°C	T=-40°C
Panel A	RMS Error (μm)	6.9	5.5	9.5	14.1
	Focal length(mm)	2004.6	1996.2	1985.1	1978.2
Panel B	RMS Error (μm)	8.1	7.3	7.9	8.8
	Focal length(mm)	2000.1	1994.8	1988.7	1980.9

It can be seen from Table IIthat at the lower temperatures the surface error of Panel B increases only slightly after removing the defocusing component. Instead, the surface error of Panel A increases considerably with the decreasing temperature which mainly concentrated on the edges of the panel, as shown in Fig. 6. It indicates that the large surface error of panel A is mainly caused by CFRP side plates. When the ambient temperature drops, the CFRP side plates and the aluminium core experience different amount of contraction in the panel normal direction, resulting large stress and deformation on the front skin.

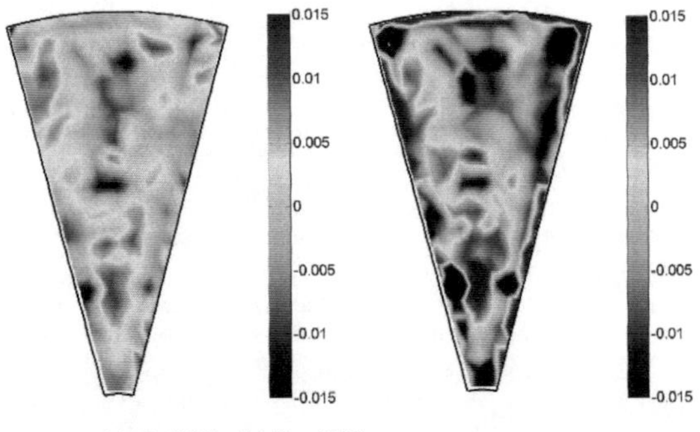

(a) T=15°C (b) T=-40°C

Fig.6Residual surface error for Panel A after best fitting with a parabolic surface.

Room temperature curing resin is used to reduce the thermal deformation in the replication process. However,the CTE of CFRP skin is still quite large (~$5 \times 10^{-6}K^{-1}$) because of the high volume ratio of resin, causinglarge variation of the focus length for both panelswhen the ambient temperature changes.We are now investigating on different carbon fiber and resin material to reduce the volume ratio of the resin.

B. Panel De-icing

The de-icing heaters are embedded between the upper skin and the aluminum coreof the panel by applying the same kind of resin system, as shown in Fig. 7. The heaters are made of a high conductive CFRP film which itself is laminated between the two isolation layers of glass fiber. High heating efficiency is realized because the distance between the heating film and the reflector surface is small.In order to heat the reflector surface in a uniform fashion, the heating film is segmented and patterned, connected by metal wire, as shown in Fig. 8.Little thermal stress will be generated when the ambient temperature varies due to the material uniformity of the heater and the skin material.

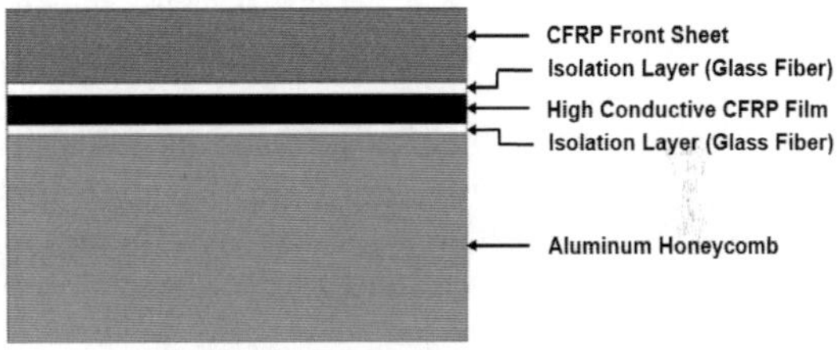

Fig. 7De-icing system of CFRP panel.

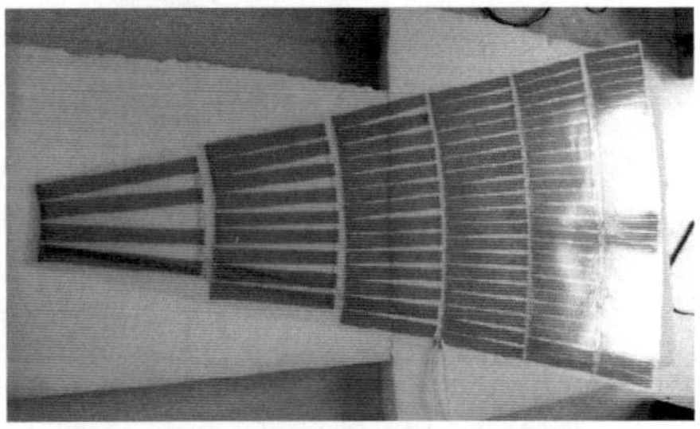

Fig. 8Pattern of the de-icing film heaters.

An input heat flux of 45W/m^2 is applied to the panel, the temperature distribution of the panel front surface obtained by an infrared camera is shown in Fig.9. The average temperature increase is measured to be 2.0°C and the peak-to-valley temperature difference across the panel is less than 0.5°C.

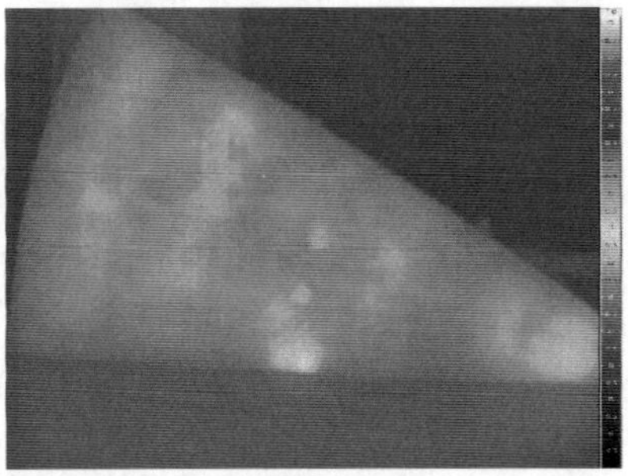

Fig. 9Temperature distribution of CFRP panel for a input heat flux of 45W/m^2.

EXPERIMENTS ON NEAR-FIELD HOLOGRAPHIC ANTENNA MEASUREMENTS

A dual-channel holographic receiver has been developed to demonstrate the near-field holography techinique. Experiment at 92 GHz have been made on a 1.45m test antenna which is a single-piece parabolic dish made from aluminum-honeycomb-sandwiched CFRP. The transmitter is located 63 m away from the antenna on a tower, with an observing elevation around 19 degrees, as shown in Fig. 10.The frontend unit is mounted in the vicinity of the primary focus of an antenna under test, with a signal horn illuminating the reflector and a reference horn, mounted back to the signal horn, pointing to a near-field transmitter. The local oscillator is also mounted inside the frontend unit.

The experiments show that, during the night time at which the ambient temperature doesn't vary rapidly, a 75-minute repeatability (repeating measurement 3 times) of ~2.3 microns rms has been achieved, with an aperture spatial resolution of 46 mm. In order to measure a known surface change, we attached a piece of aluminum foil with a thickness of 43~47 microns to the reflector. By making difference between the holographic measurements before and after the foil attached we obtained a surface change of approximately 45 microns at the foil position, which agrees with the foil thickness, as shown in Fig. 11.

Fig. 10Experimental setup of the near-field holographic measurement

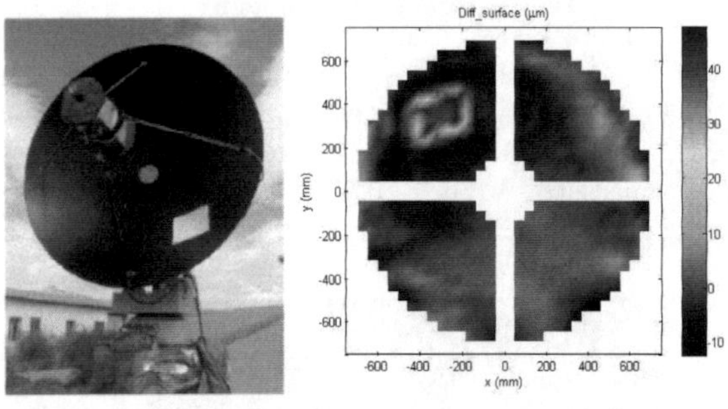

Fig. 11Surface profile measured with two layers of foil

Random errors of the experimental system, such as the pointing error, the amplitude and phase variations of the correlation receiver, have been evaluated and their contributions to the derived surface error have been simulated, indicating that the relatively poor pointing accuracy of the test antenna pedestal is the major contribution to the repeatability, and better repeatability will be expected if the pointing accuracy improves.

During the holographic data processing, we need to fit and remove 6 phase terms (constant, 2 linear gradients in the horizontal and vertical directions, 3 focus translations) from the aperture field. These terms account for a phase offset, an antenna pointing error and a small vector displacement of the signal horn relative to the nominal position. After long-time repeated measurements we have observed regular variations of the fitted terms, which we believe is correlated with the variation of the ambient temperature and solar direction. This implies that if the temperature distribution of the antenna structure is measured simultaneously, we can by this means obtain the relation between the temperature distribution and the variation of pointing and optimum signal horn position. This is very helpful for a telescope to improve its pointing accuracy and aperture efficiency, and also for verifying the related finite element analysis.

CONCLUSIONS

Both aluminium and CFRP panels have been fabricated and tested to evaluate their feasibility for DATE5. No significant additional surface errors of aluminum panel at the low temperatures are found.The structure and process of CFRP panel are still in progress. The performances of CFRP panel at low temperature strongly depends on the panel structure design.The power required for anti-icing for the entire main reflector is less than 1kW for both two types of de-icing systems developed for aluminium and CFRP panels, and the temperature field produced by the de-icing systems have trivial effects on the surface accuracy of the panel.A repeatibility accuracy of 3 microns rms is achieved for the near-field radio holography measurements.Our experiments also shows thatan intentionally introduced surface perturbation can be precisely measured by this technique.

ACKNOWLEDGMENT

This work has been supported by the National Natural Science Foundation of China (Grant Nos.11190014, 11373073 and11503093).

REFERENCES

[1] H. Yang, C. Kulesa, C. K. Walker, *et al.*, "Exceptional Terahertz Transparency and Stability above Dome A, Antarctica,"*PASP*, vol.122, pp. 490-494, April 2010.

[2] J. Yang, Y. X. Zuo, L. Zheng, et al., "Conceptual design studies of the 5m terahertz antenna for DomeA, Antarctica,"*Research in Astronomy & Astrophysics*, vol. 13, no. 12, pp. 1493-1508, 2013.

[3] C. Xiao, Y. Li, andA. Ian, et al.,"Surface characteristics at Dome A, Antarctica:first measurements and a guide to future ice-coring sites,"*Annals of Glaciology*, vol. 48, no. 1, pp. 82-87, 2008.

[4] Y.-X. Zuo,Z. Lou, J. Yang, and J.-Q. Cheng, "Design Study onNear-Field Radio Hologrphy of the 5-Meter Dome A Terahertz Explorer," *2013International Symposium on Antennas and Propagation(ISAP2013)*, Nanjing, Oct. 23~25, 2013, pp. 25-28.

[5] T. Moon and W. Hwang,"Design of Composite Multilayer Surface Antenna Structure and Its Bending Fatigue Characteristics," *Advanced Composite Materials*, vol.17, no.3,pp. 215-224,2008.

Reconfigurable Beam Measurement System andUse for ALMA Band 11 (1.25-1.57 THz)

A. Gonzalez*, T. Kojima, Y. Fujii and S. Asayama

National Astronomical Observatory of Japan, Mitaka, Tokyo 181-8588, Japan*
* Contact: Alvaro.Gonzalez@nao.ac.jp

A new reconfigurable beam measurement system has been recently established at NAOJ. The aim of such a system is to be able to measure beam patterns for any ALMA band, just by changing the RF source, receiver and a few microwave components used for the phase lock. This beam measurement system will be used from ALMA band 1 (35-50 GHz) to ALMA band 11 (1.25-1.57 THz). The possibility of reconfiguration has been tested and proven by measuring beam patterns at two different ALMA bands, band 10 (787-950 GHz) and 11. Measurements at band 1 frequencies are also scheduled for early 2016. Results show good quality amplitude and phase beam patterns with dynamic ranges as good as 60 dB, even at 1.5 THz.

This paper will describe the general ideas behind the concept of re-configurability, and show measurements performed at different RF bands with different configurations. We will also present some of the first amplitude and phase measurements at ALMA band 11 frequencies, including probe compensation measurements.

Air Liquide Cryogenic Space Coolers for Science Applications – Past, Present and Future

T. Wiertz, P. Crespi, Y. Pennec

Air Liquide Advanced Technologies, BP 15, 38360 Sassenage, France

Air Liquide Advanced Technologies (AL-AT) is an industry leader in space and ground based cryogenics for scientific and industrial applications thanks to its long standing heritage. It roots in the early development of all insulated and equipped cryogenic tanks for the Arianne launcher. Its business expanded in the last two decades into scientific and commercial Satellite applications with cryogenic devices. Recent achievements include:

- MELFI, a -80°C turbo Brayton freezer onboard the International Space Station. This cooler is achieving more than 80,000 hours and almost 10 years operations and proving to be one very reliable asset of ISS.
- HERSCHEL, a superfluid helium cryostat supplying 1.8K cold fluid to the far infra red instruments for more than 3 years
- Planck, the world's first Helium dilution cooler functioning in space, achieving flawless operation at 0.1K for 2.5 years.

Space qualified cryocooler development is a challenging endeavor and can be of great concern for an instrument development team. Thanks to its wide range expertise, Air Liquide Advanced Technology not only provides a trouble free cooling solution, but brings additional scientific and technological know-how to the developers to integrate the instrument at higher system level.

Air Liquide Advanced Technology looks forward and remains at the fore-front of spatial scientific exploration. We are already engaged in the development of a 15K pulse tube which will provide the foundation of cryogenic cooling chains. We are collaborating closely with the research team working on the next generation closed loop dilution cooler for sub-50mK applications.

The presentation will showcase our experience and the lessons learnt from past programs. The presentation will also introduce the development and performance of our novel 15K cooler.

Air Liquide is committed to work with the science community to offer high performance, reliable and tailored cooling solutions.

F1 Session: THz projects & Instruments (III)

Antarctic Observatory at Chinese Kunlun Station

J. Yang

Purple Mountain Observatory, Chinese Academy of Sciences, Nanjing 210008, China
contact: jiyang@pmo.ac.cn

Dome A has been selected as one of the goals of development for ground-based observational astronomy for the next decade. Major Instruments, including a 2.5m opt/NIR telescope and a 5m THz telescope, are proposed as Chinese National Mega-Science Facility. Conceptual designs have been carried out during the past several years. Significant progress has been achieved in R&D for the two telescopes. Here we introduce the progress of Dome A observatory preparations and some technical details on the R&D of THz telescope will be presented.

4.7-THz Quantum-Cascade Laser for the upGREAT Array Heterodyne Spectrometer on SOFIA

H.-W. Hübers[1,2*], H. Richter[1], M. Wienold[1,2,3], K. Biermann[3], L. Schrottke[3],and H. T. Grahn[3]

[1]*German Aerospace Center (DLR), Institute of Optical Sensor Systems, Rutherfordstr. 2, 12489 Berlin, Germany*
[2]*Humboldt-Universität zu Berlin, Department of Physics, Newtonstr. 15, 12489 Berlin, Germany*
[3]*Paul-Drude-Institut für Festkörperelektronik, Hausvogteiplatz 5–7, 10117 Berlin, Germany*
* Contact:heinz-wilhelm.huebers@dlr.de

The upGREAT instrument is an array heterodyne spectrometer on SOFIA. It consists of two frequency channels. The low-frequency channel covers the 1.9–2.5 THz range using two 7-pixel waveguide-based hot-electron bolometer (HEB) mixer arrays and a multiplier-based local oscillator (LO). The high-frequency channel is designed for observations of the [OI] atomic fine structure line at 4.745 THz using a 7-pixel waveguide-based HEB mixer array and a quantum-cascade laser (QCL) as the LO [1]. The upGREAT spectrometer has significantly more rigorous requirements as compared to the GREAT heterodyne spectrometer, which comprises a single-pixel detector at 4.7 THz pumped with a QCL LO [2].

We report on the performance of the 4.7-THz LO for the upGREAT heterodyne spectrometer. The LO combines a QCL with a compact, low-input-power Stirling cooler. The QCL is based on a hybrid design and has been developed for continuous-wave operation, high output powers, and low electrical pump powers [3]. Efficient carrier injection is achieved by resonant longitudinal optical phonon scattering. The QCL has a single-plasmon waveguide and a lateral distributed feedback grating. The LO is a significant improvement over its predecessor, which is in routine operation in the GREAT heterodyne spectrometer on SOFIA since 2014. One key challenge was to increase the output power from less than 200 μW to about 1.5 mW while keeping the operating temperature at about 50 K and providing a frequency tunability of at least 5 GHz around the OI rest frequency. Another key issue is the beam profile. Because the beam emitted by the QCL will be split into seven parts by using a Fourier grating, a high-quality fundamental Gaussian beam is required. With an M_2 value of about 1.2, the new LO complies with this requirement. The design of the new LO and its performance in terms of output power, frequency accuracy, frequency stability, and beam profile will be presented, and potential applications in future space-based missions will be discussed.

References

1. C. Risacher, R. Güsten, J. Stutzki, H.-W. Hübers, D. Büchel, U. U. Graf, S. Heyminck, C. E. Honingh, K. Jacobs, B. Klein, T. Klein, C. Leinz, P. Pütz, N. Reyes, O. Ricken, H.-J. Wunsch, P. Fusco, and S. Rosner, "First supra-THz heterodyne array receivers for astronomy with the SOFIA observatory," to appear in: IEEE Trans. Terahertz Sci. Technol., March 2016.
2. H. Richter, M. Wienold, L. Schrottke, K. Biermann, H. T. Grahn, and H.-W.Hübers, "4.7-THz local oscillator for the GREAT heterodyne spectrometer on SOFIA," IEEE Trans. Terahertz Sci. Technol. **5**, 539–545 (2015).
3. L. Schrottke, M. Wienold, R. Sharma, X. Lü, K. Biermann, R. Hey, A. Tahraoui, H. Richter, H.-W.Hübers, and H. T. Grahn, "Quantum-cascade lasers as local oscillators for heterodyne spectrometers in the spectral range around 4.745 THz," Semicond. Sci. Technol. **28**, 035011, 5 pages (2013).

This work was supported in part by the German Federal Ministry of Economics and Technology (grant no. 50 OK 1104) and by the Deutsche Forschungsgemeinschaft.

Fast Terahertz Imaging using a Quantum Cascade Amplifier up to 20,000 pps

Y. Ren, R. Wallis,D. S. Jessop, R. Degl'Innocenti, A. Klimont, H. E. Beere, D. A. Ritchie
Cavendish Laboratory, University of Cambridge, Cambridge, CB3 0HE, UK
* Contact: yr235@cam.ac.uk

The drive to develop ever better imaging techniques at terahertz (THz) frequencies holds tremendous promise in advancing a diverse range of fields, including biomedicine, security sensing, quality control and spectroscopic mapping. Recently, a technique has been demonstrated which employs a quantum cascade laser (QCL) to act as a source and detector simultaneously, via the self-mixing (SM) effect. Radiation reflected back from a target is coupled into the laser cavity, interfering with the intra-cavity field, and producing measureable perturbations in bias voltage containing spatially dependent information about the target. However, current limitations exist in the amount of radiation capable of being back-coupled to the sub-wavelength sized laser facet, as well as the speed at which resolved information can be recorded.

In this work, we present a novel and powerful approach to SM imaging using a 2.9 THz single plasmon QCL, converted into a quantum cascade amplifier (QCA), and operated in pulsed mode at 20 kHz. This was achieved via the use of an anti-reflection coatedsilicon lens mounted on the laser facet to fully suppress lasing action. Radiation was collected and focused via off-axis parabolic mirrors onto a target, with resulting reflections coupled back into the QCA. The induced voltage perturbations were fed into a lock-in amplifier after being differentially amplified with respect to a pulsed reference voltage of equal frequency to the QCA drive current.

The use of an anti-reflection coated lens presents numerous advantages over a standard SM setup. Not only does the impedance matching of the lens increase the power output of the laser, but a much higher proportion of returning radiation is focused directly into the facet. Coupled with the reduction of the reflectivity of the Si/air interface to <5%, these considerations act in tandem to dramatically increase the amplitude of measureable signal. More significantly however, due to the suppression of lasing, the QCA can be biased at the point of maximum alignment, rather than at threshold, which unlocks the entire gain of the structure. The resulting signal to noise (S/N) at 6% duty cycle was demonstrated to be 55 dB, up to 6 times that of reported values of systems operating at threshold, and was shown to be ~1.4 times greater still at 10% duty cycle, the limit of the particular biasing setup employed.

The increase in signal allowed continuous scanning of the target object, dramatically increasing the acquisition time compared to discrete step scanning methods, without sacrificing image quality. A 0.7 mm aperture was also included in the focal point of the second parabolic mirror to increase the resolution of the system, which was estimated to be ~300 μm, obtained through knife edge tests. The sensitivity was high enough to enable acquisition of 20,000 points per second (pps), with a gold coin as the target. In conclusion, this fast THz imaging system we developed represents significant progress in the field of THz sensing, where potential applicability is extremely far-reaching.

The Sardinia Radio Telescope Front-Ends

A. Navarrini[1]*, A. Orfei[2], R. Nesti[3], G. Valente[1], S. Mariotti[2], P. Bolli[3], T. Pisanu[1], J. Roda[2], L. Cresci[3], P. Marongiu[1], A. Scalambra[2], D. Panella[3], A. Ladu[1], A. Cattani[2], L. Carbonaro[3], E. Urru[1], A. Cremonini[4], F. Fiocchi[2], A. Maccaferri[2], M. Morsiani[2], M. Poloni[2]

[1]*INAF–Osservatorio Astronomico di Cagliari, 09047 Selargius, Italy*

[2]*INAF-Istituto di Radioastronomia, 40129 Bologna, Italy*

[3]*INAF-Osservatorio Astrofisico di Arcetri, 50125 Firenze, Italy*

[4]*SKA-Organisation, Jodrell Bank Observatory, Cheshire, United Kingdom*

*Contact: navarrin@oa-cagliari.inaf.it, phone +39 - 070 711 80 237

Abstract— **The 64 m diameter Sardinia Radio Telescope (SRT) has recently started an early science program using three cryogenic front-ends covering four bands: P-band (305-410 MHz), L-band (1.3-1.8 GHz), high C-band (5.7-7.7 GHz), K-band (18-26.5 GHz). The L- and the P-bands can be observed simultaneously with a single coaxial receiver installed at the primary focus, while a seven-beam K-band receiver and a mono-feed high C-band receiver are installed, respectively at the secondary and beam waveguide focus. Additional front-ends are under construction to further expand the telescope observing capabilities. We report on design and performance of the front-ends already installed on SRT and give an overview of the new ones to be completed in the near future.**

INTRODUCTION

The Sardinia Radio Telescope (SRT, www.srt.inaf.it), a challenging scientific project of the Italian National Institute for Astrophysics (INAF), is a new general purpose fully steerable 64 m diameter radio telescope designed to operate with high efficiency across the 0.3-116 GHz frequency range [1]-[2]. The telescope is located 35 km North of Cagliari, Sardinia, Italy, at about 600 m above the sea level. The technical and scientific commissioning of the telescope was completed and an early science program has started in February 2016.

The SRT optical design is based on a quasi-Gregorian configuration (Fig. 1) with shaped 64 m diameter primary (M1) and 7.9 m diameter secondary (M2) reflectors to minimize spillover and standing waves. The primary active surface consists of 1008 aluminium panels (with a panel manufacturing RMS<70 μm) and of 1116 electromechanical actuators under computer control that compensate the gravitational deformation of the backup structure. The primary reflector was aligned to an RMS of ~290 μm using photogrammetry. Work is in progress to improve the total optics surface accuracy down to an RMS of ~150 μm using microwave holography that will allow high efficiency observations up to the highest frequencies (~100 GHz, 3 mm band).

SRT has been designed to host up to twenty receivers installed in six focal positions: Primary focus (F1), Gregorian focus (F2) and Beam-Wave Guide foci (F3&F4 and F5&F6), respectively with focal length to diameter ratio (F/D) and frequency ranges equal to 0.33 (0.3-20 GHz), 2.34 (7.5-115 GHz), and 1.38 & 2.81 (1.4- 35 GHz).

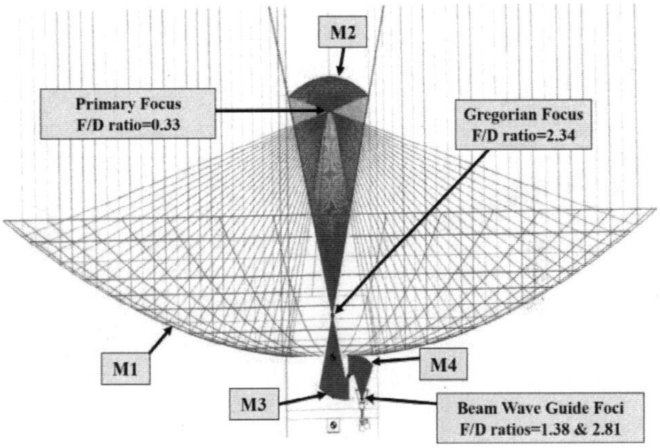

Fig. 1 Optical configuration and ray tracing of the Sardinia Radio Telescope showing the 64-m diameter primary (M1), the 7.9-m secondary (M2), and two additional Beam Waveguide (BWG) mirrors (M3 and M4). Three out of six possible focal positions (primary, Gregorian and BWG) are shown together with corresponding focal ratios.

Table I gives a summary of the receivers installed, under test and under construction for SRT.

FRONT-ENDS INSTALLED ON SRT

Currently, four receivers developed by INAF are installed on SRT: *a)* a primary focus dual linear polarization cryogenic coaxial receiver, which simultaneously covers the 305-410 MHz (P-band) and the 1.3-1.8 GHz (L-band), and is mostly used for Pulsars observations; *b)* a secondary focus dual circular polarization seven-beam cryogenic receiver operating in 18-26.5 GHz (K-band); *c)* a Beam-Wave Guide dual circular polarization mono-feed cryogenic receiver for 5.7-7.7 GHz (High-C-band); *d)* a primary focus X-Ka room temperature coaxial receiver, which simultaneously cover the 8.2-8.6 GHz (X band) and the 31.85-32.25 GHz (Ka band). While the P-, L-, C- and K-band receivers are permanently installed on SRT and used for radioastronomy science, the X-Ka band receiver is temporarily installed and used for testing the spacecraft downlink capabilities of the SRT.

The K-band and C-band use a heterodyne down-conversion to the final 0.1-2.1 GHz IF bands that are delivered to the various backends. The results of commissioning of such front-ends on the SRT antenna are described in dedicated SRT

TABLE I: RECEIVERS INSTALLED, UNDER TEST AND UNDER CONSTRUCTION FOR SRT

Front-End	Freq. range [GHz]	Focal position, F/D	Feeds × pols	Measured / Expected* receiver noise temp. [K]	Measured / Expected* antenna gain [K/Jy]	Measured / Expected* system temperature at zenith [K]	Derived / Expected* SEFD [Jy]	Status
L- and P- band coaxial feed	P: 0.305-0.410	Primary, 0.33	1 × 2	17-22	0.47-0.59	50-80	123	Commissioned
	L: 1.3-1.8		1 × 2	10-13	0.50-0.60	17-23	36	Commissioned
High C-band	5.7-7.7	BWG I, 1.38	1 × 2	6.5-9	0.64-0.70	24-28	39	Commissioned
K-band	18-26.5	Gregorian, 2.34	7 × 2	20-40[a]	0.60-0.66	40-70	89	Commissioned
S-band	3-4.5	Primary, 0.33	7 × 2	15*	0.76*	54*	71*	Under construction
Low C-band	4.2-5.6	BWG II, 2.81	1 × 2	8*	0.62-0.70*	30-35*	49*	Under construction
Q-band	33-50	Gregorian, 2.34	19 × 2	25*	0.45-0.56*	45-120*	163*	Under construction
W-band	84-116	Gregorian, 2.34	1 × 1	30-45	0.34*, b	115*	338*	Under construction
X- and Ka-band coaxial feed	8.2-8.6	Primary, 0.33	1 × 1	150	0.64	180	281	Under testing
	31.8-32.3		1 × 1	130	0.57	190	333	Under testing

Summary of the receivers installed, under test and under construction for the Sardinia Radio Telescope. The measured receiver noise temperatures are obtained in the laboratory using the Y-factor method. [a]The noise temperature of the 2×7 channels of the K-band receiver refers to the measured value across ≈18-24 GHz. [b]With a surface accuracy of ≈180 μm and the metrological systems in operation.

technical reports as well as in [3].

L-P Band cryogenic coaxial receiver for primary focus

Fig. 2 shows a cross-section of the L-P band cryogenic coaxial front-end. Details of the receiver design and performance are described in [4]-[5]-[6]-[7].

The room temperature illuminators are arranged in coaxial configuration with an inner circular waveguide for L-band (diameter of 19 cm) and an outer coaxial waveguide for P-band (diameter of 65 cm). Choke flanges are used outside the coaxial section for improving the cross-polarization performance and the back-scattering of the P-band feed. The geometry was optimized for compactness and high antenna efficiency in both bands. Four probes arranged in symmetrical configuration are used in both the P- and the L-band feeds to extract dual-linearly polarized signals and to combine them, through phase-matched coaxial cables, into 180 deg hybrid couplers. A vacuum vessel encloses the two P-band hybrids and the two L-band hybrids that are cooled respectively at 15 K and 77 K by a commercial closed-cycle cryocooler. For the P-Band, four low loss coaxial feedthroughs are used to cross the vacuum vessel, while for the L-Band a very low loss large window is employed. The RF signals are amplified by coaxial low noise amplifiers (LNAs) thermalized at 15 K inside the cryostat. Two High Temperature Superconductor (HTS) band-pass filters [8], one per polarization channel, are located in front of the LNAs to mitigate the Radio Frequency Interference (RFI) in P-band. Room temperature noise calibration sources can be injected, under remote control, through coaxial cables and coupled into the RF signal paths, also located in front of the LNAs. The noise temperatures at the input flanges of the receiver measured in the laboratory (see [9] for details) are in the range 17-22 K in P-Band and in the range 10-13 K in L-Band for both linear polarization channels. The receiver was installed on the SRT primary focus in May 2013. It is mounted on the primary focus positioner (PFP), a robotic arm that brings that front-end at its focal position during primary focus observations (see also Fig. 5).

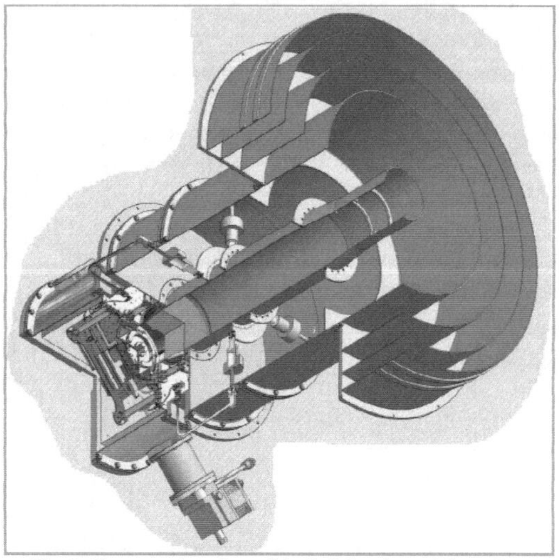

Fig. 2. Cross-section view of the L-P band coaxial receiver.

High C-band cryogenic monofeed receiver for BWG focus

SRT detected its first radio source in the summer 2012 using the high C-band (5.7-7.7 GHz) receiver that was temporarily mounted in the Gregorian focus at the time. Later, that receiver was moved to the beam waveguide focus, where is now permanently installed. The desired feed-horn illumination of the SRT antenna from the two different F/D ratios of 2.34 and 1.38, available respectively at the Gregorian and BWG foci, was obtained by adding a corrugated adapter to the front-end feed-horn when the instrument was used at the Gregorian focus. Details on the design and characterization of the high C-band corrugated monofeed receiver are given in [10]-[11]-[12]. The passive waveguide chain cascaded to the feed-horn consists of a calibration noise injector connected to a DPS (Differential Phase Shifter) and OMT (Orthomode Transducer). The DPS

Fig. 3. Fully assembled high C-band front-end installed on the BWG focus.

and OMT are thermalized at 15 K inside a cryostat. A photo of the high C-band front-end installed at the BWG, given in Fig. 3, shows the room temperature (corrugated) feed horn, the remotely controlled vacuum pump as well as the front-end downconversion and electronics control modules arranged in racks. The front-end is hold in place inside the antenna receiver cabin, which moves in elevation (and azimuth) as the antenna tracks the astronomical sources, through a mechanical frame. The frame is aligned inside the antenna receiver cabin by means of internal mechanical references. Laboratory measurements of the front-end, followed by tests in open-air, allowed accurate determination of the noise temperature of the receiver before its installation in SRT. The receiver noise was derived by the Y-factor method, which utilized room temperature and cold loads respectively given by microwave absorber and cold sky (the method also allowed to characterize the noise calibration source). The measured receiver noise temperature at the feed-horn input was below 9 K for both circular polarization channels (RHCP and LHCP).

K-Band multifeed cryogenic receiver for Gregorian Focus

A detailed description of the K-band (18-26.5 GHz) seven-beam multibeam receiver is given in [13]. The seven beam system allows to increase by a factor of seven the mapping speed of extended radio astronomy sources when compared to a monofeed. The feed-horns are arranged in hexagonal configuration plus a central one. The cryogenic part of each of the seven receiving chains includes a corrugated feed-horn, a circular waveguide directional coupler for noise calibration injection, a differential phase shifter, an orthomode transducer and an InP MMIC (Monolithic Microwave Integrated Circuit) low noise amplifier, all cooled at a physical temperature of $\approx$20 K. A photo of the instrument showing the inner part of the cryostat is given in Fig. 4.

The room temperature parts of the receiver include down-conversion and local oscillator (LO) distribution chains. Two down-conversions are used. The first LO is variable from 12 to 18.5 GHz to allow selecting a 2 GHz band across the 18-26.5 GHz RF range from each polarization channel. The input band of the second down-conversion is 6-8 GHz and utilizes a fixed local oscillator at 5.9 GHz to downconvert to the final 0.1-2.1 GHz IF band. The receiver delivers a total IF bandwidth of 28 GHz (7 feeds x 2 pols x 2 GHz).

A mechanical rotator is used in the receiver to compensate for the Earth's rotation and maintain the parallactic angle thus avoiding the astronomical field derotation caused by the SRT altitude-azimuth movement when tracking sources in the sky. The cryostat is attached to a rack cabinet containing the down-conversion modules, the local oscillator distribution chains, the bias electronics for the LNAs, and the monitoring and control electronics for the valve of the vacuum pump of the cold head, of the calibration noise source as well as of the LNAs.

The noise temperature at the vacuum window of the receiver measured in the laboratory for the 14 channels (seven dual polarization beams) is in the range 20-40 K across most of the 18-24 GHz band. The noise increases at frequencies above 24 GHz to values in the range 40-80 K.

Fig. 4. Photo of the K-band seven-beam front-end showing the assembly of the components inside the cryostat (external shields removed).

X-Ka band coaxial receiver for Primary Focus

A coaxial dual-frequency X-Ka band (X: 8.2-8.6 GHz, Ka: 31.85-32.25 GHz) room temperature receiver was installed on the primary focus of the SRT to perform an experimental verification of the antenna potential capabilities in Space Science [14]-[15]. The single circular polarization receiver was originally developed for the INAF 32-m diameter Noto radiotelescope (Italy) in order to track the Cassini spacecraft. The receiver delivers two 400 MHz wide IF bandwidths (one per band) using two internal local oscillators. The measured noise temperatures are 150 K for the X-band receiver and 131 K for the Ka-band receiver. The huge collecting area of SRT compensates the relatively high system temperatures of these two uncooled receiver with respect to other Deep Space antennas.

A view of the X-Ka band receiver installed at the SRT primary focus is visible in Fig. 5.

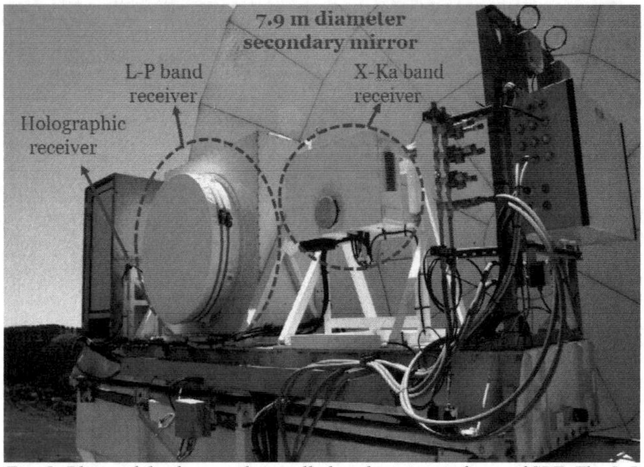

Fig. 5. Photo of the front-ends installed at the primary focus of SRT. The L-P band receiver is located between the X-Ka band and the holographic receivers. All front-ends are mounted on the primary focus positioner (PFP).The 7.9-m diameter ellipsoidal secondary mirror is visible in the back.

FRONT-ENDS UNDER CONSTRUCTION FOR SRT

Four additional cryogenic front-ends for radio astronomy application are currently being constructed for SRT (see also Table I): an S-band (3.0-4.5 GHz) seven-beam dual-linear polarization receiver for primary focus [16]-[17]-[18]; a Q-band (33-50 GHz) 19-element dual-circular polarization receiver for Gregorian focus [19]; a low C-band (4.2-5.6 GHz) monofeed receiver for the BWG focus at F/D=2.81; a W-band (84-116 GHz) monofeed receiver for the Gregorian focus [20].

CONCLUSIONS

We described the design and performance of the front-ends installed on the 64-m diameter Sardinia Radio Telescope and gave a short overview of the ones under construction. The SRT has started an early science program in February 2016 and is currently producing remarkable scientific results.

REFERENCES

[1] G. Grueff, G. Alvito, R. Ambrosini, P. Bolli, A. Maccaferri, G. Maccaferri, M. Morsiani, L. Mureddu, V. Natale, L. Olmi, A. Orfei, C. Pernechele, A. Poma, I. Porceddu, L. Rossi, G. Zacchiroli, *Sardinia Radio Telescope: The new Italian project*, Proc. SPIE Ground Based Telescopes, Vol. 5489, p. 773, 2004.

[2] G. Tofani, G. Alvito, R. Ambrosini, P. Bolli, C. Bortolotti, L. Bruca, F. Buffa, A. Cattani, G. Comoretto, A. Cremonini, L. Cresci, N. D'Amico, G. Deiana, A. Fara, L. Feretti, F. Fiocchi, E. Flamini, F. Fusi Pecci, G. Grueff, G. Maccaferri, A. Maccaferri, F. Mantovani, S. Mariotti, C. Migoni, F. Messina, J. Monari, M. Morsiani, M. Murgia, J. Musmeci, M. Nanni, V. Natale, A. Navarrini, M. Negusini, R. Nesti, L. Olmi, A. Orfei, A. Orlati, F. Palla, D. Panella, C. Pernechele, S. Pilloni, T. Pisanu, A. Poddighe, M. Poloni, A. Poma, S. Poppi, I. Porceddu, I. Prandoni, J. Roda, M. Roma, P. Sarti, A. Scalambra, F. Schillirò, A. Tarchi, G. Vargiu, G. Zacchiroli, *Status of the Sardinia Radio Telescope Project*, Proc. SPIE 5489, Ground-based Telescope, 773, 2008.

[3] P. Bolli, A. Orlati, L. Stringhetti, A. Orfei, S. Righini, R. Ambrosini, M. Bartolini, C. Bortolotti, F. Busa, M. Buttu, A. Cattani, N. D'Amico, G. Deiana, A. Fara, F. Fiocchi, F. Gaudiomonte, A. Maccaferri, S. Mariotti, P. Marongiu, A. Melis, C. Migoni, M. Morsiani, M. Nanni, F. Nasyr, A. Pellizzoni, T. Pisanu, M. Poloni, S. Poppi, I. Porceddu, I. Prandoni, J. Roda, M. Roma, A. Scalambra, G. Serra, A. Trois, G. Valente, G. P. Vargiu and G. Zacchiroli, *Sardinia Radio Telescope: General Description, Technical Commissioning and First Light*, Journal of Astronomical Instrumentation, Vol. 4, No 3&4, 1550008, 2015.

[4] T. Pisanu, P. Marongiu, A. Navarrini, and G. Valente, *A compact L-band Orthomode Junction*, Proc. SPIE, vol. 7741, Millimeter, Submillimeter, and Far-Infrared Detectors and Instrumentation for Astronomy V, Jul. 2010.

[5] G. Valente, T. Pisanu, P. Bolli, S. Mariotti, P. Marongiu, A. Navarrini, R. Nesti, A. Orfei, *The dual-band LP feed system for the Sardinia Radio Telescope prime focus,* in Proc. SPIE Millimeter, Submillimeter, and Far-Infrared Detectors and Instrumentation for Astronomy V, Vol. 7741, 774126, 2010.

[6] G. Valente, G. Montisci, T. Pisanu, A. Navarrini, P. Marongiu, A. Casula, *A compact L-Band Orthomode Transducer for Radio Astronomical Receivers at Cryogenic Temperature*, IEEE Trans. Micr. Theory and Tech., Vol. 63, Issue 10, pp. 3218, Oct. 2015.

[7] G. Valente, G. Montisci, S. Mariotti, *High-performance microstrip directional coupler for radio-astronomical receivers at cryogenic temperature*, Electronics Letters 50, 449 – 451, (2014).

[8] P. Bolli, F. Huang, *Superconducting filter for radio astronomy using interdigitated, capacitively loaded spirals*, Experimental Astronomy, March 2012, Volume 33, Issue 1, pp 225-236.

[9] S. Mariotti, G. Valente, Calibrazione della Marca di Rumore del Ricevitore *Coassiale L P,* IRA Internal Report n. 26, June 12, 2013.

[10] L. Cresci, P. Curioni, V. Natale, R. Nesti, A. Orfei, D. Panella, J. Roda, *C-band Circular Corrugated horn for the SRT Beam Waveguide Focus*, Technical Memo GAI04-TM-13.0, May 7th, 2011.

[11] O. Peverini, A. Virone, G. Addamo, R. Tascone, *Development of passive microwave antenna-feed systems for wide-band dual-polarisation receivers*, IET Microwaves, Antennas & Propag. Vol. 5, Issue 8, pp. 1008–1015, June 2011.

[12] A. Orfei, A. Cattani, M. Poloni, J. Roda, S. Mariotti, A. Maccaferri, A. Orlati, A. Scalambra, *Caratterizzazione del rivcevitore nella banda 5.7-7.7 GHz*, SRT Memo Series GAI04-FR-5.0 May 5th, 2011.

[13] A. Orfei, L. Carbonaro, A. Cattani, A. Cremonini, L. Cresci, F. Fiocchi, A. Maccaferri, G. Maccaferri, S. Mariotti, J. Monari, M. Morsiani, V. Natale, R. Nesti, D. Panella, M. Poloni, J. Roda, A. Scalambra, G. Tofani, *A Multi-Feed Receiver in the 18 to 26.5 GHz Band for Radio Astronomy,* IEEE Antennas Propag. Mag. 52(4), 62, 2010.

[14] A. Scalambra, J. Roda, S. Mariotti, R. Ambrosini, *Doppio Ricevitore Coassiale per le Bande X-Ka dedicato alla missione Cassini*, INAF-Institute of Radio Astronomy, Internal Report N. 331/03, 2004.

[15] T. Pisanu, R. Ambrosini, E. Egron, E. Flamini, N. Iacolina, S. Mariotti, G. Mariotti, P. Marongiu, A. Orlati, P. Ortu, A. Pellizzoni, M. Pili, A. Scalambra, P. Tortora, J. Roda, E. Urru, G. Valente, *Installation and characterization of an X-Ka receiver on the Sardinia Radio Telescope*, Proceedings of 36th ESA Antenna Workshop, ESA-ESTEC, The Netherlands, 6-9 October 2015.

[16] G. Valente, G. Serra, F. Gaudiomonte, A. Ladu, T. Pisanu, P. Marongiu, A. Corongiu, A. Melis, M. Buttu, D. Perrodin, G. Montisci, G. Mazzarella, E. Egron, N. Iacolina, C. Tiburzi, V. Vacca, *A MultiFeed S-Band Cryogenic Receiver for the Sardinia Radio Telescope Primary Focus*, Proc. SPIE 9153 Millimeter, Submillimeter and Far-Infrared Detectors and Instrumentation for Astronomy VII, July 2014.

[17] G. Valente, P. Marongiu, A. Navarrini, A. Ladu, T. Pisanu, M. Pili, A. Saba, G. Montisci, S. Dessi, A. Uccheddu, M. Buttu, E. Egron, N. Iacolina, A. Melis, D. Perrdin, C. Tiburzi,V. Vacca, *The 7-Beam S-Band Cryogenic Receiver for the Sardinia Radio Telescope Primary Focus: Status of Project*, to be published in Proc. SPIE Astronomical Telescope and Instrumentation, 2016.

[18] A. Ladu, G. Valente, G. Montisci, G. Mazzarella, *A wideband quadruple-ridged horn antenna for the multifeed S-band receiver of the Sardinia Radio Telescope*, accepted for publication in Journal of Electromagnetic Waves and Applications.

[19] A. Orfei, A. Cattani, A. Cremonini, L. Cresci, A. Maccaferri, S. Mariotti, R. Nesti, D. Panella, M. Poloni, J. Roda, A. Scalambra, *Elementi per la realizzazione di un multifeed per SRT nella banda 33-50 GHz*, SRT Memo TM-19.1, 2013.

[20] A. Ladu, T. Pisanu, A. Navarrini, P. Marongiu, G. Valente, *A 3 mm band SIS Receiver for the Sardinia Radio Telescope*, Proc. SPIE 9153 Millimeter, Submillimeter and Far-Infrared Detectors and Instrumentation for Astronomy VII, July 2014.

Multi-Gbit/s Data Transmission in Sub-Terahertz Range

Zhe Chen[1*], Bo Zhang[1], Yong Zhang[1], Guangrong Yue[2], and Yong Fan[1]
[1] *School of Electronic Engineering, University of Electronic Science and Technology of China (UESTC), Chengdu, 611731, China*
[2] *National Key Laboratory of Science and Technology on Communications, University of Electronic Science and Technology of China (UESTC), Chengdu, 611731, China*
*Contact: zhechen_uestc@126.com

Abstract— **Two multi-Gbit/s wireless communication experiments based on Schottky technology in sub-terahertz frequency range are presented. 120GHz on-off keying (OOK) transmission is achieved within lab distance with data rate up to 12.5Gbit/s. This proof-of-concept demonstration shows the capacity and potential of terahertz communication. 220GHz data transmission is further achieved in outdoor environment by the use of quadrature phase shift keying (QPSK) modulation.**

INTRODUCTION

It is becoming highly attractive to exploit terahertz (THz) and sub-THz frequency band (100 GHz~10 THz) for high speed wireless communication since THz communication has a number of important benefits including better environment flexibility compared with the optical communication, the availability of large absolute bandwidth and small antenna aperture size. Such benefits provide increasing attractiveness in satellite cross-link in terms of space application and short-range wireless personal area networks and secure communication for ground use [1].

Although THz communication shows attractive application advantages, yet at the current stage, the enabling physical devices remain a main technical bottleneck. In order to make effort to fulfill practical cost-effective all-electronic THz communication systems, we investigated multi-Gbit/s data transmission in two frequency bands (120GHz and 220GHz) based on Schottky electronic devices, which can be readily available at a sensible price point.

Firstly, we built a simple system set-up to carry out the proof-of-concept data transmission experiment at 120GHz in short range. After this demonstration, a more complex 220GHz set-up was proposed to achieve outdoor real-time wireless transmission, which enabled operation over practical distance range.

120GHz DATA TRANSMISSION

Fig.1 shows the 120GHz transmission experiment set-up. In the experiment, on-off keying (OOK) data signals generated by a pulse pattern generator (PPG) were fed into a self-developed 120 GHz Schottky subharmonic mixer (SHM) to transmit, and the frequency down-converted data stream was analyzed using a high sampling-rate oscilloscope and bit error detector. The transmission experiment was carried out in lab environment with a distance of 0.2m (shown in Fig. 2) and the data rate could reach up to 12.5Gbit/s with eye diagrams presenting sufficient eye opening. The eye diagram at 12.5 Gbit/s is shown in Fig. 3. Good eye diagrams and bit error rate (BER) that exceeds the threshold for forward error correction (FEC) [2] were achieved, which means further digital correction can be applied to produce much lower BER which is acceptable for practical application.

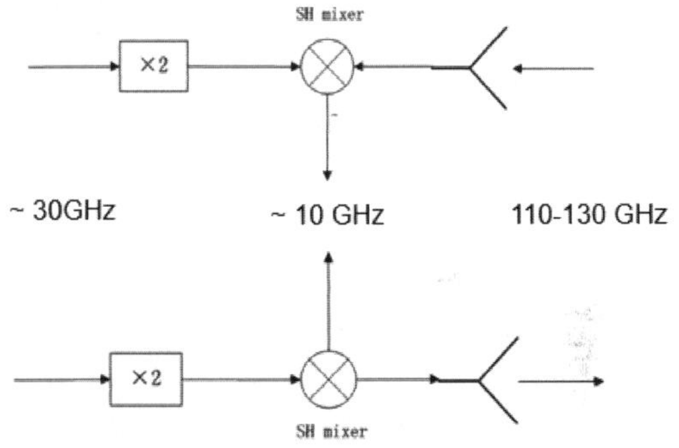

Fig. 1 120GHz transmission schematic set-up.

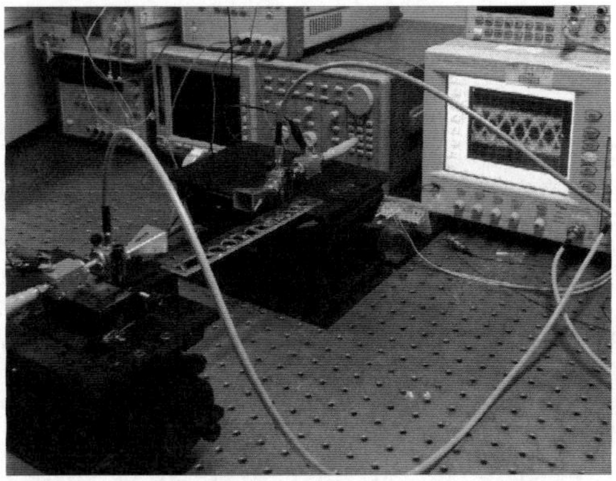

Fig. 2 Picture of 120GHz transmission experiment.

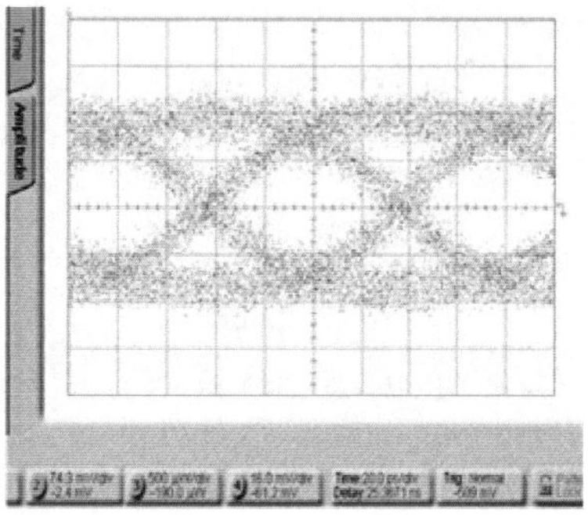

Fig. 3 Eye diagram at 12.5Gbit/s.

220GHz DATA TRANSMISSION

After the successful transmission at 120GHz, we pushed the frequency further up to 220GHz, which is an atmospheric window. Due to this merit, an outdoor transmission set-up which could enable operation over practical distance range was proposed (shown in Fig. 4). Albeit the lack of solid-state amplifiers, this drawback could be overcome by using high gain Cassegrain antennas and adopting self-developed low-noise 220GHz Schottky subharmonic mixers that could provide sufficient receiver sensitivity.

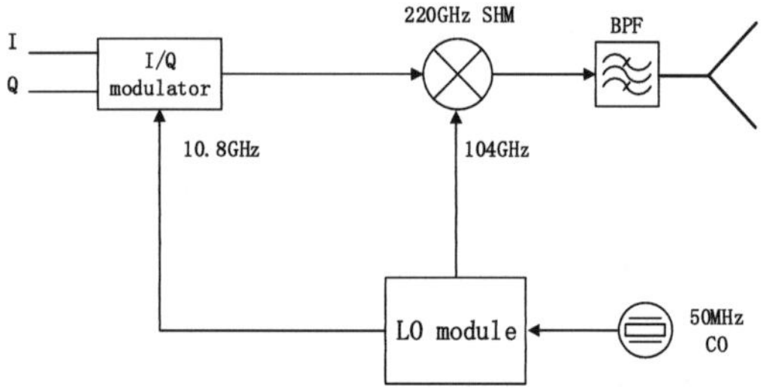

(a) transmitter.

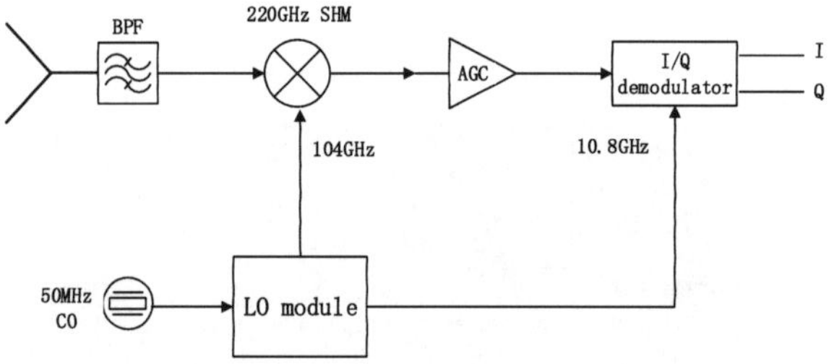

(b) receiver.

Fig. 4 220GHz transmission schematic set-up.

In the transmission set-up, Schottky subharmonic mixers (SHM) are employed as the frequency up- and down-converters respectively at the transmitter (Tx) and receiver (Rx) ends. At the Tx end, broadband base-band signals are I/Q modulated on an X-band carrier (10.8GHz) before up-converting to the transmission frequency around 218.8GHz by the subharmonic mixer which operates in the single side band mode due to the existence of a band-pass filter (BPF) between the mixer and the antenna. The Rx end consists of the same components with its Tx counterpart except for the automatic gain control (AGC) amplifier which is designed to offset the received power variation and ensure an optimum detectable power level by the I/Q demodulator. LO signals for the Tx and Rx ends are generated in the same frequency multiplication configuration based on two 50MHz crystal oscillators (CO) separately for each end.

Fig. 5 Picture of 220GHz outdoor transmission experiment.

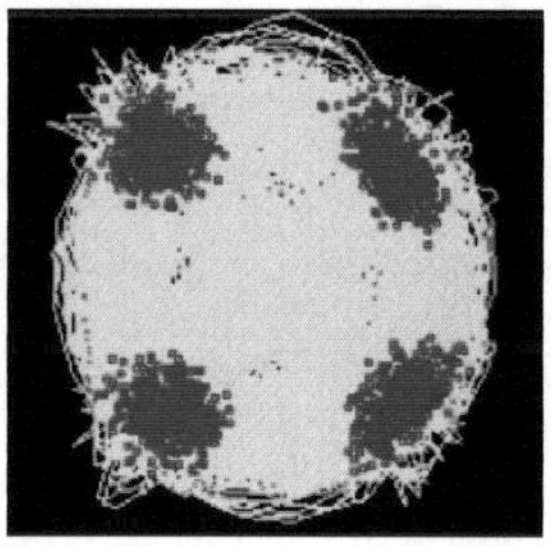

Fig. 6 Constellation diagram at 3.52Gbit/s.

The transmission experiment was conducted over a 200m distance with data rate of 3.52Gbit/s (shown in Fig. 5). The transmission quality was addressed in terms of error vector magnitude (EVM). The EVM at 3.52Gbit/s was measured to be less than 21% with the constellation diagram shown in Fig. 6. This EVM value can be translated into the signal to noise ratio (SNR) of 16dB, which exceeds the QPSK demodulation threshold. Considering the current system configuration [3], larger transmission distance and data rate are possible with proper FEC technique which can further bring 4-6 dB coding gain [2].

CONCLUSIONS

Two multi-Gbit/s transmission experiments based on Schottky technology are presented at 120GHz and 220GHz respectively. The proof-of-concept 120GHz transmission shows the capacity and potential of THz communication. The 220GHz transmission is carried out over practical distance range in outdoor environment towards future practical application. The results demonstrate the feasibility of the realization of THz communication by the use of currently available electronic technology and provide a possible technological gateway to fulfil wireless communication systems in sub-THz and THz frequency range.

ACKNOWLEDGMENT

The work is supported by the National High Technology Research and Development Program of China. The authors wish to thank all the collaborators who contribute their time and expertise to make this work done.

REFERENCES

[1] H. J. Song and T. Nagatsuma, "Present and future of terahertz communications," *IEEE Trans. Terahertz Sci. Technol.,* vol. 1, no.1, pp.256-263, Sept. 2011

[2] F. Chang, K. Onohara and T. Mizuochi, "Forward error correction for 100 G transport networks," *IEEE Com. Magazine.,* vol. 48, no.3, pp.s48-s55, Mar. 2010

[3] Z. Chen, B. Zhang, Y. Zhang, G. Yue, Y. Fan, Y. Yuan, "220 GHz outdoor wireless communication system based on a Schottky-diode transceiver", *IEICE Electron. Express*, vol. 13, no.9, pp.20160282, May 2016

F2 Session: THz Mixer & Detectors (III)

Study of Image Rejection Ratio of 2SB SIS receiver

A. Khudchenko[1,2,*], R. Hesper[2], F.P. Mena[3], J. Barkhof[2], A.M. Baryshev[1,2]

[1]*SRON Netherlands Institute for Space Research, Groningen, the Netherlands*
[2]*NOVA/Kapteyn Astronomical Institute, University of Groningen, Groningen, the Netherlands*
[3]*Electrical Engineering Department, Universidad de Chile, Santiago, Chile*
* Contact: A.Khudchenko@sron.nl

Image rejection ratio (IRR) of a sideband separating (2SB) mixer based on superconductor-insulator-superconductor (SIS) junctions is studied. By analyzing the IRR pattern and using the SIS junctionproperties it is possible to estimate whether the RF part or IF part of the 2SB mixer is limiting the IRR performance.In case of well-balanced IF hybrid the IRR vs. frequency curve is determined by RF waveguide structure imbalance,which is caused mainly by standing waves created by reflections from: 1) the SIS mixers, 2) the RF load and 3) theRF hybrid structure. We have fabricated 2SB mixer for 600-720 GHz and measured standing wave pattern using SISmixer pumping properties. In addition, 3D simulations of the complete RF structure were made and reproduced thestanding wave pattern in experiment. The expected IRR curve was calculated for 3D model and shows goodqualitative agreement with measurement.

A Zero-Bias Ultrasensitive THz Hot-Electron Direct Detector with Large Dynamic Range

B. S. Karasik* and T.J. Reck
*Jet Propulsion Laboratory, California Institute of Technology *, Pasadena, CA 91109, USA*
* Contact: boris.s.karasik@jpl.nasa.gov

As more powerful instruments are being planned for the next generation of submillimeter telescopes (e.g., the Far-IR Surveyor mission), the need for better detectors is becoming more urgent. Several advanced concepts have been pursued in the recent years with the goal to achieve a detector Noise Equivalent Power (NEP) of the order of 10^{-20} - 10^{-19} W/Hz$^{1/2}$ that corresponds to the photon noise limited operation of the future space borne far-IR spectrometers under an optical load ~ 10^{-19} W. Our recent work was focusing on the hot-electron nanobolometer (nano-HEB), a Transition-Edge Sensor (TES) where a very low thermal conductance was achieved due to the weak electron-phonon (e-ph) coupling in a micron- or submicron-size device. Using this approach, the targeted low NEP values have been confirmed recently via direct optical measurements. The kinetic inductance detector and the quantum capacitance detector demonstrated recently a similar sensitivity.

We see nevertheless the possibility to advance the state-of-the-art even further. Increase of the operating temperature and the saturation power, and simplification of the array architecture are believed to be the important areas of improvement not only for the ultrasensitive detectors but also for far-IR detectors intended for use in photometers and polarimeters where the background is higher (NEP = 10^{-18} - 10^{-16} W/Hz$^{1/2}$). Our recent paper (Karasik et al, IEEE Trans. THz Sci.&Technol. 5, 16 (2015)) analyzed the sensitivity of a normal metal nano-HEB, which uses the Johnson Noise Thermometry (JNT) to read an increase of the electron temperature caused by the absorbed far-IR radiation. Such a detector does not require any bias lines and just needs to be connected to a low-noise microwave amplifier (LNA) via a narrowband filter defining the noise bandwidth. By using a filter bank channelizer, a ~ 1000 detectors can be multiplexed using a single amplifier. The use of normal metal eliminates the need in detector material development (like, e.g., for TES) and provide a ~100 dB dynamic range. The NEP depends on the noise temperature of the readout amplifier and is not very sensitive the e-p coupling strength. An NEP < 10^{-19} W/Hz$^{1/2}$ can be achieved with commercially available LNA.

In this paper, we will present an initial experimental study of a normal metal HEB made from a 1-square-micron normal metal patch coupled to a planar twin-slot microantenna. A SQUID-rf LNA with the noise temperature T_A< 1 K followed by a HEMT LNA with $T_A \approx 5$ K and a large gain are for readout. Electrical NEP is measured by sending a dc current through the device and measuring a change of output noise power caused by the heating. At 50 mK, the NEP is ~ 10^{-19} W/Hz$^{1/2}$. The data are obtained as function of bath temperature and the filter passband and compared with the model. An on-going effort to design a 1000-element bank of narrowband (few MHz) filters needed to array multiplexing will be presented too. We will also discuss various options of LNA (HEMT, parametric superconducting amplifiers, etc.) and associated sensitivity and dynamic range trade-offs.

This research was carried out at the Jet Propulsion Laboratory, California, Institute of Technology, under a contract with the National Aeronautical and Space Administration.

Room-Temperature Direct and Heterodyne Detectors Based on Field-Effect Transistors

H. Qin[1,2*], J.D. Sun[1,2], M.C. Luo[3], S.X. Liang[4], C. Yu[4], and Z.H. Feng[4]

[1]Suzhou Institute of Nano-tech and Nano-bionics, CAS, Suzhou, Jiangsu215123, China
[2]Key Lab. of Nanodevices and Applications, CAS, Suzhou, Jiangsu 215123, China
[3]Chongqing Institute of Optoelectronics Technology, Chongqing, Sichuan 400065, China
[4]National Key Lab. of ASIC, Hebei Semiconductor Research Institute, Shijiazhuang, Hebei 050051, China
* Contact: hqin2007@sinano.ac.cn

Field-effect transistors (FETs) based on high-electron-mobility two-dimensional electron gas provide a unique nonlinear element for frequency mixing and hence detection of terahertz electromagnetic wave. To achieve high detection frequency far beyond the cut-off frequency of the field-effect transistor, on-chip antennas have to be integrated with the field-effect gate so that the terahertz electromagnetic wave is transferred to the gate-controlled electron channel. Such a near-field effect is essential for making high-sensitivity FET-based terahertz mixers without suffering from the parasitic device parameters such as the drain-ground capacitance, drain-gate capacitance or the resistance of the leads.

In this talk, we present progresses on the direct (homodyne) detectors and the heterodyne detectors based on either GaN/AlGaN heterostructure or single-layer graphene. The direct detectors are developed for four frequency bands centering at 220 GHz, 340 GHz, 650 GHz and 850 GHz, respectively. The noise equivalent power is below 50 pW/Hz$^{1/2}$ for room-temperature operation and about 1 pW/Hz$^{1/2}$ for operation at 77 K. Heterodyne detection based on GaN/AlGaN FETs and graphene FETs have been successfully demonstrated at 216 GHz and 648 GHz. Sub-harmonic mixing has also been achieved at 432 GHz and 648 GHz with the local oscillator frequency at 216 GHz. The heterodyne mixing is currently realized in a quasi-optic configuration, however, FET detectors integrated in waveguides can be foreseen. Detailed characterization on the sensitivity and the noise performance of the heterodyne detectors will be carried out in the near future.

This work is supported by the National Natural Science Foundation of China under Project No. 61271157 and 61401456.

Photon Counting Detector as a Mixer with Picowatt Local Oscillator Power Requirement

M. Shcherbatenko[1,2], Y. Lobanov[1,2] V. Kovalyuk[1], A. Korneev[1,2] and G.N. Gol'tsman[1,3]

[1]*Moscow State Pedagogical University, Moscow*
[2]*Moscow Institute of Physics and Technology, Dolgoprudny*
[3]*National Research University Higher School of Economics, Moscow*
* Contact: scherbatenko@rplab.ru

At the current stage of the heterodyne receiver technology, great attention is paid to the development of detector arrays and matrices comprising many detectors on a single wafer. However, any traditional THz detector (such as SIS, HEB, or Schottky diode) requires quite a noticeable amount of Local Oscillator (LO) power which scales with the matrix size, and the total amount of the LO power needed is much greater than that available from compact and handy solid state sources. Substantial reduction of the LO power requirement may be obtained with a photon-counting detector used as a mixer. This approach, mentioned earlier in [1,2] provides a number of advantages. Thus, sensitivity of such a detector would be at the quantum limit (because of the photon-counting nature of the detector) and just a few LO photons for the mixing would be required leading to a possible breakthrough in the matrix receiver development. In addition, the receiver could be easily tuned from the heterodyne to the direct detection mode without any loss in its sensitivity with the latter limited only by the quantum efficiency of the detector used.

We demonstrate such a technique with the use of the Superconducting Nanowire Single Photon Detector(SNSPD)[3] irradiated by both 1.5 μm LO with a tiny amount of power (from a few picowatts down to femtowatts) facing the detector, and the test signal with a power significantly less than that of the LO. The SNSPD was operated in the current mode and the bias current was slightly below its critical value. Irradiating the detector with either the LO or the signal source produced voltage pulses which are statistically evenly distributed and could be easily counted by a lab counter or oscilloscope. Irradiating the detector by the both lasers simultaneously produced pulses at the frequency f_m which is the exact difference between the frequencies at which the two lasers operate. $f_m$could be deduced form either counts statistics integrated over a sufficient time interval or with the help of an RF spectrum analyzer.

In addition to the chip SNSPD with normal incidence coupling, we use the detectors with a travelling wave geometry design [4]. In this case a niobium nitride nanowire is placed on the top of a nanophotonic waveguide, thus increasing the efficient interaction length. Integrated device scheme allows us to measure the optical losses with high accuracy. Our approach is fully scalable and, along with a large number of devices integrated on a single chip can be adapted to the mid and far IR ranges.

This work was supported in part by the Ministry of Education and Science of the Russian Federation, contract no. 14.B25.31.0007 and by RFBR grant # 16-32-00465.

1. Leaf A. Jiang and Jane X. Luu, —Heterodyne detection with a weak local oscillator, *Applied Optics Vol. 47, Issue 10, pp. 1486-1503* (2008)
2. Matsuo H. —Requirements on Photon Counting Detectors for Terahertz Interferometry *J Low Temp Phys* (2012) 167:840–845
3. A. Semenov, G. Gol'tsman, A. Korneev, "Quantum detection by current carrying superconducting film", Physica C, 352, pp. 349-356 (2001)
4. O. Kahl, S. Ferrari, V. Kovalyuk, G. N. Goltsman, A. Korneev, and W. H. P. Pernice, —Waveguide integrated superconducting single-photon detectors with high internal quantum efficiency at telecom wavelengths., Sci. Rep., vol. 5, p. 10941, (2015).

Development of a 2 THz Solid-state Radiometer for Atmospheric Sounding.

Jeanne Treuttel[1,2], Erich Schlecht[1], Choonsup Lee[1], J-V. Siles[1], Alain Maestrini[2], B. Thomas[3], Robert Lin[1]and Imran Mehdi[1]

[1]Jet Propulsion Laboratory, USA
[2]Observatory of Paris, LERMA, France
[3]Radiometer Physics GmBh, Germany

Both the interstellar medium and planetary atmospheres are incredibly rich in molecular species with spectral rotational and vibrational signatures that lie in the 1-10 THz frequency range. The atomic oxygen (OI) emission at 2.06 THz (145.525 um) is one of the two brightest emission lines in the terrestrial thermosphere and has been observed from balloon, sounding rocket and orbital platforms [1].

Schottky diode front-end receivers have been demonstrated up to 2.5 THz [2] with a CO2-pumped methanol gas laserlocal oscillator source.However, recent developmentsin Schottky multiplier sources show thatsufficient power can be obtained at 1 THz to drive a 2 THz sub-harmonic mixer. This makes possible the development of a2-THz all solid state front-end heterodyne receiver that can be deployed on CubeSat or similar miniature platforms.

Firstly we will present preliminary development of the 2THz front-end receiver, with a first circuit iteration that features a balanced sub-harmonic mixer similar to previous studies [3], along withnoise temperature measurement system. Secondly we will discuss further circuit development for a second iteration,including a novel bias-able sub-harmonic mixer. This mixerfeatures an anti-parallel pair of diodes that favors a better trade-off between available power and line losses, and was partially addressed in [4].

References

[1] K. U. Grossmann, M. Kaufmann, and E. Gerstner, A global measurement of lower thermosphere atomic oxygen, *Geophys. Res. Lett.*, Vol. 27, No. 9, 1387-1390, 2000.

[2] P. Siegel, R. Smith, M. Gaidis, and S. Martin, "2.5-THz GaAs monolithic membrane-diode mixer," *IEEE Trans. Microw. Theory Techn.l*, vol. 47, no. 5, pp. 596–604, May 1999.

[3] E. Schlecht, Siles, J.V., Lee, C., Lin, R., Thomas, B., Chattopadhyay, G., Mehdi, I."Schottky diode based 1.2 THz receivers operating at room-temperature and below for planetary atmospheric sounding" *IEEE Trans. Terahertz Sci. Tech*, Vol 4 , No. 6, Nov. 2014.

[4] Jeanne Treuttel, B. Thomas, A. Maestrini, J. V.-Siles, C. Lee, I. Mehdi " A Novel 330 GHz Sub-Harmonic Mixer with Independently Biased Schottky Diodes", *International Symposium on Space Terahertz Technology*, April 2012, Tokyo, Japan.

T4: Poster Session

Broken-step Phenomenon in SIS Mixers

A. Ermakov [1, 2*], V. Belitsky[1], P. Ahgdam[1], V. Desmaris[1], S.-E.Ferm[1], M. Fredrixon[1], S. Krause[1], I. Lapkin[1], D. Meledin[1], A. Pavolotsky[1], H. Rashid[1], S. Shafiee[1], M. Strandberg[1], E. Sundin[1]
Kamaljeet S. Saini[3], E. Bryerton[3]

[1] *Group for Advanced Receiver Technology, Chalmers University of Technology, SE41296, Gothenburg, Sweden*
[2] *Institute of Radio Engineering and Electronics, IRE RAS, Mokhovaya str. 11-7, Moscow 125009, Russia*
[3] *National Radio Astronomy Observatory (NRAO), NTC, 1180 Boxwood Estate Road, Charlottesville, VA 22903-4608, USA*

*Contact: andreye@chalmers.se

Abstract— **In this paper, we discuss a "broken step" phenomenon in an SIS mixer. This phenomena was observed in the production version of the SIS mixers, designed for the 159-211 GHz RF band, being used for the construction of the ALMA Band 5 receiver The broken step typically appears at LO frequencies above 180 GHz and manifests itself as a sharp onset in the DC current at the middle of the quasiparticle step. Correspondingly, this affects the mixer IF response in a way that is similar to the Josephson step but is however of a different nature. Such behaviour affects the SIS mixer dynamic range and complicates the tuning of the 2SB mixer to optimize its performance, for both the receiver noise as well as the sideband rejection. In this paper, we describe results of a few experiments which were performed to understand this undesirable phenomenon.**

INTRODUCTION

The broken-step was observed in the SIS mixers used for ALMA Band 5. When studying this phenomenon, we used the production version of the ALMA Band 5 2SB mixer, optimized for the 159-211 GHz RF band and largely based on the 2SB SIS mixer presented in [1] and data-acquisition software [2] specifically modified for the ALMA Band 5 project. The broken step appears typically at LO frequencies above 180 GHz and manifests itself as a sharp break in the DC current at the middle of the quasiparticle step. Correspondingly, this affects the mixer IF response in a way that is similar to a Josephson step but however is of a different nature. Such behaviour affects the SIS mixer dynamic range and complicates the tuning of the 2SB mixer for achieving optimum performance, for both the receiver noise as well as the sideband rejection. In this paper, we describe several diagnostic measurements performed on the receiver in order to isolate the root cause and physical origin of this phenomenon.

SHAPIRO VS. FISKE VS. PHOTON STEP

The mixer was operated under different conditions (of bias voltage, LO pumping level, magnetic field), and physical temperatures in an attempt to establish if this phenomenon was related to the Josephson effect (Shapiro steps), or to internal resonances of the SIS mixer tuning circuitry (Fiske steps) or if it was a manifestation of some other mechanism. By varying the magnetic field and achieving accurate suppression of the Josephson effect, we ensured that this phenomenon has no coupling to the Cooper-pair current. For the given ALMA Band 5 mixer, which uses twin-junction tuning structure [3], the standard sin(x)/x pattern is modulated by SQUID-type dependence of critical current vs. magnetic field of the twin-junction. This calls for careful and precise tuning of the magnetic coil currents for each mixer of the 2SB assembly. Fig. 1 illustrates the IV characteristic (IVC) and the IF response (the IF power was the power integrated over 600 MHz bandwidth centred at 6 GHz IF ***in all figures below***) for the mixer with adjusted magnetic field to suppress Josephson effect. The "broken step" effect is clearly visible at the first photon step (marked by dashed line arrow) while the mixer IF response demonstrates no signs of the Josephson steps.

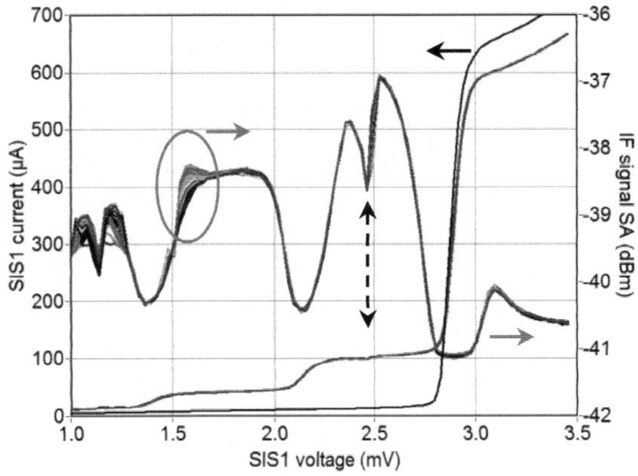

Fig. 1. Illustration of the magnetic field tuning (no visible signs of the Josephson steps affecting the IF response). The mixer IF response curve(s) indicate the IF power vs. DC bias voltage. The blue oval shows the area on the IF response where the Josephson step affects the performance and curves with different colours correspond to different magnetic field. The black and red lines show autonomous and pumped measured IVC. The "broken step" phenomenon at IVC and in the mixer IF response is marked by a dashed line arrow.

In order to completely exclude the effect of the Josephson steps or a Fiske step (the latter is unlikely as the corresponding frequency of the resonance would be around 2.2 THz), the mixer measurements were performed at several physical temperatures. The measurement setup allowed for a controlled variation of the physical temperature of the mixer between 3.4 K and 5.5 K and stabilisation at any selected temperature in this range. Fig. 2 shows the mixer IF response and the SIS mixer IVC at different physical temperatures as a function of the SIS junction gap voltage. The movement of the broken step feature is correlated with the junction gap voltage, This is indicative of the quasiparticle nature of the phenomenon and rules out internal resonance (Fiske step). This also excludes Shapiro step related origin of the observed broken step phenomenon, as it would not be dependent on temperature and as the Shapiro step-voltage counts from zero voltage.

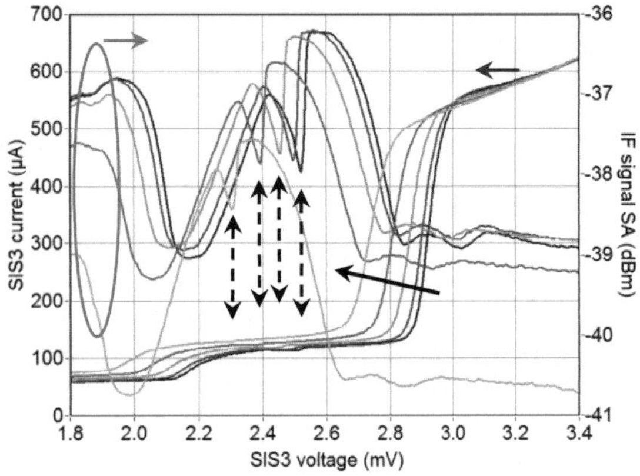

Fig. 2. The mixer IF response curves marked by blue oval shows the vs. DC bias voltage and the SIS IVC measured at different physical temperatures that is seen by the change of the SIS junction gap voltage (black arrow around 2.8 mV bias). The broken step feature moves along with the junction gap voltage, indicated by the black dashed arrows around 2.4 mV bias.

Our experiments strongly indicate that the broken step phenomenon is not coupled to Josephson effect (steps and internal resonances) but must be related with the quasiparticle tunnelling.

PHOTON STEPS FROM TWO STRONG RF SIGNALS

Presence of a series of such broken steps observed on the tested SIS mixers lead us to conclusion that a fairly strong spurious RF signal must be incident on the SIS junction along with the LO frequency. Fig. 3 shows the measured IVC and the IF response of the SIS mixer with multiple "broken step" phenomenon that manifests itself at the first and second photon steps of the LO signal with the period of the half the LO photon step.

The above measurements indicate presence of two signals withdifferent frequencies. One of them is the LO while another could be half the LO frequency. This would yield two sets of the photon steps, as observed in the experiment. Simulations indicate that such dual-frequency operation resembles the observed "broken-step" phenomenon. Interestingly, the broken-step feature also appears in the pumped IVC for the case when signals with frequencies equal to the LO and 3/2 of the LO frequency are applied to the junction. To illustrate this, simulations were carried for the case of combinations of frequencies applied to the mixer SIS junctions. By varying the normalized pumping amplitudes, e.g., $\alpha_{91.5}$ & α_{183} and α_{183} & $\alpha_{274.5}$, it was possible to reproduce the "broken-step" feature with the predicted characteristic very close to what was observed during the measurements.

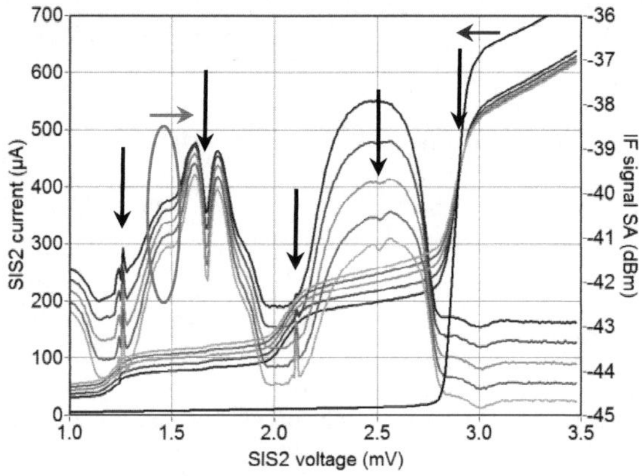

Fig. 3. The measured IVC and the IF response of the SIS mixer (marked by blue oval) with multiple "broken step" phenomenon that manifests itself at the first and second photon steps of the LO signal. It is clear that the scale of the broken-step phenomenon is half of the photon steps induced by the LO signal.

Fig. 4 and 5 show the simulated dual-frequency pumping: the experimentally measured IVC was used and signals were applied in sequence, e.g., obtain photon steps from one frequency and then use the already "stepped" IVC to apply the next frequency. The standard expression for calculating SIS junction IVC with applied RF signal, equation (1) below from [4] was used. The simulations were made using the data acquisition and calculation software IRTECON [2].

$$I(V) = \sum_n \sum_k J_n(\alpha_1) J_k(\alpha_2) I_{dc}(V + \frac{n\hbar\omega_1}{e} + \frac{k\hbar\omega_2}{e}), \text{ where } \alpha = \frac{eV_\omega}{\hbar\omega} \quad (1)$$

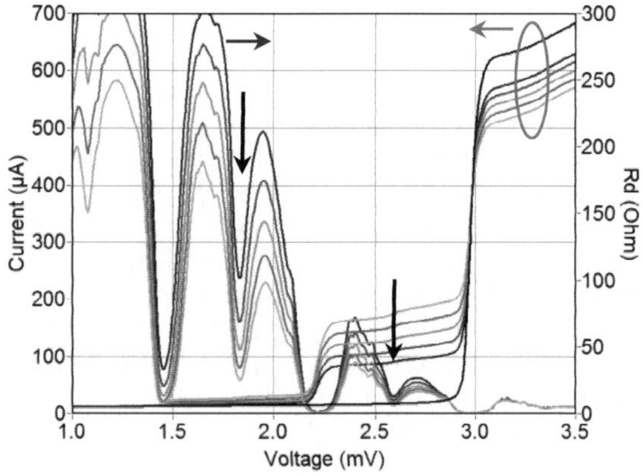

Fig. 4. Simulation results: The SIS mixer IVCs were measured through IRTECON, and were then modelled with the applied pumping signals with frequencies of 91.5 GHz and 183 GHz having different normalized pumping amplitudes, $\alpha_{91.5}$ and α_{183} (marked by blue oval). The differently coloured lines correspond to the different LO pumping, α_{183} =0.7…1.1. The IVC differential resistance, R_d was used as an indicator for the "broken-step" appearance. The "broken step" phenomenon at IVC and the R_d is marked by arrows.

The figures above, show the non-pumped SIS mixer IVC measured through IRTECON, which was then modelled to have different values of the pumping RF signal applied to have photon steps of 91.5 & 183 GHz (Fig. 4) and 183 & 274.5 GHz signals (Fig. 5). The values for parameter $\alpha_{91.5}$ & α_{183} and α_{183}

& $\alpha_{274.5}$ were adjusted to get the best fit. The first derivative of the pumped IVCs (differential resistance) was used as an indicator for the "broken-step" appearance.

It should be noted that the actual value of the pumping parameter α usually depends on the bias voltage for well-tuned SIS mixer. This is a function of the built-in on-chip integrated tuning circuitry of the SIS mixer and the embedding impedance, which is result of, e.g., the accuracy of the mixer chip placement inside the mixer block and the mixer block fabrication accuracy. In the simulations, we ignored the dependence of the parameter α on the DC bias voltage, which results in a slight difference between the simulated and measured IVC (Figure 3 vs. 4, 5) and its relative pumping levels at the second and the first photon steps.

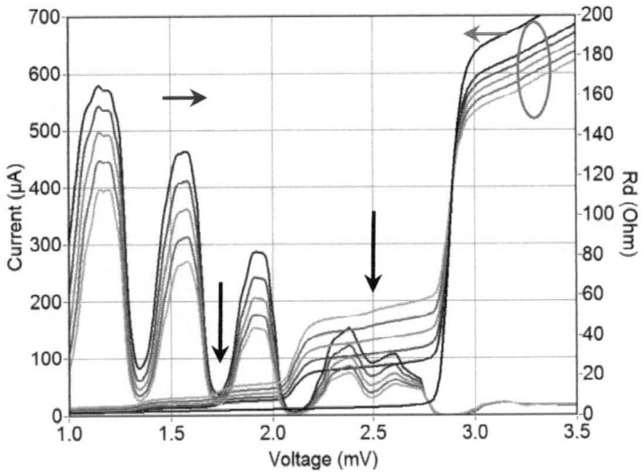

Fig. 5. Simulation results: The SIS mixer IVCs were measured through IRTECON, and were then modelled with the applied pumping signals with frequencies of 183 GHz and 274.5 GHz having different normalized pumping amplitudes, α_{183} and $\alpha_{274.5}$ (marked by blue oval). The differently coloured lines corresponds to the different LO pumping, α_{183} =0.7...1.1. The IVC differential resistance, R_d was used as an indicator for the "broken-step" appearance. The "broken step" phenomenon at IVC and the R_d marked by arrows.

DISCUSSION

From the described measurements and simulation, it can be concluded that the "broken-step" phenomenon is explained by the appearance of spurious RF signals in addition to the LO. The spurious signals could be half of the LO frequency at the SIS junction mixer or a frequency that is 3/2 times the applied LO frequency. If we consider the last multiplier stage of the LO, the frequency doubler, then the simulations indicate a combination of frequencies across the SIS mixer which are the doubler pumping tone (x1), x2 and x3 the pumping frequency.

It is extremely unlikely that a strong signal with 90-95 GHz appears directly at the Band 5 SIS mixer junction since this frequency is substantially below the cut-off frequency of the mixer LO port waveguide (dimensions are close to WR-5). Besides, the 2SB mixer has a relatively complicated waveguide configuration at its RF port (3 dB RF hybrid) and the LO port with a power divider and the LO injection coupler, all employing the WR5 waveguide circuitry [1]. Consequently, it is difficult to visualize a path for significant amount of the LO sub-harmonic to propagate to the mixer. Additionally, this frequency is unlikely to be leaking from the LO: it corresponds to the pump frequency of the final doubler in the LO chain, and cannot couple into the WR5 waveguide output of the doubler.

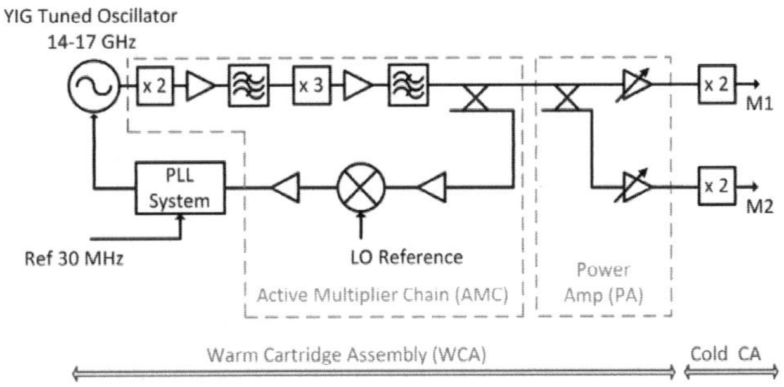

Fig. 6. Simplified block diagram of the ALMA Band 5 LO system that includes the warm cartridge assembly (WCA) and the cold cartridge assembly (CCA). M1 and M2 outputs in Fig. 6 indicate LO output to the Pol0 and Pol1 mixer assemblies.

Another possibility, which could be invoked to explain the appearance of the 90-95 GHz frequency signal is via the down-conversion of the second and third harmonic products generated at the final x2 multiplier stage of the LO source, i.e., the third harmonic around 270-285 GHz of the pump signal and the LO frequency around 180-190 GHz. Simulations indicate that the IVC is sensitive to the pumping level of the low-frequency signal, for example in the Fig.4, the $\alpha_{91.5}$ is only 24% of the α_{183} value and this is sufficient to produce well-developed "broken-step" phenomenon. This possibility however requires that the SIS mixer should be also tuned for frequencies 90-95 GHz, a frequency range, which would otherwise be nearly short-circuited by the SIS junction intrinsic capacitance.

If we now consider the presence of the signals in the range 180-190 GHz (LO) and 270-285 GHz, this combination could produce the "broken-step" directly (Fig, 5) but also provides conditions conducive to enable the down-conversion as discussed above, contingent on being complemented by "sensitivity band" of the SIS mixer itself. Both the down-conversion option and direct interaction with the third harmonic of the doubler pumping frequency in the presence of the LO (x2 pumping frequency) requires that SIS mixer is sensitive in this frequency band, 270-285 GHz.

One of the main factors affecting the SIS mixer sensitivity band is the integrated on-chip tuning circuitry. In usual design optimization procedure, the SIS mixer's built-in tuning circuitry is constructed to provide optimum SIS junction matching within a specified frequency band, which, in the case of ALMA Band 5 is 158-211 GHz, with some margins, typically 10-15% at the band edges. Simulations of the Band 5 SIS mixer tuning circuitry above 250 GHz indicate that some sensitivity, about -9 dB less than of the operational band, is possible and analysis points towards the RF/IF&DC isolation circuitry on the mixer chip. However, this simulation does not take into account the E-probe coupling with the waveguide. An accurate simulation is rather complicated by the fact that at the frequency above 225 GHz the mixer mount waveguide is not operated in the single-mode.

Another factor to consider is the relative strength of the third harmonic at the output of the LO doubler. In the ALMA Band 5 production project, the output of the multiplier was qualified for the unwanted harmonics and the output of the balanced diode doubler demonstrated to be better than -20 dBc, with measured values being closer to -25 to -30 dBc. Fig. 7 shows the results of such a screening measurement performed at room temperature.

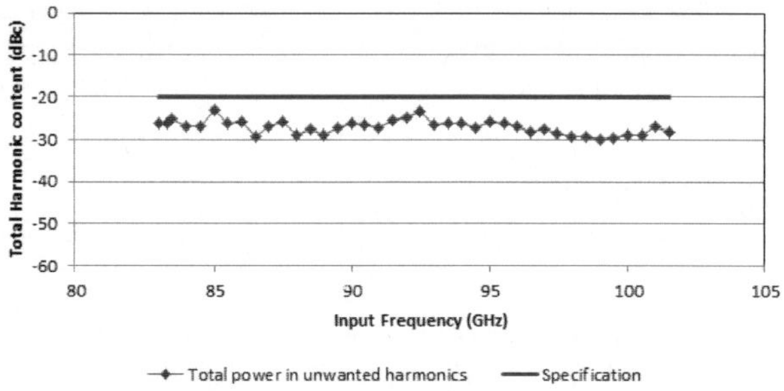

Fig. 7. Measured total power of the unwanted harmonics (x3 and above) at the output of the doubler WR5.1x2.

In light of all of the various arguments described above, interaction with the third harmonic from the LO doubler appears to be the only plausible explanation for the observed broken step phenomenon. This conclusion is supported by the results of the simulation (Fig. 5) and the modelling of the junction tuning circuitry. Further screening of the harmonic content at the output of the LO doubler and studying the correlation of the strength of the broken step phenomenon with the measured third harmonic strength in the multiplier output by measuring multiple instances of Band 5 cartridges would help definitively resolve the question..

ACKNOWLEDGMENT

This work is related to the ALMA Band 5 Full Production Project funded by European Southern Observatory under Contract No 49388/ESO/12/46561/ATR. J. Barkhof, R. Hesper and J. Adema (NOVA) and P. Yagoubov (ESO) are acknowledged for interesting and useful discussions.

REFERENCES

1. Billade, B., Nyström, O., Meledin, D., Sundin, E., Lapkin, I., Fredrixon, M., Desmaris, V., Rashid, H., Strandberg, M., Ferm, S.-E., Pavolotsky, A., Belitsky, V., "Performance of the First ALMA Band 5 Production Cartridge", *IEEE Trans. Terahertz Science and Technology*, Vol. 2, No. 2, pp.208-214, 2012.
2. A.B. Ermakov, S.V. Shitov, A.M. Baryshev, V.P. Koshelets, W. Luinge, "A data acquisition system for test and control of superconducting integrated receivers", *IEEE Trans. on Appl. Superconductivity*, Vol.11, No 1, pp. 840-843, 2001.
3. Belitsky V., Tarasov M.A., "SIS Junction Reactance Complete Compensation", IEEE Trans. on Magnetic, 1991, MAG- 27, v. 2, pt. 4, pp. 2638-2641.
4. P. K. Tien, J. P. Gordon, "Multiphoton Process Observed in the Interaction of Microwave Fields with the Tunneling Between Superconductor Films", Phys. Rev., 1963, V. 129, No 2, P. 647-651.

Development of 1.5 THz Cartridge-type Multi-pixel Receiver Based on HEB Mixers

Yen-Ru Huang*, Chuang-Ping Chiu, Hsian-Hong Chang, Yen-Yu Chiang, and Ming-Jye Wang
Institute of Astronomy and Astrophysics, Academia Sinica, Taipei 10617, Taiwan
*Contact: yenruhuang@asiaa.sinica.edu.tw, phone: +886-2-2366-5309

Abstract— **A design of 2 × 2 NbN-based hot-electron-bolometer (HEB) mixer array receiver cartridge has been demonstrated here by using multiple local oscillator (LO) beams. In our design, the 1.5 THz LO beam is split into four uniform sub-beams with a spacing of 18 mm by using a power distributor, then arrives at a four-pixel silicon lens with twin slot antenna (TSA) through a large-area beam splitter. An additional four-pixel HDPE lens is located at 120 mm in front of the silicon lens to increase the size of beam waist for fitting to the aperture parameter of sub-reflector of GLT. Some cryogenic tests of cartridge have been carried out. In this article, we report the design, assembly, thermal analysis, and some testing results of cartridge.**

INTRODUCTION

The superconducting NbN based hot-electron-bolometer (HEB) mixer is one of the commonest detectors for THz radio astronomy due to its near quantum limit noise performance and low local oscillator (LO) power requirement [1], and has already been applied in several astronomical observatories [2-4]. In the past years, we have successfully realized 1.4 THz NbN based HEB mixers with a receiver noise temperature of around 2000K and a bandwidth of 3.5 GHz. The receiver noise temperature can be further improved by replacing a silicon lens with anti-reflection coating. In addition, for astronomy receivers, multi-pixel receiver is more favourable because of high mapping speed. However, uniformly injecting enough LO power to each pixel is a critical issue in a THz multi-pixel receiver. To achieve the minimum power loss and uniform beam profiles, most HEB array receivers apply the polarizing beam splitters for the LO beams [5,6], including Fourier phase gratings, wire-grid polarizers, multilayer structures, etc. In our design, simple dielectric slab beam splitters are used which can suppress the variations of transmittance and reflectance below 2.2% in the frequency range from 1.45 THz to 1.55 THz.

The receiver cartridges for the Atacama Large Millimeter /submillimeter Array (ALMA) have many advantages, including high cooling efficiency and good modularity which can simplify the efforts on telescope site. In this paper, we report the designs of single-pixel and 2×2 HEB mixer array ALMA-type cartridge receiver at the frequency range of 1.45 - 1.55 THz. The pixel number in our design can be further extended to 9 based on currently available LO power. We target to deploy this multi-pixel THz receiver cartridge on the Greenland Telescope (GLT) [7]. An engineering model of cartridge which can be used for single-pixel and 2×2 array receiver has been constructed. The thermal analysis, cooling test of cartridge, and LO source performance will be discussed.

CARTRIDGE DESIGN CONCEPT

The receiver cartridge comprises three cooled stages with operation temperatures of 4K, 15K, and 110K. To boost output power, cooled frequency multipliers are used in the LO module which is provided by VDI Inc. The fundamental frequency of LO signal is generated by a synthesizer and multiplied to Q-band frequency. Then the power is amplified and fed through the vacuum-type waveguide feedthrough (WR22) at 300K base-plate to reduce power losses. The frequency is further multiplied by a cooled two-doubler module and a cooled two-tripler module to reach the target frequency. The output horn of LO module is fixed at the focal point of a 90 degree off-axis parabolic mirror on the 110K plate with an effective focal length of 30 mm to make the Rayleigh length enough. For 2×2 array receiver, a power distributor module is used to divide the LO beam into four sub-beams with a spacing of 18 mm. After that the LO and radio frequency (RF) beams are combined by a large-area beam splitter made of mylar films with 13 μm thickness, and then couple to the HEB mixer adhered on the backside of a silicon lens. The reflectance of mylar beam splitter is about 20%. In addition, another four-pixel HDPE lens is located at 120 mm in front of the silicon lens to increase the size of beam waist for matching the aperture parameter of the sub-reflector of the GLT antenna.

The design and assembly of the engineering model cartridge had been completed, and the testing of single-pixel design has been carried out first to check the performance. Fig. 1 is the schematic diagram of the single-pixel and four-pixel receiver cartridge design. In these two engineering models, the arrangements of cartridge are identical, except of the mixer block, the LO coupling module, and the number of IF channels. Fig. 2(a) shows the four-pixel power distributor module which consists of TE and TM mode polarizing beam splitters made of quartz and silicon slabs respectively due to the polarization direction of LO signals. To achieve uniform distribution patterns of the output sub-beams, the thicknesses of the quartz and silicon slabs are optimized to 135.5 μm and 130.5 μm respectively which are depicted in Fig. 2(b).

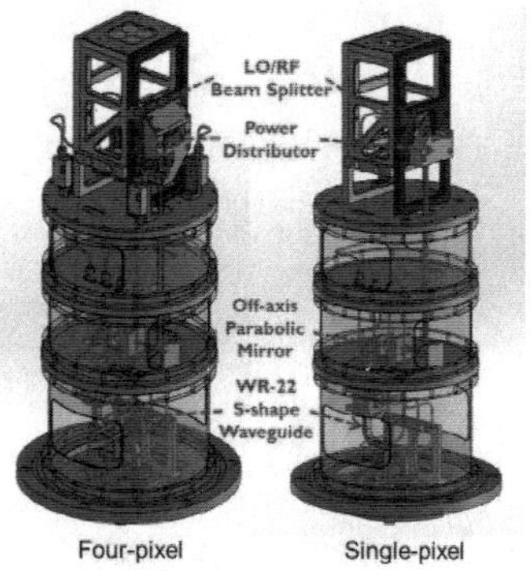

Fig. 1 The schematic diagram of the single-pixel and four-pixel cartridge design.

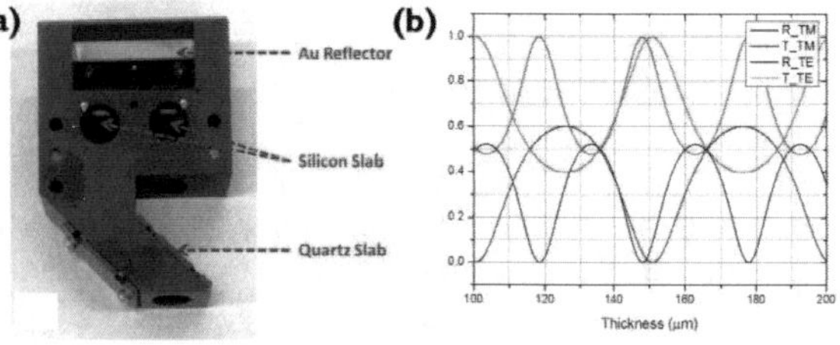

Fig. 2 (a) The four-pixel power distributor module. (b) The calculated transmittance and reflectance of polarizing beam splitters versus thickness.

COOLING TEST OF CARTRIDGE

The vacuum and cooling test of cartridge has been completed. The loading system with an ALMA-type testing cryostat is shown in Fig. 3. There are two parallel slippery tracks and a transportation plate on the system, and the cartridge can be adjusted to the target height by using a hand pulled stacker. The cartridge is mounted on the under-side of transportation plate, and then be slowly loaded into the testing cryostat along the slippery tracks. The Sumitomo RDK-3ST three stage cryocooler is used, which has cooling powers of 1.0 W at 4.4K for the 4K stage, 8 W at 18K for the 15K stage, and 33 W at 85K for the 110K stage.

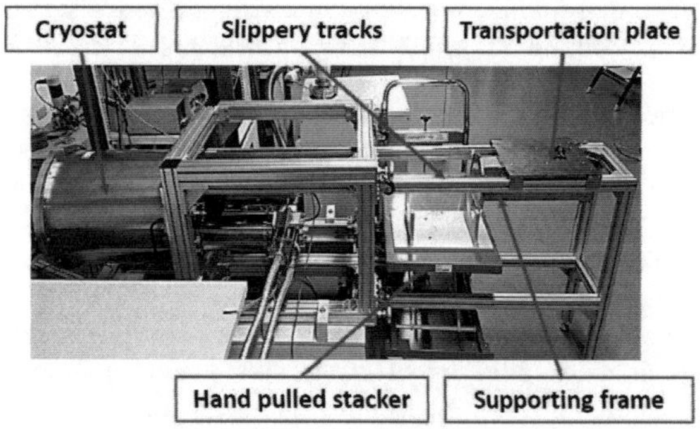

Fig. 3 The loading system and testing cryostat for cartridge.

The vacuum sealing test of 300K base-plate shows a helium leakage rate below 2.4E-6 mbar*L/s within 2000 seconds which can meet the ALMA specification, and the vacuum level of 5.7E-5 mbar at room temperature can be achieved. The temperatures of three stages during cooling procedure are shown in Fig. 4. The dashed line is the result of bare cartridge body, and the solid line is that of the engineering model cartridge. The balanced temperatures of three stages are 2.8K, 14.8K, and 79.5K for the bare cartridge, and 2.8K, 15.3K, and 80.2K for the single-pixel engineering model cartridge respectively. The DC biases for cooled multipliers of the LO module are applied before cooling. The slightly increase of temperature at 15K and 110K stages is due to extra thermal loading, such as cables, wires, and waveguides which are not included in the bare cartridge body.

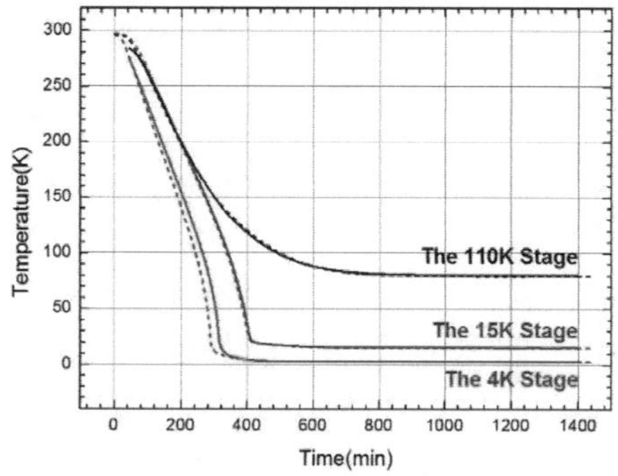

Fig. 4 The temperatures of the three cooled stages versus time during cooling procedure. The dashed line is the result of the bare cartridge, and the solid line is that of the engineering model cartridge.

To understand the thermal loading after the synthesizer of LO module is turned on, we mounted more sensors on different positions to measure the temperature variation. Fig. 5 shows the temperature evolution of each sensor when synthesizer of LO module is turned on and off. The two-doubler module is heated from 108.8K to 152.1K as the amplified Q-band LO signals are turned on, and the temperature of two-tripler module increases about 10K. However, the 4K plate and the mixer block (~ 3.4K) have a temperature increase below 0.015K. We expect that the performance HEB mixer won't be affected during operation.

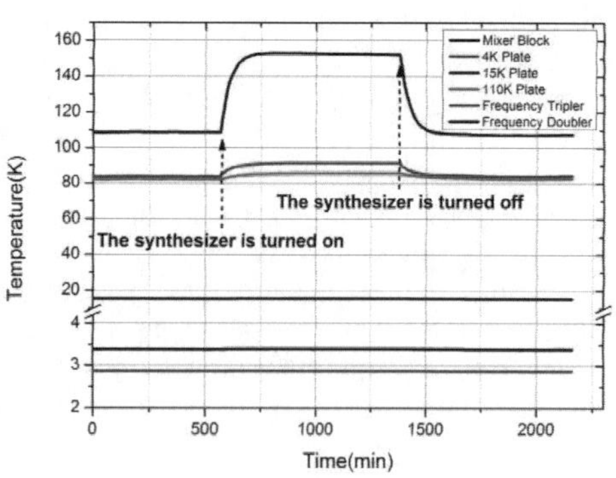

Fig. 5 The temperature distribution of the cold cartridge.

THERMAL ANALYSIS

The temperature distribution of cartridge has been simulated by ANSYS. The three cooled plates have a thickness of 10 mm and diameters of 169.0 mm at the 4K stage, 169.5 mm at the 15K stage, and 170.0 mm at the 110K stage respectively. The 4K and 15K plates are made of oxygen-free copper (OFCu) with gold plated, but the 110K plate is made of aluminum (Al). The thermal conductances of the three stages are set to 1.7 W/K, 5.6 W/K [8], and 7.0 W/K, and the heat generation of each frequency doubler of LO module are set to 0.75 W and 2.2 W when the synthesizer is turned off and on respectively. Fig. 6(a) shows the simulated temperature distribution of the bare cartridge, and Fig. 6(b) and 6(c) are that of the engineering model cartridge when the synthesizer of LO module is turned off and on respectively. The simulated balanced temperatures of the 110K plate, the two-tripler module, and the two-doubler module are 81.1K, 84.0K, and 105.2K when the synthesizer is off, and 83.3K, 91.0K, and 152.8K respectively when the synthesizer is turned on, which are consistent with our measurement results.

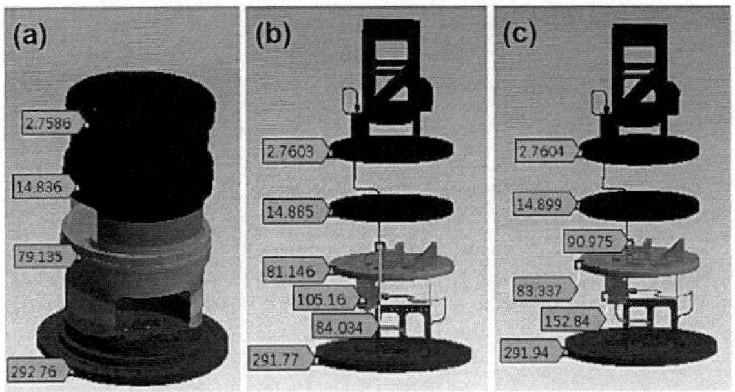

Fig. 6 (a) The simulated temperature distribution of the bare cartridge. (b) The simulated temperature distribution when the synthesizer is turned off. (c) The simulated temperature distribution when the synthesizer is turned on.

THE OUTPUT POWER OF THE LO MODULE

Cooled frequency multipliers are used in the LO module to boost the output power at the frequency range of 1.45 - 1.55 THz. The LO output power as the multipliers cooled/not cooled has been measured, as shown in Fig. 7. The LO output power with cooled multipliers is about 2.5 times higher than that at room temperature. The measurement result is much higher than the estimated value (about 40% enhancement when the two-tripler module is cooled only). The higher output power at low temperature is attributed to a higher conversion gain of the two-doubler module at low temperature, similar phenomenon to the cooled tripler. Such LO power boost could provide us more margins on our 9- or more pixels receiver in the future.

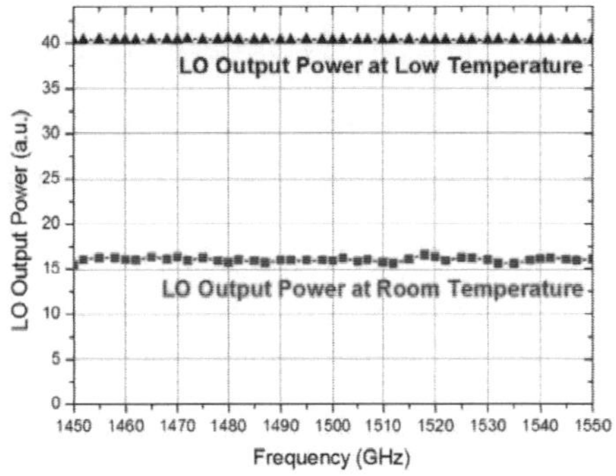

Fig. 7 The LO output power measurement at room temperature and low temperature.

SUMMARY

We present the design of engineering model of 1.5 THz single/four -pixel cartridge-type receiver based on HEB mixers. The cooling test of single-pixel cartridge has been carried out, and the measured temperature distribution of the cold cartridge is consistent with the simulation result. Mechanical analysis of cartridge will be done soon. The output power of the LO module at low temperature is nearly twice higher than the predicted value, which can provide more margins for 9- or more pixels receiver in the future. Testing with HEB mixers will be the next milestone to evaluate the performance of the cartridge receiver.

REFERENCES

F. Rodriguez-Morales, K. S. Yngvesson, E. Gerecht, N. Wadefalk, J. Nicholson, et al. "A terahertz focal plane array using HEB superconducting mixers and MMIC IF amplifiers," *IEEE Microw. Wireless Compon.*, vol. 15, pp. 199-201, Apr. 2005.

C. Risacher, R. Guesten, J. Stutzki, H. Huebers, D. Buechel, et al. "First supra-THz heterodyne array receivers for astronomy with the SOFIA observatory," *IEEE Trans. Terahertz Sci. Technol.*, vol. 6, pp. 199-211, Mar. 2016.

S. Cherednichenko, V. Drakinskiy, T. Berg, P. Khosropanah, and E. Kollberg, "Hot-electron bolometer terahertz mixers for the Herschel Space Observatory," *Rev. Sci. Instrum.*, vol. 79, 034501, Mar. 2008.

C. Walker, C. Kulesa, P. Bernasconi, H. Eaton, N. Rolander, et al. "The Stratospheric THz Observatory (STO)," *Proc. SPIE*, vol. 7733, 77330N, July 2010.

Y. C. Luo, X. X. Liu, D. J. Hayton, L. Wei, J. R. Gao, and C. Groppi, "Fourier phase grating for THz multi-beam local oscillators." in *ISSTT 2015: Proceedings of the 26th International Symposium on Space Terahertz Technology*, 2015, paper T2-4.

U. U. Graf and S. Heyminck, "Fourier gratings as submillimeter beam splitters," *IEEE Trans. Antennas Propag.*, vol. 49, pp. 542-546, Apr. 2001.

H. Hirashita, P. M. Koch, S. Matsushita, S. Takakuwa, M. Nakamura, et al. "First-generation science cases for ground-based terahertz telescopes," *Publ. Astron. Soc. Jpn.*, vol. 68, R1, Feb. 2016.

M. Sugimoto, Y. Sekimoto, S. Yokogawa, T. Okuda, T. Kamba, et al. "Thermal link for cartridge-type cryostat," *Cryogenics*, vol. 43, pp. 435-439, Mar. 2003.

Photon Noise Limited Performance over an Octave of Bandwidth of Kinetic Inductance Detectors for Sub-millimeter Astronomy

J. Bueno[1], O. Yurduseven[2], S. Yates[3], N. Llombart[2], V. Murugesan[1], D. J. Thoen[2], A. Baryshev[3,4], A. Neto[2], and J. Baselmans[1,2]

[1]SRON – Netherlands Institute for Space Research, Utrecht, The Netherlands
[2]Terahertz Sensing Group, Delft University of Technology, Delft, The Netherlands
[3]SRON– Netherlands Institute for Space Research, Groningen, The Netherlands
[4]Kapteyn Astronomical Institute, University of Groningen, Groningen, The Netherlands

We present the development of background limited kinetic inductance detectors (KIDs) for sub-millimeter (sub-mm) astronomy applications to be used in space based observatories. The sub-mm radiation is coupled to the KID via a leaky-wave antenna covering the frequency range from 1.4 to 2.8THz. We have developed a hybrid niobium titanium nitride/aluminium (NbTiN/Al) KID, fabricated on a silicon (Si) substrate, in which the leaky-wave antenna and absorbing section of the KID are fabricated on a suspended silicon nitride (SiN) membrane. The radiation is coupled to the leaky-wave antenna with a Si lens placed on top of it at a distance of 3μm. We observe photon noise limited performance both in the phase and amplitude readout simultaneously, with a good optical efficiency at a frequency of 1.55THz. The Fourier Transform Spectroscopy (FTS) measurements showing the broadband radiation coupling for an octave of bandwidth, and the beam pattern measurements at 1.55THz are also presented. In summary, we have developed a new fabrication route that assures photon noise limited performance, and a scalable assembly method that provides the 3μm gap space between the antenna and the lens. These developments assure background limited performance with a broad frequency coupling over an octave of bandwidth for sub-mm radiation. Given these promising results, hybrid NbTiN/Al leaky-wave antenna coupled KIDs will enable astronomically usable kilopixel arrays for sub-mm imaging for future space missions.

This work was supported as part of a collaborative project, SPACEKIDs, funded via grant 313320 provided by the European Commission under Theme SPA.2012.2.2-01 of Framework Programme 7.

Gap Frequency and Photon Absorption in a Hot Electron Bolometer

A. Trifonov[1*], C.-Y. E. Tong[1], Y. Lobanov[2,3], N. Kaurova[3], R. Blundell[1] and G. Goltsman[3]

[1]Harvard Smithsonian Center for Astrophysics, Cambridge, MA 02138, USA
[2]Moscow Institute of Physics and Technology, Dolgoprudny 141700, Russia
[3]Moscow State Pedagogical University, Moscow 119571, Russia
* Contact: avtrifonov@cfa.harvard.edu

The superconducting energy gap is a crucial parameter of a superconductor when used in mixing applications. In the case of the SIS mixer, the mixing process is efficient for frequencies below the energy gap, whereas, in the case of the HEB mixer, the mixing process is most efficient at frequencies above the gap, where photon absorption takes place more readily. We have investigated the photon absorption phenomenon around the gap frequency of HEB mixers based on NbN films deposited on silicon membranes. Apart from studying the pumped I-V curves of HEB devices, we have also probed them with microwave radiation, as previously described [1].

At frequencies far below the gap frequency, the pumped I-V curves show abrupt switching between the superconducting and resistive states. For the NbN HEB mixers we tested, which have critical temperatures of ~9 K, this is true for frequencies below about 400 GHz. As the pump frequency is increased beyond 400 GHz, the resistive state extends towards zero bias and at some point a small region of negative differential resistance appears close to zero bias. In this region, the microwave probe reveals that the device impedance is changing randomly with time. As the pump frequency is further increased, this random impedance change develops into relaxation oscillations, which can be observed by the demodulation of the reflected microwave probe. Initially, these oscillations take the form of several frequencies grouped together under an envelope. As we approach the gap frequency, the multiple frequency relaxation oscillations coalesce into a single frequency of a few MHz. The resultant square-wave nature of the oscillation is a clear indication that the device is in a bi-stable state, switching between the superconducting and normal state.

Above the gap frequency, it is possible to obtain a pumped I-V curve with no negative differential resistance above a threshold pumping level. Below this pumping level, the device demonstrates bi-stability, and regular relaxation oscillation at a few MHz is observed as a function of pump power. The threshold pumping level is clearly related to the amount of power absorbed by the device and its phonon cooling.

From the above experiment, we can derive the gap frequency of the NbN film, which is 585 GHz for our 6 μm thin silicon membrane-based device. We also confirm that the HEB mixer is not an efficient photon absorber for radiation below the gap frequency.

1. A. Trifonov et al., "Probing the stability of HEB mixers with microwave injection", IEEE Trans. Appl. Supercond., vol. 25, no. 3, June 2015.

Frequency Agile Heterodyne Detector for Submillimeter Spectroscopy of Planets and Comets

J. H. Kawamura, B. S. Karasik [1], M. S. Sherwin [2], S. Gulkis [1], P. Focardi [1], K. J. Stone [1], N. Chahat [1], and L. Pfeiffer [3]

[1] *Jet Propulsion Laboratory, California Institute of Technology*
[2] *University of California, Santa Barbara*
[3] *Princeton University*
* Contact: kawamura@jpl.nasa.gov

We are developing a novel heterodyne detector called a Tunable Antenna-Coupled Intersubband Terahertz (TACIT) detector, a four-terminal semiconductor device that has numerous advantages over current state-of-the-art Schottky and superconducting mixers. This detector absorbs radiation via micro-antennas which are ohmically contacted to a front and back gate biased two dimensional electron gas (2DEG) quantum well structure engineered from high mobility (>10^6 cm^2/Vs) GaAs. The photons are initially absorbed in a resonant intersubband transition of a 2DEG, which then creates a population of hot electrons that changes its resistance. The resonant frequency of the intersubband transition is electrically tunable in the 1-5 THz range, and the mixing mechanism should allow low-noise near-quantum limited operation up to a 70 K bath temperature. Because of the low density of electrons in the mixer volume, the TACIT mixer requires just a few μW of Local Oscillator (LO) power. Additionally, the technology is compatible with proven technology used to generate LO power at THz frequencies (in contrast to the > 1mW required to pump Schottky mixers). The low LO power requirement further translates into the possibility of deploying array receivers. Thus the advantages of the TACIT mixer over existing mixer technologies are as follows:

(1) Low noise operation at bath temperatures achievable using passive radiative cooling in space (50-70K).
(2) Very low LO power requirements compatible with solid-state sources with flight heritage (μW).
(3) Planar structure allows for simple and straightforward implementation of array receivers.

This mixer is ideal for long-duration planetary, astrophysics and Earth science missions, as they do not require active cooling like superconducting devices. Preliminary devices have demonstrated some salient properties of the device, in particular, by using high mobility (>10^6 cm^2/Vs) GaAs 2DEG antenna-coupled devices we have shown the key principles of electrical tuning and enhanced resonant absorption of THz radiation. This ability to electrically tune the resonant frequency over a wide range is what sets this detector apart from previous 2DEG mixers. We are currently optimizing the design as well as producing test structures to assess the performance of the antenna and the influence of the back gate.

Characterization of a free-standing membrane supported superconducting Ti transition edge sensor

W. Zhang[1,2,*], W. Miao[1,2], Z. Wang[1,2], D. Liu[1,2], J.Q. Zhong[1,2], F. Wu[1,2], Q.J. Yao[1,2], X.H. Guo[1,2], S.C. Shi[1,2]
[1]*Purple Mountain Observatory, CAS, Nanjing, 210008, China*
[2]*Key Lab of Radio Astronomy, CAS, Nanjing,210008, China*
*Contact: wzhang@pmo.ac.cn, phone +86-25-83332229

Abstract—**Superconducting transition edge sensors (TES) based on a Ti microbridge on Si substrate have demonstrated a very low noise equivalent power (NEP). Their effective response time, however, is on the order of microsecond due to relatively high transition temperature (i.e. 300-400 mK) of the Ti microbridge, making it difficult to read out the signal of a large Ti TES array with a SQUID-based multiplexer. We propose a twin-slot antenna coupled superconducting Ti microbridge separated from the antenna feed and supported by a free-standing membrane. Its resistive transition (R-T) and current-voltage (I-V) curves are measured before and after wet etching of the Si substrate underneath the Ti microbridge. The free-standing membrane supported Ti TES shows slightly lower transitions and higher normal resistance. Its thermal conductance is reduced to ~150 pW/K from ~2000 pW/K. In addition, its effective response time measured with a current pulse signal is about 30 μs.**

INTRODUCTION

The THz and FIR band contains a wealth of information about the cold universe. Observations of gas and dust can probe the earliest stages in the formation of galaxies, stars, and planets. Due to the limited atmospheric transparency, a large-pixel detector array with high sensitivity is desirable for ground-based THz/FIR telescopes [1]. Purple Mountain Observatory is leading the efforts on a 5-m THz telescope (DATE5) [2] to be constructed at Dome A, the highest position on the Antarctic plateau at an elevation of 4100 m. Dome A has been confirmed to be an ideal site on earth for terahertz astronomy. We are currently developing a terahertz superconducting imaging array (TeSIA) for DATE5 at 350 μm [3], and TES is a potential detector candidate. Superconducting Ti TESs have demonstrated a very low optical NEP at 620 GHz [4], but its response time is on the order of microsecond due to relatively high transition temperature (i.e. 300~400 mK) of the Ti microbridge [5, 6], making it difficult to read out the signal of a large Ti TES array with a SQUID-based multiplexer. Thermal conductance between Ti microbridge and Si substrate should be further lowered to increase the effective response time. One way is to suspend the Ti microbridge from the substrate with legs for thermal isolation [7]. Here we propose a twin-slot antenna coupled superconducting Ti microbridge separated from the antenna feed and supported by a free-standing membrane. We present the details of the detector design, fabrication, electric characterization, and study the back etch effect on thermal conductance and response time.

DESIGN AND FABRICATION OF FREE-STANDING MEMBRANE SUPPORTED TI TES

Our design uses a twin-slot antenna to couple THz radiation (see Fig. 1b). The combination of twin-slot antenna and silicon elliptical lens has a nearly symmetric beam and linear polarization, which have been used successfully in THz heterodyne mixers [8, 9]. The electric field from each slot antenna propagates along coplanar waveguide (CPW) transmission line, then coherently added, and terminated at the Ti microbridge. The Ti microbridge works both as absorber and thermistor. The slot length, width and separation are 246 μm, 16 μm and 140 μm, respectively. Simulation shows that the twin-slot antenna is resonant at 345 GHz, and it is well matched with Ti TES with 30 Ω normal resistance via CPW transmission line.

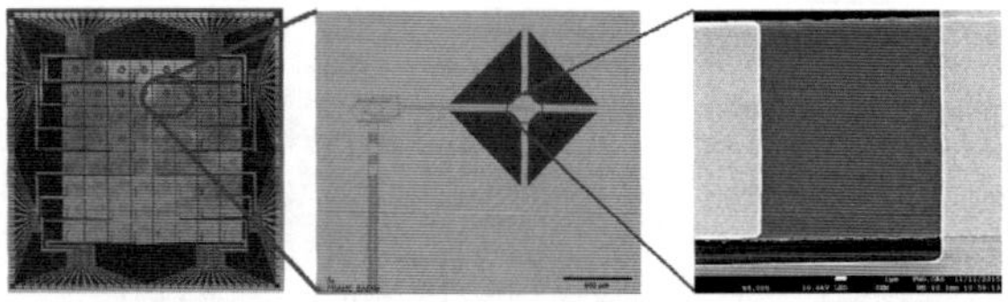

(a) (b) (c)

Fig. 1 (a) 8x8 superconducting Ti transition edge sensor array. (b) photo of the microbridge integrated with twin-slot antenna via CPW transmission line. (c) Ti microbridge with dimension of 16μm x16μm.

The superconducting TES devices are based on a Ti film, which is electron-beam evaporated on a 250-μm high-resistivity Si substrate in an ultrahigh vacuum environment. The microbridge is patterned by optical lithography. RF cleaning is used to remove the TiOx on the surface of Ti film before the deposition of 150 nm thick Nb contacts. The critical temperature of the Nb contacts is about 9K, so they serve as the Andreev reflection contact material. The Ti microrbidge is chosen to be 16 μm × 16μm, providing suitable saturation power (see Fig.1c). As shown in Fig.1a, we designed and fabricated a 8x8 superconducting Ti TES array. Fig. 1b shows a single-pixel Ti TES, including twin-slot antenna, RF choke filter, CPW transmission line, and Ti microbridge.

The fabricated Ti TES device is tested using an Oxford Triton 400 dilution cooler [10] that is able to cool the device down to 20 mK (see Fig. 2). The Ti TES wafer is anchored to the copper holder, then mounted on the mixing chamber (MC) stage. A temperature sensor is used to monitor the holder temperature. The constant voltage is realized with a 0.68-Ω shunt resistor (RSH) in parallel with the Ti TES device, which is then connected to the input coil of a Magnicon single-stage SQUID [11] on the 1K stage via twisted superconducting NbTi wires. The input inductance of the single-stage SQUID is 150 nH, and its current noise contribution is about 1 $pA/Hz^{0.5}$. The TES current is read out by the SQUID with a closed flux-locked loop.

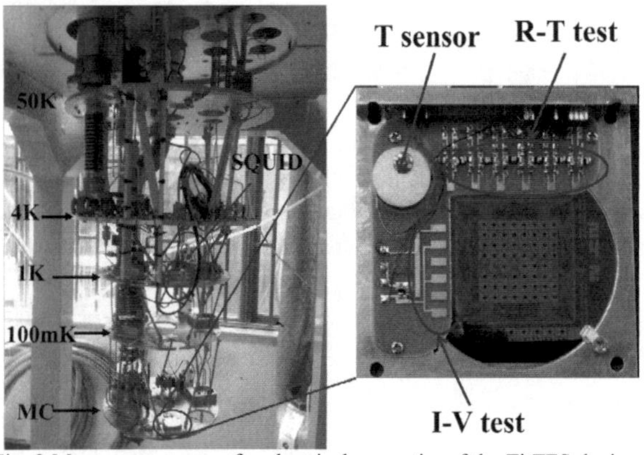

Fig. 2 Measurement setup for electrical properties of the Ti TES device. The TES device is mounted on the MC stage of dilution cooler. The TES current is read out using a single-stage SQUID operated at 1 K cold plate.

R-T CURVES

The resistance as a function of temperature is measured with an ac bridge. The results are shown in Fig.3. The resistance of Ti TES is about 1.6 KΩ at room temperature, and decreases with temperature. After Nb transition at ~9 K, the Ti microbridge shows a normal resistance of 2.3 Ω, consistent with sheet resistance of Ti film. There are two transition temperatures (i.e. 290 mK and 358 mK) from Ti microbridge and Ti/Nb contact pads, respectively. After KOH wet etch, the Ti TES device shows the same temperature dependence of its resistance, indicating that the Nb wire is not influenced by the KOH process. However, the normal resistance of Ti microbridge at 0.4 K is increased to 6.4 Ω, and the transition temperatures of Ti microbridge and Ti/Nb contacts are 266 mK and 338 mK, respectively. The superconductivity of Ti film is deteriorated by the KOH wet etch process. The reason is not clear and needs further study.

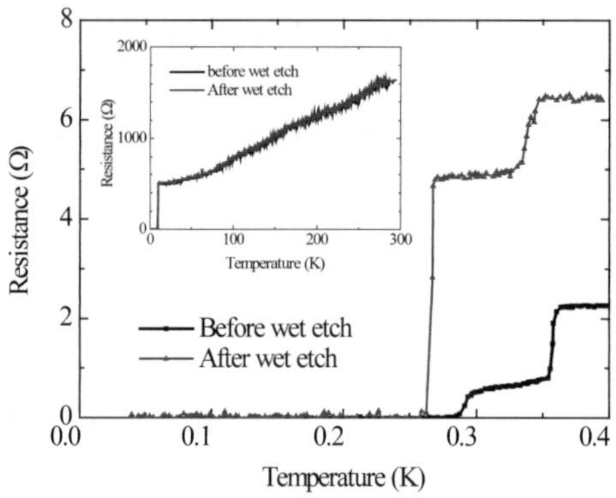

Fig.3 Measured resistance as a function of temperature of the Ti TES before and after wet etch

CURRENT-VOLTAGE CURVES

We measured the current–voltage characteristics of the Ti TES device at different bath temperatures between 50 mK and 350 mK (see Fig. 4). The current of the TES device (I_{TES}) can be directly obtained from the output voltage with the SQUID amplifier gain. Parasitic resistance (R_{PAR}) and normal resistance (R_N) are determined from its superconducting and normal branches of the I–V curves, respectively. The TES voltage (V_{TES}) is calculated through the Thevenin equivalent circuit model [12]. As plotted in Fig.4, there are several steps in the I–V curves in the transition regime, which might be due to the proximity effect. Superconductivity of the Ti film under the Nb contacts is indeed enhanced, leading to other transitions with higher critical temperatures in the measured resistive curve, which can be seen from the R-T curves in Fig.3. We choose the data points of the I–V curves at 1 Ω and plot the power level as a function of bath temperature (see Fig. 5).

The power flow from electrons to phonons follows $P_{DC} = K(T_c^n - T_{bath}^n)$, where P_{DC} is the DC bias power applied to the TES device, K a constant that depends on the geometry and material properties of the supporting structure, and n a thermal-conductance exponent depending on the dominant

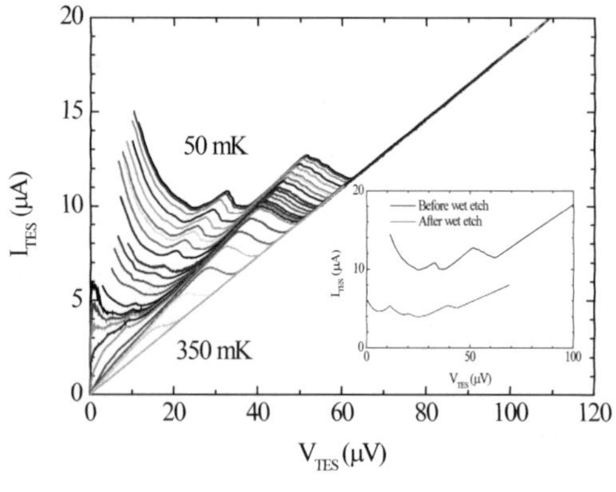

Fig. 4 I-V curves of the Ti TES at different bath temperatures before wet etch. The inset shows the I-V curves measured before and after KOH wet etch.

thermal transport mechanism. We can fit the measured DC power as a function of bath temperature to find K, n, and T_c. The best fit is obtained using $K=1.628\times10^4$ pW/K^n, n = 3.6 and T_c = 283 mK. Thermal conductance (G) between the TES device and the substrate can be calculated straightforwardly by $G = nKT_c^{n-1}$ =2219 pW/K. As n is close to 4, we indicate that the electron–phonon coupling is the dominant energy relaxation mechanism. After that, the Si substrate underneath the Ti microbridge was wet etched with KOH and the I-V curves were measured once again (see the inset of Fig. 4). The calculated thermal conductance (G) is reduced to 153 pW/K as shown in Fig. 5.

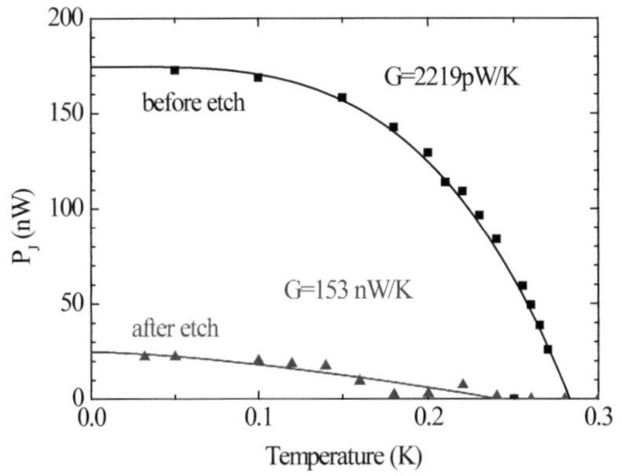

Fig.5 The power level as a function of bath temperature

As shown in the inset of Fig. 5, the two I-V curves show the similar shape. Due to the wet etch, the critical current is apparently reduced to about 5μA. This may be caused by the strain change of the TES device. After wet etch, n is reduced to 1.48 from 3.6, indicating that dominant thermal transport mechanism is diffusive phonon transport instead of electron-phonon interaction.

TABLE I
KEY PARAMETERS

	R_N (Ω)	T_c (K)	n	K	G (pW/K)	τ (μs)
Before etch	5.5	0.28	3.6	16280	2200	3
After etch	8.6	0.24	1.48	204	150	29

Effective Response Time

We then applied a current pulse with an amplitude much less than the bias current to the TES device at 2.66 μV, and measured its response. The rise and fall time of the current pulse is below 100 ns, which is much shorter than the measured response time. The decay curve can be fitted with an equation $V(t) = A_0+A_1\exp(-(t-t_0)/\tau_{eff})$, where A_0 and A_1 are the shift and the pulse height, t_0 and τ_{eff} are the current pulse incident time and response time, respectively. The best fitting value of the effective response time is τ_{eff} = 29 μs as shown in Fig.6. For comparison, the measured decay curve for a Ti TES without back etch is also plotted in the same figure. The effective response time is 3 μs, about 10 times faster than that with back etch.

Table 1 summarizes the main results of the Ti TES. After back etch, the thermal conductance is reduced from 2200 pW/K to 150 pW/K. Consequently, the effective response time is increased to about 30μs, 10 times larger than before. However, the KOH back etch has some influence on the superconductivity of Ti TES, the normal resistance becomes larger and the critical temperature shifts towards lower temperature, which should be studied further.

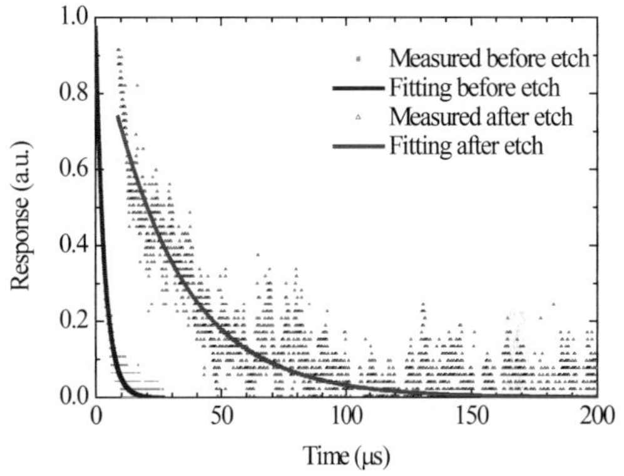

Fig.6 The measured and fitting pulse response of TES device before etch and after etch

Conclusion

We have studied the electrical performance of free-standing membrane supported superconducting Ti TES device. The thermal conductance extracted from the measured I–V characteristics at different bath temperatures is 150 pW/K as expected. As a result, the effective response time is 29 μs, ~10 times larger than that without back etch. We will further lower the thermal conductance by etching the SiN membrane into four legs, and the response time can be increased to 1ms, making it possible to use a SQUID-based time domain multiplexing to read out the signal of a 8x8 TES array.

Acknowledgment

The authors would like to thank W.Y. Duan and K. Zhang for their technical support. This work was supported in part by the NSFC under Grants 11127903, 11190012, 11473075, by the Chinese Academy of Sciences under Grant XDB04010300, and by the operation, maintenance and upgrading fund for astronomical telescopes and facility instruments.

References

[1] W. S. Holland et al., “SCUBA-2: The 10000 pixel bolometer camera on the James Clerk Maxwell Telescope,” Mon. Not. R. Astron. Soc., vol. 430, p. 2513, 2013

[2] J. Yang, Y.X. Zuo, Z. Lou, J.Q. Cheng, Q.Z. Zhang et al., Res. Astron. Astrophys. 13(12), 1493–1508 (2013).

[3] S.C. Shi, W. Zhang, J. Li, W. Miao, Z.H. Lin, Z. Lou, and Q.J. Yao, "A THz superconducting imaging array developed for the DATE5 telescope", J. Low Temp. Phys.vol. 184, 2016.

[4] B.S. Karasik, R. Cantor, "Demonstration of high optical sensitivity in far-infrared hot electron bolometer," Appl. Phys. Lett. **98**, 193503, 2011.

[5] B. S. Karasik, S.V. Pereverzev,D. Olaya, M. E. Gershenson, R. Cantor, J. H. Kawamura, P. K. Day, B. Bumble, H. G. LeDuc, S. P.Monacos, D. G.

Harding, D. Santavicca, F. Carter, and D. E. Prober, “Development of the nano-HEB array for low-background far-IR applications,” in Proc. SPIE, vol. 7741, p. 774119, 2010.

[6] W. Zhang, J.Q. Zhong, W. Miao, Z. Wang, D. Liu, Q.J. Yao, S.C. Shi, T.J. Chen, M.J. Wang, "Electrical chracteristics of superconducting Ti transition edge sensors," J. Low Temp. Phys, 184,11-16, 2016.

[7] M.J. Myers, W. Holzapfel, A.T. Lee, R. O'Brient, P.L. Richards, and H.T. Tran, "An antenna-coupled bolometer with an integrated microstrip bandpass filter," Appl. Phys. Lett. 86, 114103, 2005.

[8] W. Zhang, J.R. Gao, M. Hajenius, W. Miao, P. Khosropanah, T.M. Klapwijk, and S.C. Shi, "Twin-slot antenna coupled NbN hot electron bolometer mixer at 2.5 THz" IEEE Trans. Terahz. Sci. Tech. vol.1 (2), pp.378-382, 2011.

[9] W. Zhang, D. Hayton, J.R. Gao, M. hajenius, W. Miao, D. Liu, T.M. Klapwijk, and S.C. Shi, "Optimized sensitivity and beam pattern of a twin-slot antenna coupled NbN HEB mixer at 1.6THz," in Proc. SPIE, vol. 8452, p. 84522H, 2012.

[10] Triton TM 200/400, Cryofree dilution refrigerator, Issue 3.0 (Oxford Instruments, 2012)

[11] *High performance DC SQUID electronics, XXF-1 manual*, v3.3.11. Magnicon Phys. Res. Instrum., Hamburg, Germany (2013)

[12] K. D. Irwin and G. C. Hilton, “Transition-edge sensors, in Cryogenic Particle Detection,” in *Topics in Applied Physics*, vol. 99. Berlin, Germany: Springer-Verlag, 2005.

A HEB Waveguide Mixer Operatingwith a Waveguide QCL at 1.9 THz

D. Büchel[1*], M. Justen[1], K. Jacobs[1], P. Pütz[1], M. Schultz[1], K. Otani[2], U. U. Graf[1], C. E. Honingh[1], and J. Stutzki[1]

[1]*Kölner Observatorium für Submm-Astronomie (KOSMA), Universität zu Köln, Zülpicher Str. 77, 50937 Köln, Germany*
[2]*Institute for Quantum Electronics, ETH-Zürich, Zürich, Switzerland*
* Contact: buechel@ph1.uni-koeln.de

We present results of heterodyne measurements at 1.9 THz using a hot electron bolometer (HEB) waveguide mixer with a waveguide quantum cascade laser (QCL) and with a commercial solid-state multiplier as local oscillator(LO).

The HEB mixer is similar to the currently operating in the upGREAT low frequency array on the Stratospheric Observatory for Infrared Astronomy (SOFIA). The HEB mixer devices employ NbN microbridges integrated into an on-chip matching circuit with a waveguide probe antenna. The circuit is defined on a 2 µm thick Si membrane and is suspended and contacted with beamleads. The device is assembled into a 48 µm by 96 µm rectangular metal waveguide and uses a waveguide spline profile feedhorn (commercial) as an interface to free space. The 1.9 THz QCL is a double metal QCL, which is embedded in a waveguide with a broadband coupling structure and radiates to free space with an integrated diagonal feedhorn. The QCL operates at an ambient temperature of 12 K.

In a standard heterodyne measurement setup with aMylar beam splitter, system noise temperatures between 700 K and 1500 K are measured over a 0.2-4 GHz IF bandwidth. We study the performance ofboth types of LOs and their impact on the measured receiver noise performance.

Our HEB mixer work currently is supported by the Federal Ministry of Economics and Technology via the German Space Agency (DLR), grant 50 OK 1103 and by Collaborative Research Council 956, sub-project area D, funded by the Deutsche Forschungsgemeinschaft (DFG).

Single Junction Design for 790-950GHz SIS Receiver

KirillRudakov[3,4,*], AndreyKhudchenko[1],PavelDmitriev[3], ValeryKoshelets[3], RonaldHesper[2], AndreyM. Baryshev[1,2]

[1]*SRON Netherlands Institute for Space Research, Groningen, the Netherlands*
[2]*University of Groningen, Kapteyn Astronomical Institute, Groningen, the Netherlands*
[3]*Kotel'nikov Institute of Radio Engineering and Electronics RAS, Moscow, Russia*
[4]*Moscow Institute of Physics and Technology, Dolgoprudny, Russia*
*Contact: rudakov@phystech.edu

Two years ago, we have fabricated and tested superconductor-insulator-superconductor (SIS) mixers based on Nb/AlN/NbN twin tunnel junctions for waveguide receiver operating in frequency range of 790 - 950 GHz, which demonstrated noise temperature from 250 K at low frequencies to 500 K at the high end of the band. Due to high current density of the junctions (up to 30 kA/cm^2), a wideband design was realized.

Based on result for twin junction, a new mixer design with single SIS junction was proposed and produced. The single junction design is expected to have narrower frequency range than twin-based one, but still wide enough, and provide better coupling in the middle of the band. In addition, the critical current suppression will be much better, than for twin junction, which has an intrinsic problem of difference between areas of individual junctions. The SIS junction in new design is made by Nb/AlN/NbN technology and incorporated in a microstrip line consisting of a 300 nm thick bottom electrode (ground plane) made of NbTiN and a 500 nm thick top electrode made of Al.The top Al electrode is passivated by the SiO_2 layer for protection. The DC tects of the fabricated wafer shows, that optimization of the technology allowed to realize gap voltage of 3.2 mV even for junctions of 0.5 square micron area fabricated on the NbTiN bottom layer. The FTS and noise temperature measurement of new mixers will be done in the nearest future and will be presented in comparison with the twin junction design results at the conference.

A 1080-1280 GHz Sub-Harmonic biasable Schottky Front-end Design for Planetary Science and Remote Sensing

D. Moro-Melgar[1*], A. Maestrini[1], *Member, IEEE*, J. Treuttel[1], L. Gatilova[1&2], F. Yang[4], F. Tamazouzt[1], T. Vacelet[1], Y. Jin[2], A. Cavanna[2], J. Mateos[5], *Member, IEEE*, A. Feret[1], C. Chaumont[1], C. Goldstein[3]

[1]*LERMA, Observatoire de Paris, PSL Research University, CNRS, Sorbonne Universités, UPMC Paris06, F-75014, France*
[2]*Laboratoire de Photonique et de Nanostructures, CNRS, 91460 Marcoussis, France*
[3]*Centre National d'Etudes Spatiales (CNES), Toulouse, France*
[4]*State Key Lab, Nanjing, China*
[5] *(USAL) Universidad de Salamanca, Salamanca 37008, Spain*
* Contact: diego.moro-melgar@obspm.fr

The design and optimization of a sub-harmonic biasable frequency mixer at 1.2 THz based in the Schottky diode technology is presented in this work. The design is dedicated to the Sub-millimeter Wave Instrument (SWI) which will take part of the payload in the JUpiter ICy moon Explorer (JUICE) mission of the European Space Agency (ESA). The procedure previously followed by LERMA in the design of a front-end receiver at 600 GHz that performs an average of 1284 K DSB noise temperature, presented in [1], is applied on the design of this receiver at 1.2 THz. The methodology used in the design and optimization is based on a non-linear harmonic balance simulator (Agilent ADS) coupled with a 3D-Electro-magnetic simulator (Ansoft HFSS). Both the Monolithic Microwave Integrated Circuits (MMIC) of the local oscillator (LO) chain and the mixing stage are manufactured using the E-beam photolithography LERMA-LPN process [2], especially important for the fabrication of the smallest features in the chip, such as the Schottky anodes and air-bridges. The local oscillator chain consist of an already tested power combiner frequency doubler at 270-320 GHz, which performs up to 27% of conversion efficiency using the on board power supply configuration, and a second single chip frequency doubler at 540-640 GHz which has been designed to perform up to 17% of conversion efficiency to deliver at least 1 mW of local oscillator input power. Due to the low LO power delivered on the mixer, a bias configuration has been included in the design in order to increase the tune-ability of the receiver along the frequency band. Several technical improvements in the manufacture process of these MMIC devices have allow us the design of more sophisticated structures, such as the DC path and the edging of the GaAs-membrane to reduce transmission losses. In addition, some of the possible options related to the Schottky diode structure have been studied by considering a 2-dimensional physical simulator base on the *ensemble* Monte Carlo method for semiconductor devices [3].

Preliminary discrete Schottky components fabricated for the 1.2 THz frequency mixer, have given us a set of measurements of the I-V characteristics to be introduced in the ADS Schottky diode model. The parameters used in the Schottky diode model consist of a $C_{j0} < 1$ fF junction capacity, a $I_s = 3.65 \cdot 10^{-13}$ A saturation current, a $\eta > 1.4$ ideality factor and a $R_s > 60\ \Omega$ series resistance. The ADS test-bench includes the losses introduced by the RF antenna, the real LO and RF path dimensions and the noise figure of the low noise amplifier (LNA), in order to reproduce as well as possible the real conditions of the receiver. A DSB noise temperature under 5000 K is expected at room temperature in the entire frequency band according to our ADS-HFSS simulations. The experimental verification of this performances waits for the finalization of the current manufacture process at LPN.

References

[1] J. Treuttel, L. Gatilova, A. Maestrini, D. Moro-Melgar, F. Yang, F. Tamazouzt, T. Vacelet, Y. Jin, J. Mateos, A. Feret, C. Chaumont, C. Goldstein. "A 520-620 GHz Schottky receiver front-end for Planetary Science and Remote Sensing with 1070K-1500K DSB noise temperature at Room Temperature".*IEEE Trans.Terahertz Science and Tech.*, vol. PP, no 99, p.1-8 , 2015.

[2] J. Treuttel, B. Thomas, A. Maestrini, H. Wang, B. Alderman, J. V. Siles, S. Davis and T. Narhi. "A 380 GHz sub-harmonic mixer using MMIC foundry based Schottky diodes transferred onto quartz substrate". *20th International Symposium on Space Terahertz Technology, Charlottestville*.p. 20-22, 2009.

[3] J. Mateos, T. González, D. Pardo, V. Hoel, H. Happy and A. Cappy. "Design optimization of AlInAs-GaInAs HEMTs for high-frequency applications. *IEEE Trans. Electron Devices*, vol. 51, no 4, p. 521-528, 2004.

Development of an RF Waveguide Frequency Multiplexer for a Multiband Heterodyne System

T. Kojima[1*], A. Gonzalez[1], S. Asayama[1], and Y.Uzawa[2]

[1]*National Astronomical Observatory of Japan *, Mitaka, Tokyo 181-8588, Japan*
[2]*National Institute of Information and Communications Technology, Koganei,Tokyo, 184-8795, Japan*
* Contact: t.kojima@nao.ac.jp

The Atacama Large Millimeter/submillimeter Array (ALMA) is the most powerful ground-based radio telescope. The ALMA telescope uses ultra-sensitive cartridge-type heterodyne receivers. The instantaneous bandwidth of the ALMA for Bands 3 to 8 (except for Band 5) is currently 4 GHz per sideband and polarization, which is limited by the amplifiers in principle. Recently, although microwave low-noise cryogenic amplifiers with wide bandwidth exceeding 10 GHz have been developed, the instantaneous bandwidth is still limited if compared with the radio frequency (RF) bandwidth, especially at the higher frequency bands. The purpose of this study is to increase the instantaneous bandwidth, focusing on a front-end receiver system. The system proposed introduces the concept of a multiband receiver which consists of two multiplexers, one for dividing the full RF band into smaller bandwidths, and the other to separate the tones in a multi-frequency local oscillator (LO) signal. Each of the smaller bandwidth RF signals and the corresponding LO tone are injected into individual dedicated SIS mixers. The resulting down-converted signals will be in the same IF bandwidth and can be simultaneously amplified with dedicated similar IF amplifiers. This allows down-converting the full RF bandwidth at once, which translates into ultra-wideband operation.

This paper will describe the concept design of the multiband heterodyne receiver system in the 380-500 GHz band. We will also present the design of a waveguide RF multiplexer with 25 GHz bandwidth for each channel. The 25 GHz bandwidth has been chosen because it corresponds to the state-of-the-art bandwidth of current IF amplifiers. The designed multiplexer has no frequency gap between channels. This is done by using a hybrid-coupled multiplexer, which is composed of two identical 90 degree hybrid couplers and two identical filters. Waveguide iris coupled bandpass filters and 3-dB blanch-line couplers were used for the multiplexer. Full-wave simulation results will be compared with measurements.

ACKNOWLEDGMENTS

Part of this work was supported by JSPS KAKENHI Grant Number 15K18057, and was made possible by the Visiting Scholar Program grant supported by the Research Coordination Committee, National Astronomical Observatory of Japan (NAOJ).

Concept Design of a Dual-Polarization Sideband-Separating Multi-Pixel SIS Receiver

Wenlei Shan[1*,2], GonzalezAlvaro[1], Matthias Kroug[1], Takafumi Kojima[1], Yasunori Fujii[1], Shinichiro Asayama[1], Takashi Noguchi[1]

[1]*National Astronomical Observatory of Japan*, Mitaka, Tokyo, 181-8588, Japan,*
[2]*Purple Mountain Observatory, Nanjing,JiangSu, 210008, China*
* Contact: wenlei.shan@nao.ac.jp

ALMA is a unique facility not only because it combines high sensitivity and high angular resolution, but also because it is sensitive to the cold gas in the universe that radiates mainly in mm/sub-mm frequencies, which has not been fully explored. The weak point of ALMA is its narrow field of view (FOV) with a typical value of one arc minute. The proposed multi-pixel SIS receiver is aimed to extend the FOV of ALMA by an order of magnitude. The frontend of the receiver is an integrated type. This assembly concept will largely reduce the use of electrical and mechanical interconnections, and thus brings about robustness and reliability of the multi-pixel frontend. In this design, two linear polarizations are received in a sideband-separating (2SB) manner. Both the orthomode transducers and the quadrature hybrids are implemented with planar circuits in order to reduce the difficulty in the machining of metallic waveguides. Low power-consumption MMIC cryogenic amplifiers are considered to be integrated into the frontend module. Radio-over-fiber technique is designed to transmit tens of IF channels out of the receiver cryostat with little thermal leakage. Challenges also come from the multi-pixel optical design and the increase in computing cost. They will be also discussed.

Development of Terahertz SIS Mixers Using Nb/AlN/Nb Tunnel Junctions Integrated with All NbTiN Tuning Circuits

Y. Uzawa[1, 2*], M. Kroug[2], T. Kojima[2], K. Makise[1], A. Gonzalez[2], S. Saito[1], Y. Fujii[2], K. Kaneko[2], H. Terai[1], and Z. Wang[1]

[1]*National Institute of Information and Communications Technology, Koganei, Tokyo,184-9795, Japan*
[2]*National Astronomical Observatory of Japan, Mitaka, Tokyo, 181-8588, Japan*
* Contact: uzawa@nict.go.jp

Abstract—We are developing ultra-sensitive superconductor-insulator-superconductor (SIS) mixers at terahertz frequencies such as the Atacama Large Millimeter/submillimeter Array (ALMA) Band 10 (0.787-0.95 THz). Current SIS mixers for the Band 10 receivers employ high-quality Nb/AlOx/Nb tunnel junctions and low-loss Al/SiO2/NbTiN tuning circuits and have shown excellent noise performance compliant with the stringent ALMA requirements of less than 5 hf/k for all of our mass-produced 73 receivers. To further improve the noise performance, our approach is to replace the normal metal Al wiring in the tuning circuit with the superconducting material NbTiN. We have so far investigated the superconducting properties of NbTiN films deposited on sputtered SiO2 layers to make all NbTiN microstrip transmission lines (i.e. NbTiN/SiO2/NbTiN structure) at terahertz frequencies and confirmed that the properties of NbTiN films on SiO2 are as good as those of NbTiN films directly sputtered on quartz substrates, showing a critical temperature of 14 K and a gap frequency of 1.2 THz. However, it is known that I-V characteristics of Nb SIS junctions embedded in all NbTiN circuits are degraded because of quasi-particle trapping in the Nb electrodes of the junctions due to the superconducting gap difference between Nb and NbTiN. This brings about an effective temperature increase in the junctions and thus results in a reduction of the gap voltage. To solve this issue, it would be effective to increase the volume of the Nb electrodes. We have fabricated Nb/AlN/Nb tunnel junctions with relatively thin (~ 50 nm) and thick (~ 200 nm) counter electrodes contacting the NbTiN wirings and compared their I-V curves. We observe that the gap voltage reduction is smaller for the junctions with the thick Nb counter electrode, hence the heating effect is lessened. This indicates that SIS mixers employing Nb junctions with thick electrodes and embedded in an all NbTiN tuning circuits may work well at terahertz frequencies.

The authors would like to thank Prof. Ivan Nevirkovets of Northwestern University and HYPRES Inc. for his valuable suggestions about the device structure. This work was supported in part by JSPS KAKENHI under Grant 26420330 and by a grant for the Joint Research by the Chile Observatory, NAOJ.

Gas cell measurement using an HEBM with a phase-locked THz-QCL as a local oscillator at 3 THz band

Y. Irimajiri*, A. Kawakami, I. Morohashi, M. Kumagai, S. Nagano,
N. Sekine, S. Ochiai, S. Tanaka, Y. Hanado, Y. Uzawa, and I. Hosako
National Institute of Information and Communications Technology, Koganei, Tokyo, Japan
* Contact: irimaji@nict.go.jp, phone +81-42-327-6089

Abstract— **We have developed a 3 THz-band HEB mixer with a THz-QCL as a local oscillator. We demonstrated phase-locking of a THz-QCL using an HEBM and a THz-reference, which was generated by a frequency-comb and a UTC-PD. We have measured emission line spectra of methanol [CH_3OH] at 3.7 THz band using the HEBM with a phase-locked THz-QCL as a local oscillator. We also measured H_2O, HDO, and D_2O lines.**

I. INTRODUCTION

We are developing a superconducting low noise heterodyne receiver based on a hot electron bolometer mixer (HEBM) with a THz quantum cascade laser (QCL) as a local oscillator for astronomical and atmospheric observations at THz frequencies (2-5 THz). The observing target will be OH (1.83 THz, 3.55 THz) and O-atom (2.06 THz, 4.75 THz) etc. in the Earth's atmosphere. We may plan a balloon experiment in the future. We may consider the first balloon flight at the balloon launch sight of ISAS/JAXA in Japan (Taiki-cho, Hokkaido) mainly for demonstration of a THz receiver.

An NbN HEBM device was fabricated in our laboratory. The device was installed in a quasi-optical mixer mount with a log-spiral antenna and an AR-coated Si lens. A Fabry-Perot type metal-metal THz-QCL lasing at 3.1 THz with output power of 140 μW was also fabricated in our clean room facility. We have achieved uncorrected receiver noise temperature of 1,200 K (DSB) as a best value using a vacuum optics and a 4-μm thick Polyester beam splitter [1]. The THz-QCL was phase-locked [2] to a THz reference, which was generated by an optical-comb and a UTC-PD [3].

We have demonstrated gas cell measurement of emission line spectra of molecules using a HEBM with a phase-locked THz-QCL as a local oscillator at 3.7 THz-band.

II. MEASUREMENT SETUP AND RESULT

Fig.1 shows a gas cell system for the measurement of emission line spectra of molecules using an HEBM with a phase-locked THz-QCL. Fig. 2 shows the photograph of the measurement setup. Two HEBMs and LNAs were installed into a same cryostat of a pulse tube cooler. A THz-QCL was cooled to 45 K using a Stirling cycle cooler with active vibration cancellation (AVC) system. The temperature fluctuation of the cooler was also stabilized less than 0.01 K. A 3rd order DFB THz-QCL lasing at 3.7 THz was phase locked using an HEBM1 to a THz reference, which was generated by an optical-comb and a UTC-PD. The 3.7 THz-QCL array was fabricated by Longwave Photonics LLC. The output power is ~1 mW at an operation temperature of 45 K in CW-mode. For the spectroscopic measurement, it is important to design and fabricated the THz-QCL with the emission frequency close to the frequency of spectral line. The frequency of a THz-QCL should be close to that of the spectral line around +/-3 GHz, which is an IF bandwidth of a HEBM. At the present, we cannot fabricate a THz-QCL lasing at the required frequency with an accuracy of +/-3 GHz, however, we will try to do. The phase-locked THz signal was fed into another HEBM2 using a beam splitter. The line spectra were obtained using a digital spectrometer with a bandwidth of 1 GHz and 13 K channels. We have also successfully phase-locked the THz-QCL using a superlattice (SL) harmonic mixer. A 181-194 GHz AMC (amplifier/multiplier chain) with a typical output power of 20 mW was used as a local oscillator for the SL mixer. We can also use the SL mixer for this experiment instead of an HEBM. It is noted that the frequency of the local oscillator should be selected considering the required frequency of a THz-QCL, because the conversion efficiency of the superlattice mixer is higher for even harmonic number than odd number.

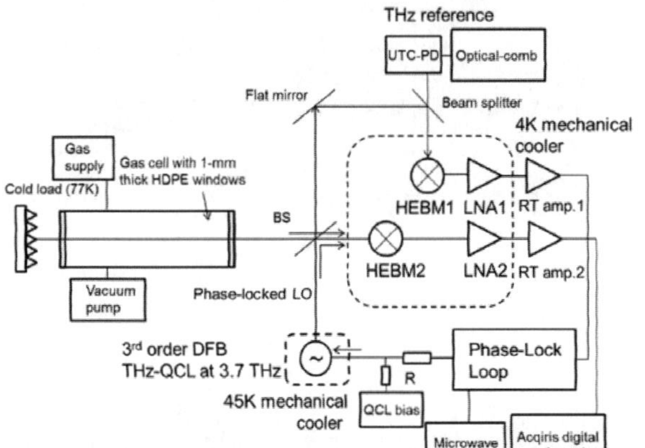

Fig. 1. A gas cell system for the measurement of emission line spectra of molecules using an HEBM with a phase-locked THz-QCL-HEBM. A 3rd order DFB THz-QCL at 3.7 THz was phase locked using an HEBM1 to a THz reference, which was generated by an optical comb and a UTC-PD. The phase-locked THz signal was fed into another HEBM2 using a beam splitter. The line spectra were obtained using a digital spectrometer with a bandwidth of 1 GHz and 13 K channels.

Fig.3 (a) and (b) show the measured emission line spectra of methanol [CH_3OH] at 3.7-THz band. The integration time was around 50 s. Because the receiver was operated in DSB mode, both upper and lower sideband lines were measured. The frequency of the THz-QCL was phase-locked to 3786.321 GHz. We confirmed the frequency of all the lines of methanol using JPL catalogue. We measured the lines at the different gas pressure at 50 Pa and 100 Pa. We see the pressure broadening of the spectral lines. Fig. 3 (c) and (d) show the measured lines of [H_2O, HDO] and [D_2O] with an integration time of 5 s and 25 s, respectively. The THz-QCL was operated in free-running for the measurement of H_2O and its isotopes. We will improve the S/N ratio of the data, and measure the center frequency of the spectral lines accurately using a hydrogen maser as a reference for the phase-locking.

For the real application of the phase-locked THz-QCL to an air-borne, balloon-borne, or satellite-borne system, it is important to stabilize the phase-locking in long term, hopefully more than a week. Currently, we have measured longest stabilization time of ~15 hours. The larger S/N ratio of a beat signal more than 30-40 dB is one of the important issues. The bias point of a THz-QCL should be at the point where the emission frequency of a THz-QCL changes linear with the bias. It may be necessary slow control of an operation temperature of a THz-QCL monitoring a correction signal for a phase-locking in addition to fast control using PLL. An automatic or remote-control recovery system might be considered when a THz-QCL becomes unlocked.

Fig. 2. A photograph of a gas cell system for the measurement of emission line spectra of molecules. A 1-m long gas cell with 1-mm thick HDPE windows at both sides was used.

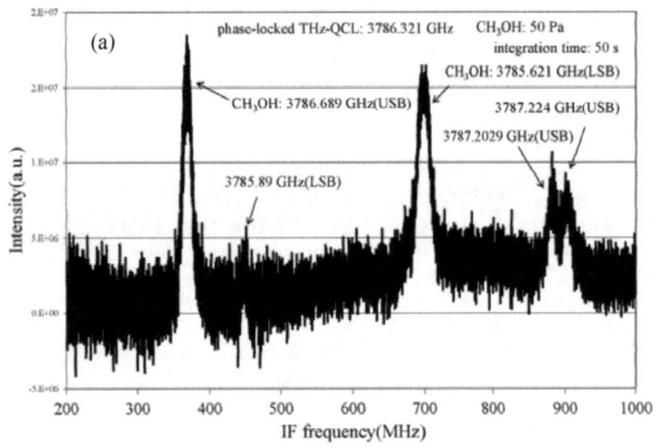

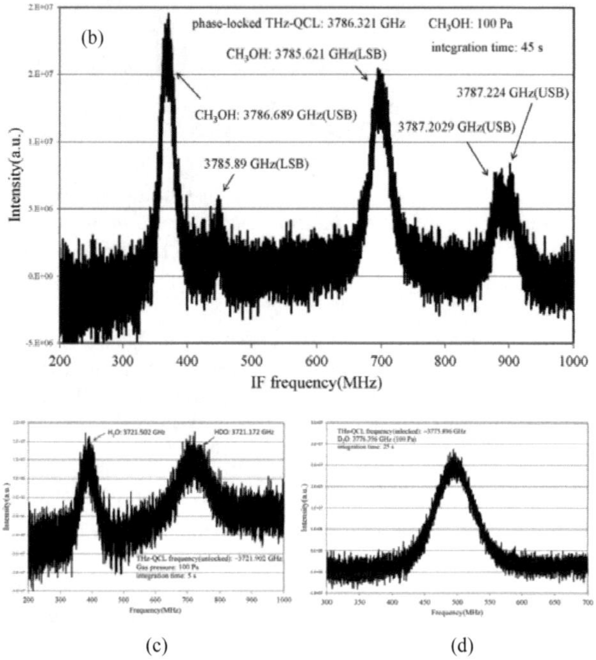

Fig. 3. Measured emission line spectra of methanol [CH_3OH] (a), [H_2O, HDO] (b), and [D_2O] (c) at 3.7-THz band. The methanol lines were measured using a HEBM and a phase-locked THz-QCL. The integration time was around 50 s for the measurement of methanol. We see pressure broadening at the different gas pressure. The spectral lines of H_2O and its isotopes were measured using a free-running THz-QCL with the integration time of 5 s and 25 s, respectively. The vertical line was not calibrated in this experiment.

III. Conclusions

We have developed a 3-THz HEB mixer with a phase-locked THz-QCL as a local oscillator. We have measured emission line spectra of methanol [CH_3OH] at 3.7 THz band using a gas cell system. We also measured H_2O, HDO, and D_2O lines. Long-term stabilization of the phase-locked THz-QCL will be an important task for the real application.

Acknowledgments

The authors would like to thank to all the staff members of the Advanced ICT Device Lab for providing the equipment and supporting the fabrication of the THz-QCL.

References

[1] Y. Irimajiri, M. Kumagai, I. Morohashi, A. Kawakami, N. Sekine, S. Nagano, S. Ochiai, S. Tanaka, Y. Hanado, Y. Uzawa, and I. Hosako, "Development of a Superconducting Low-Noise 3.1-THz Hot Electron Bolometer Receiver," *IEEE Trans. THz Sci. Technol.*, vol. 5, no. 6, pp.1154-1159, Nov. 2015.

[2] Y. Irimajiri, M. Kumagai, I. Morohashi, A. Kawakami, N. Sekine, S. Nagano, S. Ochiai, S. Tanaka, Y. Hanado, Y. Uzawa, and I. Hosako, "Precise Evaluation of a Phase-Locked THz Quantum Cascade Laser," *IEEE Trans. THz Sci. Technol.*, vol. 6, Issue 1, pp.115-120, 2016.

[3] I. Morohashi, Y. Irimajiri, T. Sakamoto, T. Kawanishi, M. Yasui, and I. Hosako, "Generation of millimeter waves with fine frequency tunability using Mach-Zehnder-modulator-based flat comb generator," *IEICE Trans. Electron.*, vol. E96-C, no. 2, pp. 192-196, 2013.

Critical Temperature Dependence of the Noise Temperature and IF Bandwidth of Superconducting Hot Electron Bolometer Mixers

W. Miao[1,2], W. Zhang[1,2], K.M. Zhou[1,2], H. Gao[1,2], K. Zhang[1,2], Q.J. Yao[1,2], W.Y. Duan[1,2], S.C. Shi[1,2,*], Y. Delorme[3], and R. Lefevre[3]
[1]*Purple Mountain Observatory, CAS, 2 West Beijing Road, Nanjing, 210008, China*
[2]*Key Lab of Radio Astronomy, CAS, Nanjing, 210008, China*
[3]*Observatoire de Paris, 61 avenue de l'Observatoire, Paris 75014, France*
*Contact: scshi@mail.pmo.ac.cn, phone +86-25-8333 2204

Abstract—**We present a study of the critical temperature dependence of the noise temperature and intermediate frequency (IF) noise bandwidth of superconducting hot electron bolometer (HEB) mixers. We simulate the noise temperature of a superconducting niobium nitride (NbN) HEB mixer with different critical temperatures (T_c) by a distributed hot spot model. The simulation shows that the mixer noise temperature is the lowest at T_c approximately equal to 7 K~9.5 K, and increases beyond this range due to the decrease of conversion gain or the increase of output noise. To verify this result, three superconducting HEB mixers of different critical temperatures (i.e., 7.5 K, 8.8 K and 10.3 K) are measured (at 3.5 K) at 0.85 THz and 1.3 THz. A good agreement is observed between simulation and measurement. In addition, we also study the dependence of IF noise bandwidth of superconducting HEB mixers on T_c. It appears that the larger T_c, the wider the IF noise bandwidth is.**

I. Introduction

Superconducting hot electron bolometer (HEB) mixers are presently the most competitive devices for low-noise heterodyne detection at frequencies above 1 THz [1, 2], and they are increasingly being used on space- and ground-based telescopes for radio astronomical observations [3, 4]. Superconducting HEB mixers are essentially a microbridge made of an ultra-thin superconducting film such as niobium nitride (NbN), with a typical thickness of a few nanometres. When a superconducting HEB mixer is driven to its operating point by applying terahertz radiation from a local oscillator (LO) and a direct current (dc) voltage, a "hot-spot" region will be formed in the microbridge centre with the electron temperature close to the microbridge critical temperature T_c. It is known that the performance of superconducting HEB mixers, such as noise temperature and IF noise bandwidth, is closely related to the electron temperature and the microbridge critical temperature T_c. However, the dependence of noise temperature and IF noise bandwidth of superconducting HEB mixers upon the microbridge critical temperature T_c is yet to be fully understood. In this paper, we study the critical temperature dependence of the mixing performance of a superconducting HEB mixer by a distributed hot spot model [5, 6]. We also measure the receiver noise temperature and IF noise bandwidth of three superconducting HEB mixers fabricated from different batches of NbN films and with different critical temperatures (i.e., 7.5 K, 8.8 K and 10.3 K) at 0.85 THz and 1.3 THz. A detailed comparison between simulation and measurement is then given.

II. Noise Temperature

We first modelled the critical temperature dependence of the noise performance of a superconducting HEB mixer (at a bath temperature of 3.5 K) by a distributed hot spot model [7]. It can be seen from Fig. 1 that the input noise temperature of the superconducting HEB mixer is the lowest at T_c approximately between 7 K and 9.5 K. We attribute the degradation of mixer noise performance to the decrease of conversion gain at low critical temperatures and to the increase of output noise at high critical temperatures.

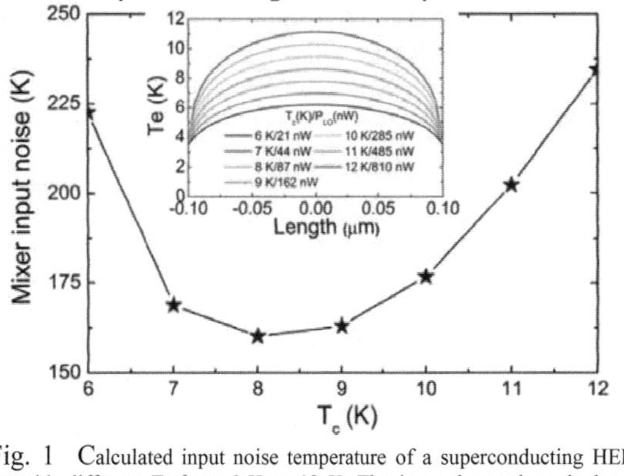

Fig. 1 Calculated input noise temperature of a superconducting HEB mixer with different T_c from 6 K to 12 K. The inset shows the calculated electron temperature profiles.

We calculated the conversion gain and output noise temperature (including Johnson noise and thermal fluctuation noise) of the superconducting HEB mixer as a function of critical temperature with a small signal model [8]. Fig. 2 shows the calculated results at the optimal bias point of the superconducting HEB mixer. Apparently, the higher critical temperature is, the higher the conversion gain and the higher the noise temperature appear. It should be pointed out that the conversion gain of the superconducting HEB mixer approaches nearly zero at critical temperatures close to the bath temperature and becomes saturated at high critical temperatures.

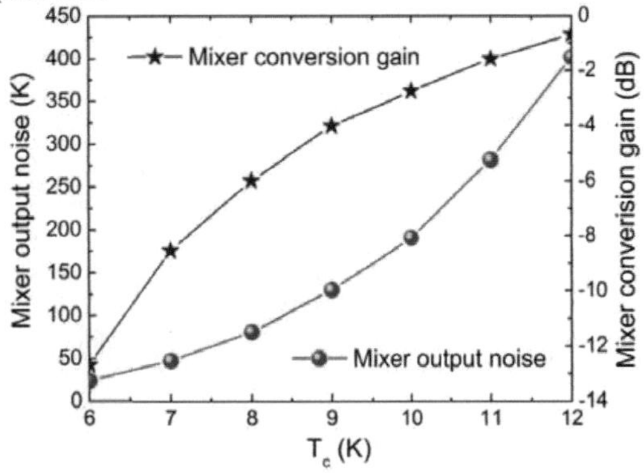

Fig. 2 Calculated output noise temperature and conversion gain of the superconducting HEB mixer as a function of critical temperature at the optimal bias point.

To compare with the predicted results, three superconducting HEB mixers with different critical temperatures (i.e., 7.5 K, 8.8 K and 10.3 K) were chosen for measurements. The inset of Fig. 3 shows the measured receiver noise temperature of the superconducting HEB mixers as a function of bias voltage. Their lowest receiver noise temperatures are about 650±50 K at 1.3 THz for mixer No. 1, 500±50 K at 1.3 THz for mixer No. 2 and 800±50 K at 0.85 THz for mixer No. 3. The uncertainty is mainly attributed to the temperature fluctuation and mechanical vibration of the 4 K cooler. In order to understand the critical temperature dependence of mixer intrinsic noise performance, we corrected the noise contributions of the quasi-optical components. Fig. 3 shows the measured input noise temperature of the superconducting HEB mixers as a function of critical temperature together with the calculated one. Clearly, the measured result is in good agreement with the calculated one. Superconducting NbN HEB mixers do have the lowest noise temperature in the critical temperature range from 7 K to 9.5 K.

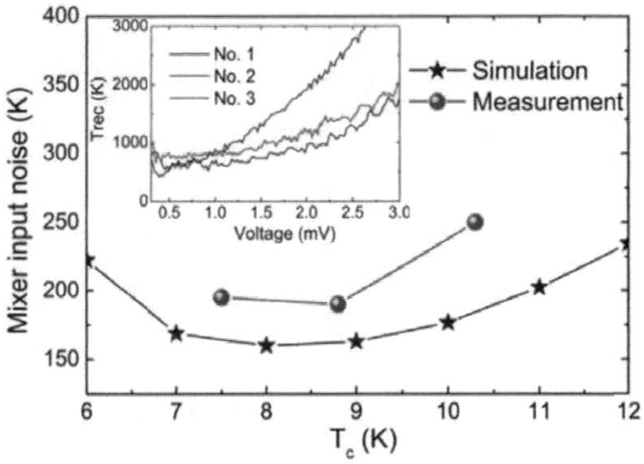

Fig. 3 Measured input noise temperature of the superconducting HEB mixers together with the calculated one. The inset shows the measured receiver noise temperature of the superconducting HEB mixers as a function of bias voltage.

III. IF NOISE BANDWIDTH

Fig. 4 shows the calculated IF noise bandwidth of the superconducting HEB mixer as a function of critical temperature. Also shown is the measured IF noise bandwidth of the three superconducting HEB mixers used above for the noise measurements. The inset of Fig. 4 shows the measured receiver noise temperature of the three superconducting HEB mixers as a function of IF frequency. It can be seen from Fig. 4 that the calculated and measured IF noise bandwidths show the same dependence on the critical temperature, i.e., the IF noise bandwidth of the superconducting HEB mixers gets wide at high critical temperatures. However, the calculated IF noise bandwidth is roughly two times higher than the measured one. We think this difference is likely due to the imperfect interface between the superconducting NbN microbridge and Au contact pads (200 nm thick) in our superconducting HEB mixers. Using a proximity effect model based on the Usadel theory [9], the interface transparency of the superconducting HEB mixers is estimated to be ~0.1 from their measured R-T curves. This estimated interface transparency is still smaller than the value of ~0.15, which is expected from the Fermi velocity mismatch using v_f=1.39×10^6 m s^{-1} [10] for Au and v_f=5.7×10^4 m s^{-1} [11] for a superconducting NbN film. In addition, the diffusion of hot electrons may be partly restricted due to the Andreev reflection since the energy gap of the superconducting microbridge becomes wide in the direction of heat diffusion.

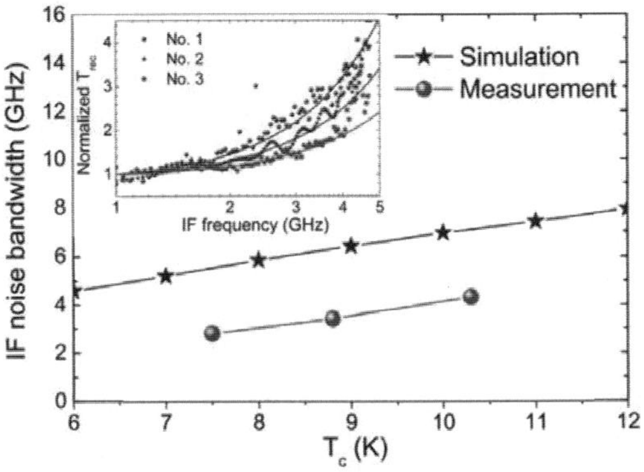

Fig. 4 Calculated and measured IF noise bandwidth as a function of critical temperature. The inset shows the measured receiver noise temperatures of the superconducting HEB mixers with different critical temperatures as a function of IF frequency. The solid lines are the fitting results using T_{rec}~$(1+(f/f_{IF})^2)$.

IV. Conclusions

In summary, we have studied the noise temperature and IF noise bandwidth of superconducting NbN HEB mixers with different critical temperatures by a distributed hot spot model. The modelling has shown that superconducting HEB mixers have the lowest noise temperature when their critical temperature is in the range from 7.0 K to 9.5 K, and the IF noise bandwidth of superconducting HEB mixers gets wide due to the reduction of phonon cooling time and diffusion cooling time at high critical temperatures. We have also measured the noise performance of three superconducting HEB mixers with different critical temperatures at 0.85 THz and 1.3 THz. A good agreement has been observed between modelling and measurement.

Acknowledgment

We acknowledge G.N. Gol'tsman and B. Voronov of MSPU for providing superconducting NbN films. This work was supported by CAS under Grant No. XDB04010300, NSFC under Grant Nos. 11127903, 11190012, and 11473075, and the CAS Joint Key Lab for Radio Astronomy.

References

[1] I. Tretyakov, S. Ryabchun, M. Finkel, A. Maslennikova, N. Kaurova, A. Lobastova, B. Voronov, and G. Gol'tsman, "Low noise and wide bandwidth of NbN hot-electron bolometer mixers," Appl. Phys. Lett., vol. 98, p. 033507, 2011.

[2] W. Zhang, P. Khosropanah, J. R. Gao, E. L. Kollberg, K. S. Yngvesson, T. Bansal, R. Barends, and T. M. Klapwijk, "Quantum noise in a terahertz hot electron bolometer mixer," Appl. Phys. Lett., vol. 96, p. 111113, 2010.

[3] S. Cherednichenko, V. Drakinskiy, T. Berg, P. Khosropanah, and E. Kollberg, "Hot electron bolometer terahertz mixers for the Herschel Space Observatory," Rev. Sci. Instrum., vol. 79, p. 034501, 2008.

[4] D. Meledin, A. Pavolotsky, V. Desmaris, I. Lapkin, C. Risacher, V. Perez, D. Henke, O. Nystrom, E. Sundin, D. Dochev, M. Pantaleev, M. Fredrixon, M. Strandberg, B. Voronov, G. N. Gol'tsman, and V. Belitsky, "A 1.3 THz balanced waveguide HEB mixer for the APEX telescope," IEEE Trans. Microw. Theory Technol., vol. 57, no. 1, pp. 89-98, 2009.

[5] W. Miao, Y. Delorme, A. Feret, R. Lefevre, B. Lecomte, F. Dauplay, J. M. Krieg, G. Beaudin, W. Zhang, Y. Ren, and S. C. Shi, "Comparison between hot spot modelling and measurement of a superconducting hot electron bolometer mixer at submillimeter wavelengths," J. App. Phys., vol. 106, p. 103909, 2009.

[6] W. Miao, W. Zhang, J.Q. Zhong, S.C. Shi, Y. Delorme, R. Lefevre, A. Feret, and T. Vacelet, "Non-uniform absorption of terahertz radiation on superconducting hot electron bolometer microbridges," Appl. Phys. Lett., vol. 104, p. 052605, 2014.

[7] R. Barends, M. Hajenius, J.R. Gao, and T.M. Klapwijk, "Current-induced vortex unbinding in bolometer mixers," Appl. Phys. Lett., vol. 87, p. 263506, 2005.

[8] F. Arams, C. Allen, B. Peyton, and E. Sard, "Millimeter mixing and detection in bulk InSb," Proc. IEEE, vol. 54, pp. 612-622, 1966.

[9] J.M. Martinis, G.C. Hilton, K.D. Irwin, and D.A. Wollman, "Calculation of Tc in a normal-superconductor bilayer using the microscopic-based Usadel theory," Nucl. Instrum. Methods Phys. Res. A, vol. 444, pp. 23-27, 2000.

[10] N.W. Ashcroft and M.N. David, Solid State Physics, International Edition, 1976.

[11] M. Hajenius, J.J.A. Baselmans, J.R. Gao, T.M. Klapwijk, P.A.J. de Korte, B. Voronov, and G. Gol'tsman, "Low noise NbN superconducting hot electron bolometer mixers at 1.9 and 2.5 THz," Supercond. Sci. Technol., vol. 17, pp. S224-S228, 2004.

An 8×8 CPW MKIDs Developed at 0.35THz

Jing Li, Qing Shi, Jin-Ping Yang, Dong Liu, Zhen-Hui Lin, Wei Miao, Wen Zhang and Sheng-Cai Shi
Purple Mountain Observatory, Chinese Academy of Sciences
Key Lab of Radio Astronomy, Chinese Academy of Sciences
2 West Beijing Road, Nanjing, JiangSu 210008, China
lijing@pmo.ac.cn

***Abstract*—Microwave Kinetic Inductance Detectors (MKIDs) are rather promising for THz direct detector arrays of large size, particularly with simple frequency-division multiplexing. Purple Mountain Observatory is developing a terahertz superconducting imaging array (TeSIA) for the DATE5 telescope to be constructed at Dome A, Antarctica. Here we report on the development of a prototype array for the TeSIA project, namely an 8×8 CPW MKIDs array at 0.35THz. The array consists of 64-pixel superconducting resonators in the frequency range of 4-5.575 GHz with an interval of 0.025 GHz and 64-pixel twin-slot antennas at 0.35 THz for coupling THz radiation to the detector array. Based on our design, we fabricate the MKIDs array using TiN superconducting film with Tc about 4.5 K. And the performance is characterized at different temperatures. Detailed results and analysis will be presented.**

Keywords—MKIDs; TiN; Superconducting resonator; CPW; TeSIA

I. Introduction

China is planning to construct an observatory at Dome A, Antarctica, which has been found to be the best site (with low perceptible water vapor and low atmospheric boundary layer) on the earth for THz and Optical/IR astronomy. One of the telescopes to be built there is a 5-m THz telescope (DATE5 [1]) targeting at 350um and 200um atmospheric windows. One science case for the DATE5 is to observe extreme starburst galaxies at different redshifts to better understand the nature and evolution of these enigmatic and important objects. Therefore, we are going to develop an imaging camera with 1024-pixel. We are currently developing a detector array demonstrator to meet such a requirement in the future [2].

The detector array demonstrator has 8×8 pixels and will work at 0.35 THz, which is chosen to demonstrate the performance of the detector array on a small sub-millimeter telescope (POST, with a diameter of 30cm) situated at Delingha, China. As is well known, microwave kinetic inductance detectors (MKIDs [3]) use frequency domain multiplexing that allows thousands of pixels to be read out over a single microwave transmission line followed by a cryogenically cooled low noise amplifier. Furthermore, a large number of MKIDs can be integrated like a filter bank to realize on-chip spectrometers such as DESHIMA and SuperSpec [4, 5]. Hence we choose MKIDs to develop our detector array demonstrator. The MKIDs make use of TiN superconducting films with a critical temperature of approximately 4.5 K, which can be operated at temperatures of ~0.3K [6]. Furthermore, TiN MKIDs can reach a noise equivalent power (NEP) below 1x10-19 W/Hz0.5, which is a sufficiently high sensitivity for ground-based astronomical observations [6]. The 8×8 TiN MKIDs will be integrated with a micro-lens array of 8×8 0.95-mm hyper-spherical Si lens with a separation of 2 mm between individual lens. The readout for this MKIDs detector array demonstrator is similar to others, but adopting a commercial arbitrary wave-function generator to generate 64-tone input signal. The 0.3K refrigerator for the demonstrator is a He3 two-stage sorption cooler (CRHe7 [7]) based on a liquid helium cryostat, which offers over 10 hours continuous cooling with a total heat load of up to 20 μW. In this paper, we mainly introduce the design, fabrication, and characterization of the 8×8 TiN MKIDs.

II. MKIDs Development

We adopted the coplanar-waveguide (CPW) type resonator to design our 8×8 TiN MKIDs array because it has a relatively simple architecture of only one thin-film layer on the substrate. This kind of resonator has a quarter-wavelength transmission line with one end capacitively coupled to a feed line and the other matched to a planar antenna (twin-slot antenna, for example) [8].

The 8×8 MKIDs array was fabricated in the cleanroom of RIKEN Center for Advanced Photonics (Japan). As introduced before, we chose TiN superconducting films for this MKIDs detector array [9, 10]. MKIDs based on TiN superconducting films have a few advantages including a) T_c can be controlled between ~0-5 K by the components of Ti and N_2 [6]; b) quasi-particle lifetime is around 10-200 μs [6]; c) low loss in the superconducting state [6]. Firstly, a 100-nm thick TiN film was deposited on a high resistivity Si wafer in a DC magnetron sputtering system. Secondly, the CPW lines were defined in contact lithography by a mask aligner. Thirdly, the etching course was done in an ICP machine. The fabricated MKIDs have a T_c approximately equal to 4.5 K. Hence they can be operated in a 0.3 K low temperature environment.

For the fabricated 8×8 TiN MKIDs array, as shown in Fig. 1, we mainly measured its Q factors and the dependence of Q factors upon the bath temperature and the input power for the multiplexing readout. The measurements were done simply by a scalar network analyzer to the 8×8 TiN MKIDs array and a following 0.1~12 GHz cryogenically cooled low-noise amplifier, which has an equivalent noise temperature of 5 K and gain of approximately 35 dB.

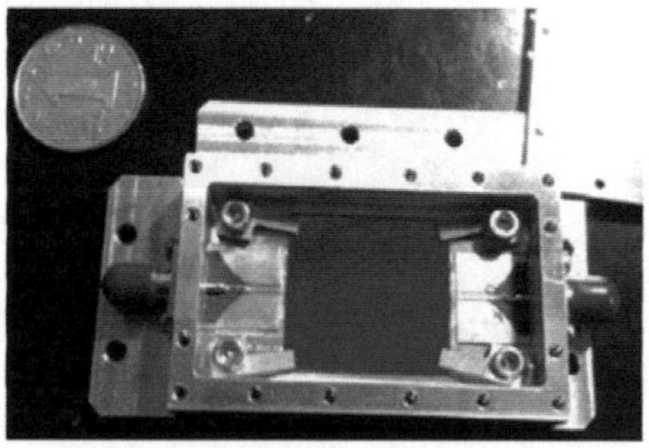

Fig. 1. Measured MKID 8×8 TiN MKIDs array, with its input and output of a CPW-to-microstrip transition (both in 50 Ω).

The transmission characteristic of the 8×8 TiN MKIDs array was firstly measured at a temperature of approximately 600 mK in the frequency range from 5.2-7.7 GHz. The result is exhibited in Fig. 2. By checking the resonance dips individually, we found that they all survived. The inset in Fig. 2 shows the result zoomed in for a 100-MHz frequency interval.

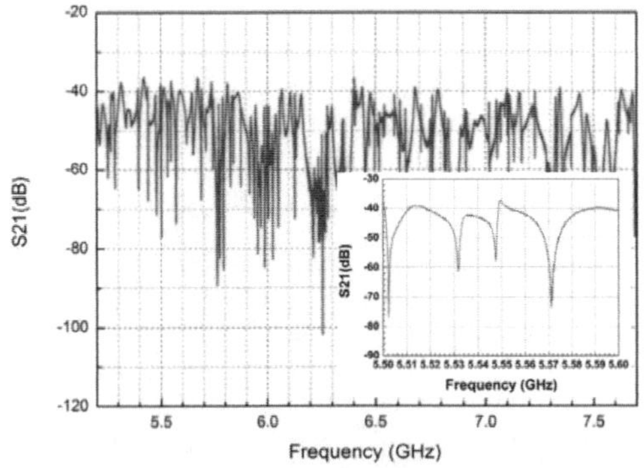

Fig. 2. Measured transmission characteristic of the 8×8 TiN MKIDs array at a temperature of approximately 600 mK in the frequency range from 5.2-7.7 GHz. The inset shows the part zoomed in for a 100-MHz interval, with four resonance dips seen clearly.

We then studied the transmission characteristic of the 8×8 TiN MKIDs array with increasing the bath temperature (600 mK to 950 mK). The measured results are shown in Fig. 3 for the frequency range of 5.5~5.6 GHz. Obviously, the four resonances demonstrate the same behavior, namely the higher the temperature, the lower the resonance frequency and the shallower the resonance dip. We also plotted the dependence of the resonance frequency upon temperature, as shown in the inset in Fig. 3.

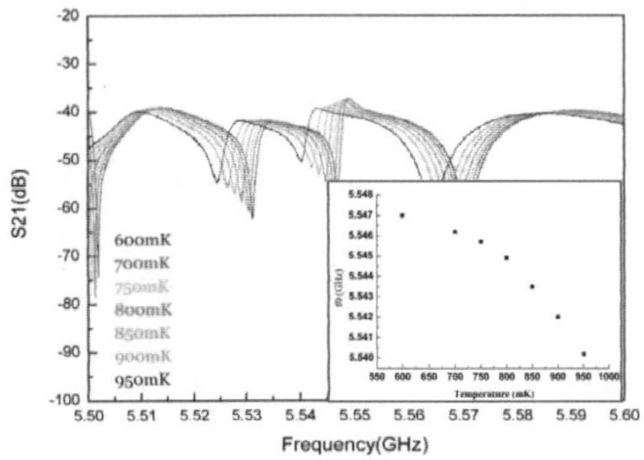

Fig. 3. Measured transmission characteristic of the 8×8 TiN MKIDs array as a function of the bath temperature, shown in the frequency interval of 5.5~5.6 GHz. The inset presents the dependence of then resonance frequency upon temperature.

III. Conclusion

We have designed and fabricated an 8×8 TiN MKIDs array. Its transmission characteristic has been measured with respect to the bath temperature and input power. The temperature dependence follows that predicted by the Mattis-Bardeen theory, while the resonance frequency and the quality factor are both insensitive to the input power when it is below -40 dBm.

Acknowledgment

The authors are grateful to Zheng Lou of Purple Mountain Observatory for the designs of the 3.5 THz twin-slot antenna and the 8x8 micro-lens array. Jing Li thanks Dr. C. Otani and S. Mima for the help during the device fabrication process.

References

[1] J Yang, et al., "China's Antarctic observatory," project proposal, 2013.

[2] S.C. Shi and TeSIA team, "Development of a terahertz superconducting imaging array (TeSIA)," ISSTT2014, Moscow, Russia, Apr. 2014.

[3] Peter K. Day, Henry G. LeDuc, Benjamin A. Mazin, Anastasios Vayonakis and Jonas Zmuidzinas "A broadband superconducting detector suitable for use in large arrays," *Nature* 425, 817-821, 2003.

[4] A. Endo, J.J.A. Baselmans, P.P. van der Werf, B. Knoors, S.M.H. Javadzadeh, S.J.C. Yates, D.J. Thoen, L. Ferrari, A.M. Baryshev, Y.J.Y. Lankwarden, P.J. de Visser, R.M.J. Janssen and T.M. Klapwijk, "Development of DESHIMA: A Redshift Machine Based on a Superconducting On-Chip Filterbank," *Proc. SPIE*, Vol. 8452, 2012.

[5] Erik Shirokoff, et al., "MKID development for SuperSpec: an on-chip, mm-wave, filter-bank spectrometer," *Proc. SPIE,* 8452, 84520R (2012).

[6] Henry G. Leduc, Bruce Bumble, Peter K. Day, Byeong Ho Eom, Jiansong Gao, Sunil Golwala, Benjamin A. Mazin, Sean McHugh, Andrew Merrill, David C. Moore, Omid Noroozian, Anthony D. Turner and Jonas Zmuidzinas, "Titanium Nitride Films for Ultrasensitive Microresonator Detectors," 2010.

[7] 0.3K He7 refrigerator, Chase Research Cryogenics, UK.

[8] R. M. J. Janssen, J. J. A. Baselmans, A. Endo, L. Ferrari, S. J. C. Yates, A. M. Baryshev, and T. M. Klapwijk, "High optical efficiency and photon noise limited sensitivity of microwave kinetic inductance detectors using phase readout," *APL*, 103, 203503, 2013.

[9] Kensuke Koga, Seiichiro Ariyoshi, Mitsuhiro Yoshida, Noboru Furukawa, Massashi Hazumi, and Chico Otani, "Development of TiN MKIDs for CMB polarization observations," *23rd International Symposium on Space Terahertz Technology*, 2012.

[10] K. Koga, S. Ariyoshi, C.Otani, M. Yaoshida, H. Watanabe, M. Hazumi, "Development of TiN-MKIDs for cosmic inflation exploration," 2011.

Shot Noise in NbN Distributed Superconducting Tunneling Junctions

Dong Liu[1, 2, 3], JieHu[1, 2, 3], Ming Yao[1, 2, 3], Jing Li[1, 2], and Sheng-Cai Shi[1, 2*]

[1]PurpleMountain Observatory, CAS, 2 West Beijing Road, Nanjing, 210008, China
[2]Key Laboratory of Radio Astronomy, CAS, 2 West Beijing Road, Nanjing, 210008, China
[3]GraduateSchool of ChineseAcademy of Sciences, 19 Yuquan Road, Beijing, 10040, China
*E-mail: scshi@pmo.ac.cn

With sensitivity approaching the quantum limit, superconductor-insulator-superconductor (SIS)mixers play an important role in radio astronomy and atmosphericresearch at millimeter and sub-millimeter wavelengths. As one of the intrinsic noise sources in superconducting tunneling junctions, shot noise is still not well understood, particularly for SIS junctions of relatively high energy gap (e.g., NbN/AlN/NbN). In this paper, we mainly study the shot noise of three different NbN junctions (i.e., parallel connected twin junctions, distributed junction array and long junction) as well as its temperature dependence. It has been found that the shot noise increases withtemperature in general, while the fraction due to the MAR effect is inversely proportional to temperature. In addition, the tunnel barrier transmission in superconducting junctions is found to be nearly independent of temperature.Detailed measurement results and analysis will be presented.

A 4.7 THz HEB QCL Receiver for STO2

Darren J. Hayton[1*], Jian-rong Gao[1,2], Wouter Laauwen[1], Behnam Mirzaei[2], Andrey Khudchenko[1], Jenna L. Kloosterman[3], Dmitry Paveliev[4], Qing Hu[5] Wilt Kao[6], Vladimir Vax[7], Chris K. Walker[8],

1 SRON Netherlands Institute for Space Research, Groningen, 9747 AD,The Netherlands
2 Delft University of Technology, Delft, 2628 CJ, The Netherlands
3 Jet Propulsion Laboratory, 4800 Oak Grove Drive, Pasadena, CA 91109 USA
4 Lobachevskii State University of Nizhny Novgorod, Nizhny Novgorod, 603950 Russia
5 Dept. of Electrical Engineering and Computer Science, MIT,Cambridge, Massachusetts 02139, USA
6 Longwave Photonics LLC, Mountain View, CA 94043, USA
7 Institute for Physics of Microstructures, Russian Academy of Sciences, Nizhny Novgorod, 603950 Russia
8 Steward Observatory, University of Arizona, Tucson, Arizona 85721, USA

*Contact: D.J.Hayton@sron.nl

Abstract— **We report on a 4.7 THz heterodyne receiver designed for high resolution spectroscopy of the astronomically important neutral oxygen (OI) line at 4.745 THz. The receiver is based around a hot electron bolometer (HEB) mixer and quantum cascade laser (QCL) local oscillator. This receiver has been developed to fly on the Stratospheric Terahertz Observatory (STO-2), a balloon-borne 0.8 m telescope observing from an altitude of 44 km for 14 days or more. We measure a double sideband receiver noise temperature of 815 K (~ 7 times quantum noise) with a noise temperature IF bandwidth of 3.5 GHz. We describe the receiver performance expected in flight and outline novel approaches to QCL amplitude and frequency stabilization.**

INTRODUCTION

The fine structure line of neutral oxygen (OI) at 4.7448 THz offers astronomers a valuable tool with which to study the lifecycle of star forming regions within giant molecular clouds in the Milky Way. Large scale surveys with extremely high spectral resolution and sensitivity are required to determine large scale kinematics within these clouds prompting the development of a super-THz heterodyne receiver near this frequency. Due to strong water absorption in the atmosphere, it is not possible to observe the OI line from Earth. Therefore, this receiver requires a compact high powered local oscillator (LO) that can be operated in fight from an aircraft or higher.

We report on a 4.7 THz heterodyne receiver designed specifically for high resolution spectroscopy of the OI line. This receiver has been developed to fly on the 2016 flight of the Stratospheric Terahertz Observatory (STO-2) balloon craft. Following from STO [1], the STO-2 platform consists of a balloon-borne observatory operating a 0.8 m diameter telescope at an altitude of > 40 km for 14 days or more. Heterodyne receivers are used to detect the brightest of the fine structure lines, namely those of, ionized nitrogen [NII] at 1.4 THz, ionized carbon [CII] at 1.9 THz, and neutral oxygen [OI] at 4.7 THz.

STO-2 is primarily aimed at improving understanding of the life cycle of stars in our Galaxy by observation of interstellar clouds and star forming regions and by attempting to further understand the relationship between star formation and the life cycle of interstellar clouds. STO-2 proposes to specifically address the following points:

1. Determine the life cycle of Galactic interstellar gas.
2. Study the creation and disruption of star-forming clouds in the Galaxy.
3. Determine the parameters that affect the star formation rate in a galaxy.
4. Provide templates for star formation and stellar/interstellar feedback in other galaxies.

STO-2 will make 3-dimentional maps of the dynamics, structure, energy balance, turbulence and pressure of the Milky Way's Interstellar Medium (ISM).

SUPER-THZ RECEIVER

The 4.7 THz receiver is based on a single pixel NbN hot electron bolometer (HEB) mixer [3] pumped by a 4.7 THz 3rd order distributed feedback quantum cascade laser (QCL) as local oscillator ([4]-[7]). The QCL is operated at 44 K using a commercial (Sunpower CT) Stirling cryocooler. The QCL, provided by MIT [8], emits ~ 150 μW at 4.7 THz and with tuning coefficients of 3 GHz/V and -160 MHz/K. The receiver has been characterized in the lab to determine sensitivity, IF bandwidth, stability, and beam pattern. We measure a double sideband receiver noise temperature of 815 K (~ 7 times quantum noise) with a noise temperature IF bandwidth of 3.5 GHz [2]. The output intensity of the QCL is stabilized using an auto gain control (AGC) loop [9] in which a voice coil shutter is driven to maintain the optimum bias conditions of the HEB, mitigating the effects of the 60 and 120 Hz modulation that are induced by the Stirling cooler free piston. The resulting increase in spectroscopic Allan variance time greatly improves the efficiency of on-the-fly mapping by increasing the period between calibration scans. In addition, the frequency of the QCL is stabilized using a superlattice harmonic mixer ([10], [11]) in which the IF output of the 24th harmonic of a 197.7 GHz source forms the input to a PID based frequency lock loop. These two control loops have been demonstrated to operate in parallel and independently. We describe the expected receiver performance in flight and outline the novel approaches to QCL amplitude and frequency stabilization.

MIXER PERFORMANCE

The HEB mixer used for the STO-2 4.7 THz receiver is a single pixel HEB in a quasi-optic configuration. The detector element is a 0.2 x 2 μm NbN superconducting bolometer (Tc = 9.1 K) coupled to a tight wound spiral antenna capable of operation from 0.6 to 6 THz. The HEB/antenna is coupled to a 10 mm diameter anti-reflection coated Si lens. Prior to integration into the instrument, the performance of the HEB mixer and QCL LO was characterized in the lab. For the HEB mixer, a standard hot/cold Y-factor technique was used. The experimental setup consists of a mixer mounted in a lHe cryostat with a 3 μm Mylar beamsplitter to combine the LO power with the vacuum hot and cold blackbody surfaces. For the LO source in these tests we used both a FIR gas laser at 4.3 and 5.3 THz and the flight QCL at 4.7446 THz. As indicated in Figure 1 below, we measure a DSB receiver noise temperature of 826 K at 4.7 THz.

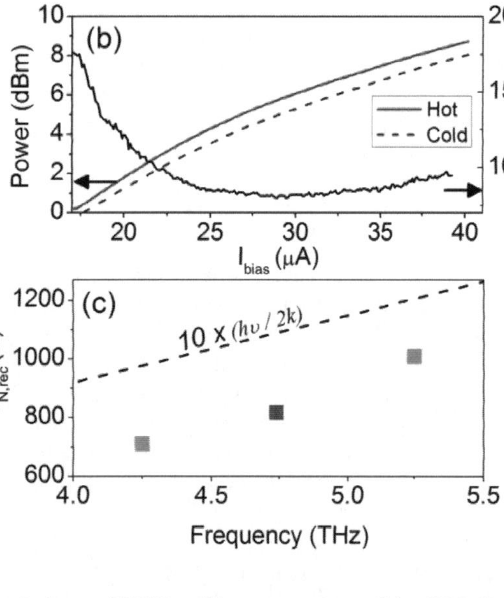

Fig. 1 (upper) DSB noise temperature of the STO-2 4.7 THz HEB as characterized in the lab using the flight QCL at optimum bias. (lower) Noise temperature versus frequency for the flight mixer at 4.3, 4.7 and 5.3 THz where green indicates the gas laser points and red indicated a QCL point.

LO PERFORMANCE

The 4.7 THz LO unit comprises a 120W Stirling cooler for 1 W heat lift from the QCL stage at 45 K. Special attention is given to the minimization of cooler vibration. This is achieved using an additional active balancer for the Sunpower CT cooler in which the vibration is sensed and actively compensated for by an additional inductive motor. This, coupled with frequency and amplitude stabilization schemes, results in an ultra-stable LO source. The thermal performance of the LO unit is as expected from the calculations reaching a base temperature of 44 K under a 0.7 W load from the QCL. Thermal stability is better than ± 0.1 K and cooling from room temperature to 44 K takes in the order of 45 minutes with payload coolant at 35 C. Radiation from the QCL is coupled to the HEB via a single parabolic metal mirror. The beam pattern plot in Fig. 2 shows the distribution of THz power at the equivalent position of the HEB Si lens, 750 mm away from the QCL. The majority the THz power falls within the diameter of the Si lens and is therefore easily possible to fully pump the mixer over a wide bias range of the QCL.

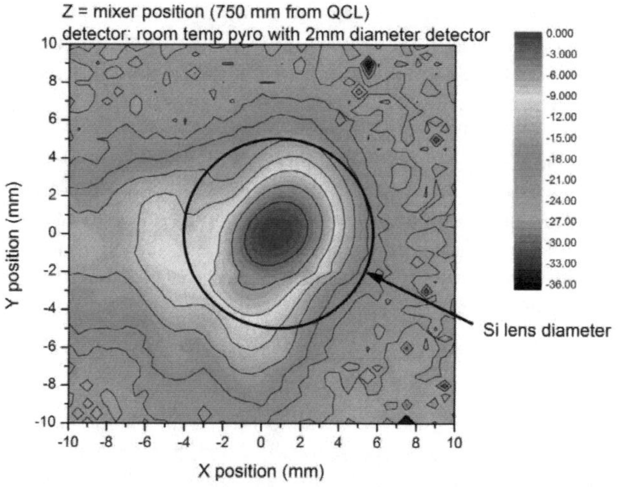

Fig. 2 Beam pattern measurement using a room temperature pyro-electric detector scanned in X and Y plane at the position of the HEB.(signal in dB)

RECEIVER PERFORMANCE

The performance of the complete 4.7 THz receiver was evaluated in-situ on the STO-2 payload. Fig. 3 shows an image of the installed LO hardware. The receiver noise temperature was measured at 1600 K @ 1.5 GHz IF and with a 12 μm Mylar beamsplitter. Initially, this thicker beamsplitter is used during alignment of the LO to the mixer so that pumping of the mixer may be more easily obtained. For flight, the beamsplitter will be replaced with a thinner membrane (6 μm) so that the system noise temperature will be reduced towards a predicted value of 1200 K. Receiver testing also demonstrated effective use of both the AGC and the superlattice frequency locking loops. In the latter, both the small residual 60/120 Hz modulation from the cooler and any slow drift in LO power could be reduced to a level that is no longer visible in the mixer IF. In addition, the effect of a small

(< 1 mm) movement in the alignment of the LO unit during elevation changes of the telescope from stow position (vertical) to horizontal was fully removed by the AGC loop.

The superlattice based frequency lock loop was also assessed whilst in-situ on the instrument. Frequency locking could be established using a 400 MHz IF tone produced by the 24th harmonic of a197.7 GHz AMC based LO. Free funning frequency stability was already shown to be excellent when running the QCL on the payload battery supply. With this favourable free running condition, the frequency locked line could be reliably established.

Fig. 2 The 4.7 THz QCL based LO as installed and ready for flight on the STO-2 payload

In addition to the FLL, the superlattice was also used to carefully characterize the frequency of the QCL versus bias voltage and QCL bath temperature. With this data, even in free running mode, the recorded values of bias and temperature could be used to determine the frequency of the QCL to within an estimated 10 MHz.

CONCLUSIONS

We demonstrate a QCL based 4.7 THz local oscillator and HEB mixer receiver for the STO-2 THz telescope. The receiver performs as predicted in terms of noise temperature, stability and optic coupling. The QCL based hardware is designed around a Sterling cooler to produce a turn-key LO system suitable for operation on a long duration stratospheric balloon. STO-2 is scheduled to fly in mid-December 2016.

REFERENCES

[1] Walker, C. K., et al. “The Stratospheric Terahertz Observatory (STO),” Proc. SPIE 7733, 77330N (2010).

[2] Kloosterman, J. L., et al.” Hot electron bolometer heterodyne receiver with a 4.7-THz quantum cascade laser as a local oscillator,” Appl. Phys. Lett. 102, 011123 (2013)

[3] Zhang, W., Khosropanah, P., Gao, J. R., Bansal, T., Klapwijk T. M., Miao, W., and Shi, S. C., “Noise temperature and beam pattern of an NbN hot electron bolometer mixer at 5.25 THz,” J. Appl. Phys. 108, 093102 (2010).

[4] Köhler, R., et al., “Terahertz semiconductor-heterostructure laser,” Nature, 417, 156-159 (2002)

[5] Williams, B. S., “Terahertz quantum-cascade lasers,” Nat. Photonics, 1, 517 (2007).

[6] Ren, Y., et al. “Terahertz heterodyne spectrometer using a quantum cascade laser” Appl. Phys. Lett. 97, 161105 (2010)

[7] Amanti, M. I, Scalari, G., Castellano, F., Beck, M., and Faist, “Low divergence Terahertz photonic-wire laser,” J. Opt. Express 18, 6390 (2010).

[8] Kao, T.-Y., Hu, Q., and Reno, J. L., “Perfectly phase-matched third-order distributed feedback terahertz quantum-cascade lasers,” Opt. Lett. 37, 2070 (2012).

[9] Hayton, D. J., Gao, J. R., Kooi, J. W., Ren, Y., Zhang, W., and de Lange, G., “Stabilized hot electron bolometer heterodyne receiver at 2.5 THz,” Appl. Phys. Lett. 100, 081102 (2012).

[10] D.G. Paveliev et al. “Experimental study of frequency multipliers based on a GaAs/AlAs semiconductor superlattices in the terahertz frequency range” Semiconductors 46, 125 (2012).

[11] D.J. Hayton, A. Khudchenko, D. G. Pavelyev, J. N. Hovenier, A. Baryshev, J. R. Gao, T. Y. Kao, Q. Hu, J. L. Reno, and V. Vaks, “Phase locking of a 3.4 THz third-order distributed feedback quantum cascade laser using a room-temperature superlattice harmonic mixer”, *Appl. Phys. Lett.*, vol. 103, 051115, (2013).

Room Temperature Terahertz SubHarmonic Mixer Based on GaNNanodiodes

F. Yang[1*], L.Q. Chen[2], J. Torres[3], J. Mateos[4]
1 State Key Lab of Millimeter Waves, Southeast University, Nanjing, China
2 School of Information Science and Engineering, Southeast University, Nanjing, China
3Institut d'Electronique du Sud, Université Montpellier, France
4Departamento de FisicaAplicada, Universidad de Salamanca, Spain

*Contact: yangfei@seu.edu.cn, phone +86-139-51855309

Abstract—**GaN Unipolar Nanochannels is fabricated by etching in the AlGaN/GaNheterojunction. Adjusting the GaNnanochannel width results in a nonlinear, quadrantal symmetry, current voltage (I/V) characteristic. It means GaNNanochannel is quite suitable for subharmoicemixing(fIF=|fRF-2fLO|). Here, we will present the DC and RF performance prediction of a novel terahertz sub-harmonically pumped mixer that uses the GaN Unipolar Nanochannels.**

INTRODUCTION

Two dimensional Electron Gas (2-DEG) unipolar nanodiodesknown as self-switching diode(SSD) is fabricated by etching two symmetrical L-shaped trenches in the semiconductor heterojunction[1]. Asymmetric nanodiodebased on different semiconductor heterojunctions has been proposed to obtain planar devices with non-linearcurrent voltage (I-V) characteristic [2]. Devices using InGaAs/GaAs or AlSb/AlGaSbheterojunction have been demonstrated as room temperature direct detectors at millimeter[3]and terahertz frequencies already[2][4]. GaNNanodiodes is fabricated by etching in the AlGaN/GaNheterojunction. Firstly, the possibility of GaNnanodiodes as direct and heterodyne detectors in terahertz frequency range has been foundin the numerical Monte Carlo (MC) simulations[5]. Then, GaNnanodiodesutilized as the direct detector and the fundamental mixer have been realized in the laboratory [6][7].

A rigorous device's model which describes the physical mechanism or the electronic characteristic is very important for the circuit nonlinear simulation. Åberg et al. later modelledthe I/V characteristic of silicon-based SSDs based on FET-equations, and their model is brieflyreviewed in [9]. Schottky diode equation has also been used to describe theI/V characteristic of SSDs, however it is a challenge still to deal with the GaNNanodiodesquadrantal symmetryI/V characteristic. Here we introduced the phenomenologicalunified diode equation for GaNnanodiodes and the mixing conductivity g of the device and the expression is as the following,

$$g = \frac{di_d}{dv_d} = \left[\frac{1}{2}\left(e^{v_d} + e^{-v_d}\right)\right]^{-2} \quad (1)$$

Then, Taylor series expansion to the diode's mixing conductivity g around its operating point v_0。Following the taylor expansion,$v_d$in the Eqn(1) was substituted bythe local signal $v_{LO}e^{j\omega_{LO}t}$and then the curves of each series expansion parameters depend of the operating point v_0 wereplotted in @Mathematica and were shown in Fig1. If the operating point $v_0$was set to zero, which means the device was zero biased, the linear, cubic and quantic terms are all zero atthis case. And the constant and even order terms (quadratic and quartic) are at their peak position. With the RF signal $v_{RF}e^{j\omega_{RF}t}$ exciting, only the even order terms can be obtained, final result contains the $|2n\omega_{LO} - \omega_{RF}|$ termsafter the low pass filtering ($n = 1,2,3 \cdots$).

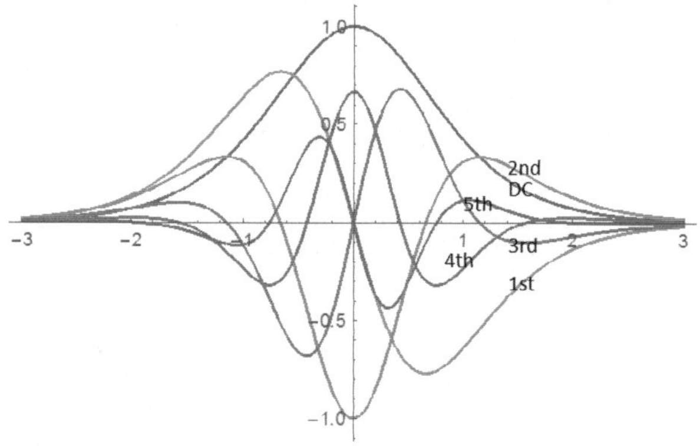

Figure 1.Curves of each series expansion parameters depend of the operating pointv_0.

DESIGN METHODOLOGY

Fig.2 depicts the basic mixer circuit,consisting of a mixer chip placed in a channel across the inputwaveguide. The mixer are using the silicon-on-insulator (SOI) substrate comprisedof the antenna, filter and the SSDs, issuspended at the channel, plus a waveguidebackshort which is the critical part for the optimization of a better LO/RF coupling.

SSDs model is implemented together with the ideal input and output matching networks to optimize the electrical parameters of the anode for certain input power using the Harmonic-balance simulator in ADS.

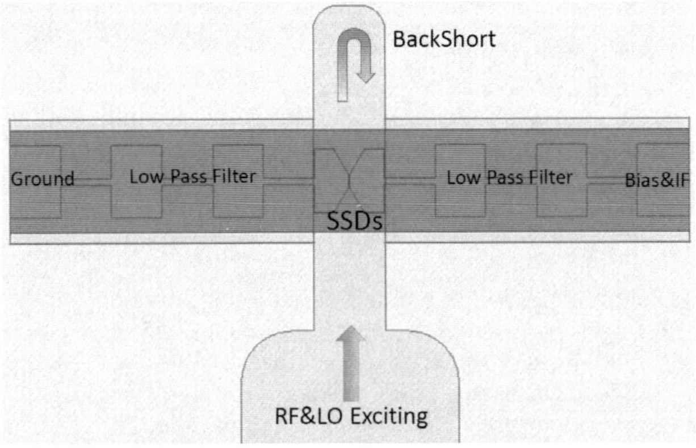

Figure 2.Configuration of mixertopology.

Passive networks around the diodes are simulated in HFSS to get the S-parameter matrices. These matrices were then utilized to determinate initialdimensions of each section using the linear simulator and nonlinear simulator in ADS step by step. Iteration is required to run between the linear and the nonlinear steps until the emergence of the acceptable performance. Finally, the values of each part from the nonlinear step are feedback to reconstruct the mixer in HFSS, then the SSDs models and the S-parameter matrices of the whole mixer circuit are combined to check the performance using the nonlinear harmonic balance simulator in ADS.

SIMULATION RESULTS

The design is designed with the procedure mentioned in the last section and is optimized for available input pump power in practice. lowerconversion lossis obtained as a consequenceof the wider bandwidth, for the mixer working at 300 GHz the relative bandwidth is higher than 12%. It is important toremark that the design owns the character with high port performance, as shown in Fig.3.

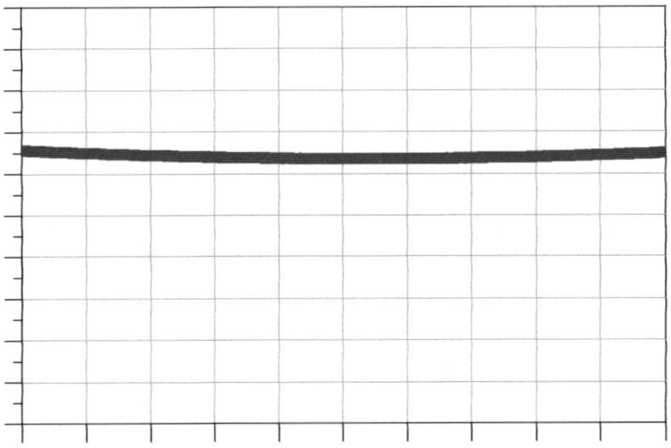

Figure3.Predicted conversion loss as a sub-harmonical mixer.

ACKNOWLEDGMENT

The work proposed in this paper is supported by the National Natural Science Foundation of China (Grant No. 11303004, Grant No. 11573007)and the National Natural Science Foundation of Jiangsu,China (Grant No. BK20130637)..

REFERENCES

P. Sangaré et al., Experimental demonstration of direct terahertzdetection at room-temperature in AlGaN/GaN asymmetricnanochannels, J. Appl. Phys., vol. 113, no. 3, p. 034305, 2013.

A. Westlund, G. Moschetti, H. Zhao, P. A. Nilsson, J. Grahn, Fabrication and DC characterization of InAs/AlSb Self-Switching Diodes, IEEE 24th International Conference on Indium Phosphide & RelatedMaterials, IPRM, pp. 65-68, Aug 2012.

Balocco C, Song A M, Åberg M, Forchel A, González T, Mateos J, Maximov I, Missous M, Rezazadeh A A, Saijets J, Samuelson L, Wallin D, Williams K, Worshech L and Xu H Q, Microwave detection at 110 GHz by nanowires with broken symmetry, Nano Letters, 5, 1423, 2005.

C. Balocco, S. R. Kasjoo, X. F. Lu, L. Q. Zhang, Y. Alimi, S. Winnerl,and A. M. Song, Room-temperature operation of a unipolar nanodiode at terahertz frequencies, Appl. Phys. Lett. 98, 22, 3501, 2011.

I. Íñiguez-de-la-Torre et al., "Operation of GaN planar nanodiodes asTHz detectors and mixers," IEEE Trans. Terahertz Sci. Technol., vol. 4,no. 6, pp. 670–677, Nov. 2014.

Daher, C.; Torres, J.; Iniguez-de-la-Torre, I.; Nouvel, P.; Varani, L.; Sangare, P.; Ducournau, G.; Gaquiere, C.; Mateos, J.; Gonzalez, T., "Room Temperature Direct and Heterodyne Detection of 0.28–0.69-THz Waves Based on GaN 2-DEG Unipolar Nanochannels," in Electron Devices, IEEE Transactions on , vol.63, no.1, pp.353-359, Jan. 2016, doi: 10.1109/TED.2015.2503987.

S. Paul, D. Guillaume, G. Bertrand, B.Virginie,and et al.,Experimental demonstration of direct terahertz detection at room-temperature in AlGaN/GaN asymmetric nanochannels, Journal of Applied Physics, 113, 034305 (2013), DOI:http://dx.doi.org/10.1063/1.4775406.

P. B. Shah and H. A. Hung. 2012. Critical design issues for high-power GaN/AlGaN anti-serial Schottky varactor frequency triplers. Microelectron. J. 43, 6 (June 2012), 410-415. DOI://dx.doi.org/10.1016/j.mejo.2012.03.003.

M. Åberg, J. Saijets, A. Song, and M. Prunnila, "Simulation and Modelingof Self-switching Devices," Phys. Scr., vol. T114, pp. 123–126, Jan. 2004.

A.Maestrini, J.Ward, C. Tripon-Canseliet, J. Gill, C. Lee, H. Javadi, G. Chattopadhyay, and I. Mehdi, "In-Phase Power-Combined Frequency Triplers at 300 GHz," IEEE Microwave and Wireless Component Letters, Vol. 18, No.3, pp.218-220, March 2008.

J.Treuttel, F.Yang, M.Benzazaa, A.Maestrini, J.V-Siles, H.Wang, H.Sanghera, B.Alderman."Solid State Frequency Multipliers at Sub-Millimeter Wavelength Using European Schottky Technology," 22TH ISSTT 2012, Tokyo, April 2012.

J.Treuttel, Gatilova, L., F.Yang, A.Maestrini, "A 330 GHz frequency doubler using European MMIC Schottky process based on e-beam lithography", General Assembly and Scientific Symposium (URSI GASS), 2014 XXXIth URSI, Beijing, pp.1-4, April 2014.

Development of Wideband 100-GHz SIS Mixers foraNew Multi-beam Receiver

Y. Kozuki[1,2*], Y.Hasegawa[1], T.Onishi[1],H.Ogawa[1],
T.Minamidani[2], T.Takahashi[2], Y.Iizuka[2], Y.Niizeki[2], Y.Fujii[2], S.Asayama[2]

[1]*Osaka Prefecture University *, Osaka, Japan*
[2]*National Astronomical Observatory of Japan,Tokyo, Japan*
* Contact: s_y.kozuki@p.s.osakafu-u.ac.jp

We are developing wideband SIS mixers at 100-GHz band. As the first step of this development, we are developing SIS mixers for the*"FOREST"*(FOur beamREceiver System on 45m-Telescope*)* installed on the45-m millimeter-wave telescope located at Nobeyama Radio Observatory (NRO), Japan. Each beam of *FOREST*consists ofone ortho-mode transducer and two sideband-separating (2SB) mixers, and then, the *FOREST*needs eight 2SB mixers in total.The target receiver noise temperature is 40 K or lower over the IF range of 4-12 GHz and the RF range of 80-116GHz. We newly designed SIS mixersthat have a series-array junction in order to avoid saturation, acoplanar inductor loaded microstrip impedance transformer for wideband operation, and hammer filter as RF choke. In this design, we quantify and correct the differences between the model and actual transmission-lines, which occur from structural discontinuity in the circuit. These mixers were evaluated in our laboratory and showed good performances that met the present specifications of *FOREST*. These mixers will be installed into the *FOREST* from the next observation season. Based on these experiences, the next step of this development is to achieve a wider IF bandwidth of ~20GHz with covering the RF frequency range of 67-116GHz. We will also describethis future plan.

Development of NbN-based Hot Electron Bolometer Mixers fabricated by standard UV lithography

C. Chaumont*, F. Boussaha, R. Lefèvre, G. Gay, T. Vacelet, F. Reix
Observatoire de Paris, 77 avenue Denfert-Rochereau, 75014 Paris
*Contact: christine.chaumont@obspm.fr, phone +33 1 40 51 22 59

Abstract— **We present the fabrication of NbN-based hot electron bolometer mixers for THz spectral line astronomy applications. Superconducting HEB nanobridges are defined from ultrathin NbN layers sputtered on Si substrates. The thin films of 3-5 nm exhibit critical temperatures (T_c) of ~ 8-10K so far. We propose to define the whole HEB mixer structures, including nanobridges, by the conventional UV lithography technique with a limited number of steps.**

I. Introduction

Nowadays, to detect and resolve atomic and molecular transitions in THz frequency range, typically beyond 1 THz, the hot electron bolometer (HEB) mixer is the key element to build ultrasensitive heterodyne receivers [1]. The HEB device which features a nanoscale superconducting strip on the order of few hundred nm, typically 100-300nm, is usually defined by the e-beam writing technique [2]. However, this may not be available and usually requires high-cost machines.

In this paper, we present the current development of HEB devices from ultrathin superconducting NbN films deposited by means of a new sputtering machine at Paris Observatory. In order to fabricate HEB mixers, these films are patterned using the standard 320nm UV lithography technique to realize HEB nanobridges.

II. Optimisation of Ultrathin NbN deposition

The NbN layers are deposited using the new sputtering machine Plassys MP700S which works at a base pressure of 3×10^{-8} mbar. It is equipped with a 6-inch diameter Nb target and 3-inch silicon carbide substrate heater designed to go up to 1000 °C. The distance between the target and the substrate is 8 cm. In order to find out the reasonable starting range for NbN deposition parameters, we first recorded the complete current-voltage characteristics of the Nb dc magnetron cathode using current stabilization in pure argon Ar gas then in a N_2/Ar mixture. Fig. 1 shows the cathode I-V characteristics obtained with a N_2/Ar flow ratio and Ar flow rate set at, respectively, 0.12 and 50sccm. Depending on the current between electrodes, the curve shows three slopes corresponding to three states of the system during the sputtering when the N_2 gas is injected along with Ar gas into the chamber. At low currents, a small amount of nitrided Nb atoms are sputtered from the target. The increase of the current leads to enhancing the reaction between N_2 and Nb atoms. This leads to the decrease of N_2 partial pressure in the chamber causing a drop in the discharge voltage resulting in a negative resistance region observed in the I-V curve [3]. Thus, the best conditions for deposition of NbN thin films should be found in this region. At high currents, the curves with and without N_2 are joined indicating likely the vanishing of the N_2 amount as it reacts with Nb atoms. The quality of the NbN films can be assessed by measuring the difference discharge voltage ΔV between the discharge voltage with Ar gas only and its value when N_2 is injected into the deposition chamber [3].

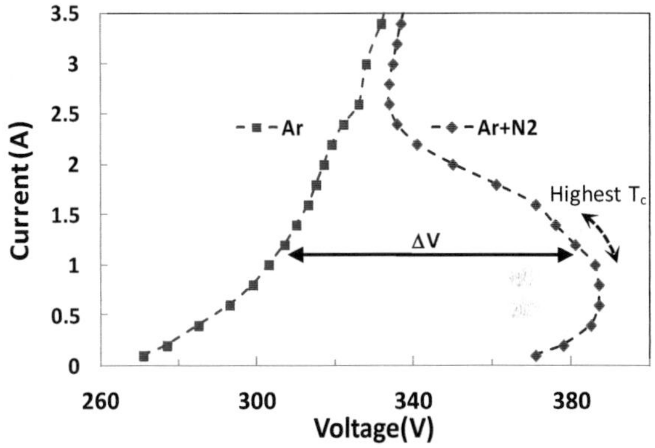

Fig. 1. Current-voltage characteristic of the Nb magnetron sputtering cathode using current stabilization in Ar/N_2 and in pure argon gas mixture.

Ultrathin NbN Film deposition

As it is not possible to efficiently measure a thickness below 10nm using the existing contact profilometry technique, the targeted thickness of 4-5nm would be achieved by adjusting the sputtering time. This is deduced from the thickness versus sputtering time curve which was recorded for measurable thicknesses, above 20nm. NbN films are deposited on 3-inch Si substrates which are heated between 600 and 700°C during the deposition. Using the atomic force microscopy (AFM) technique as well as RIE end-point detection compared to a known sample, the thickness of NbN films is estimated to be 3-7nm thick.

Measurement of critical temperatures

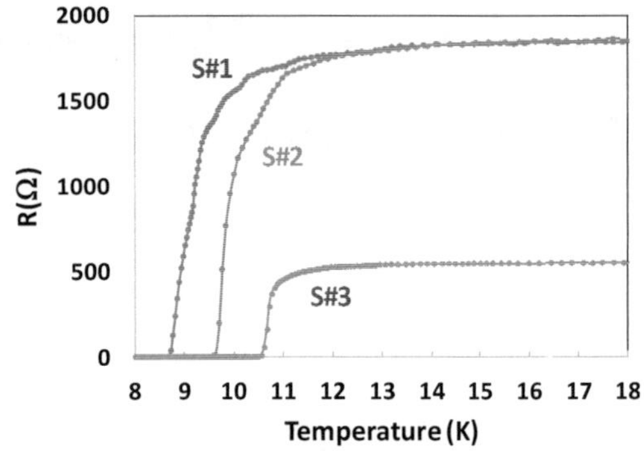

Fig. 2. Resistance versus temperature curves of ultrathin NbN films deposited on 3-inch Si substrates for a substrate temperature of 700°C during deposition. Critical temperatures are measured at the mid temperature transition. Using AFM technique and RIE end-point detection, the thickness is estimated to be 4-5 nm for films 1 and 2 and 6-7 nm for films 3.

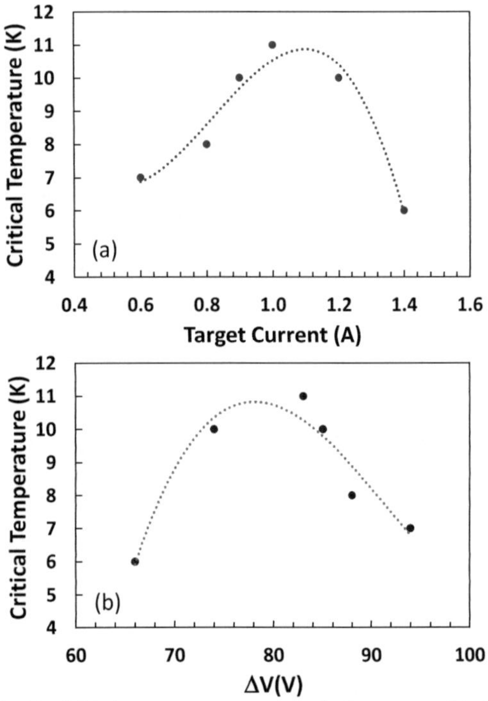

Fig. 3. (a) Critical temperature versus cathode current. (b) Critical temperature versus ΔV.

Rectangular samples are cut off and mounted on a dipstick which is dipped into liquid helium to measure the critical temperature (T_c). Fig. 2 shows typical measured resistances as a function of temperature R(T) curves we performed. In Figs. 3a and 3b, we show curves of respectively the cathode current and ΔV as a function of measured T_c. The highest critical temperatures varies between 9 and 10.7 ± 1K and are obtained with a cathode current in 0.9-1.2A range and ΔV at around 80V. As expected, the currents belong to the negative resistance region with the largest value of ΔV as shown in Fig. 1. These critical temperatures would allow to achieve high quality HEB mixers. Furthermore, even a higher T_c is expected as it is possible to heat the substrate temperature up to 1000°C. As depicted in [3], the optimal reaction between Nb atoms and N_2 gas takes place within this negative resistance region.

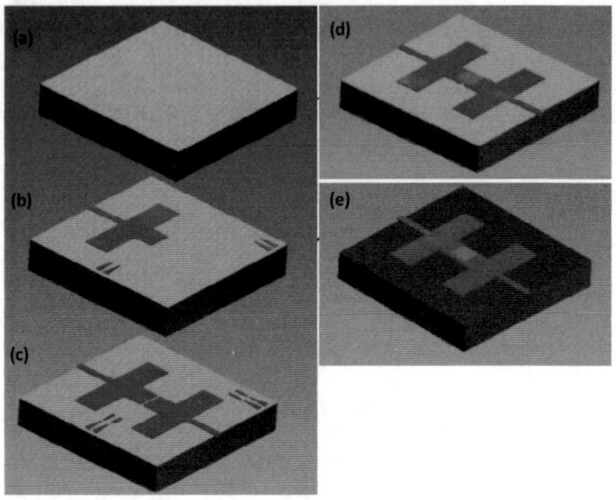

Fig. 4 Fabrication process sequence using standard 320nm UV lithography.

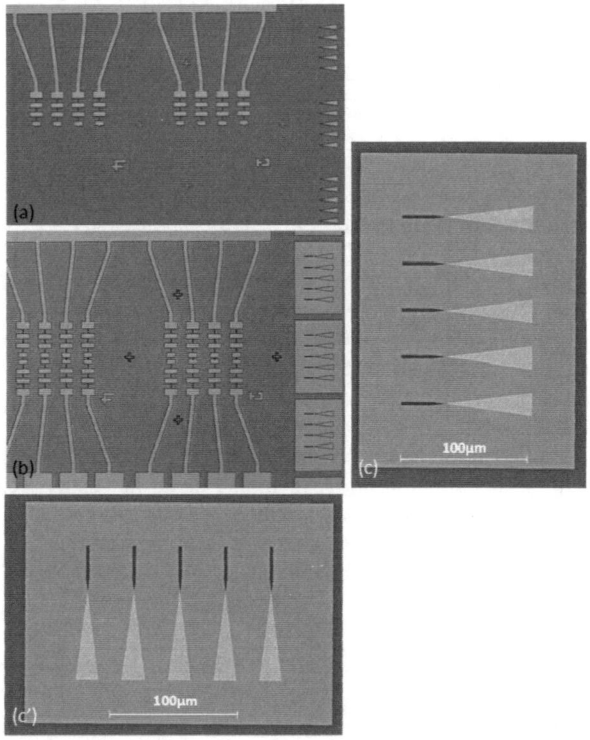

Fig. 5 Fabrication process sequence using standard 320nm UV lithography. (a) Definition of the first Au electrode, (b) definition of the second Au electrode which is closely and precisely aligned to the first one using horizontal and vertical arrow mark alignment.

III. Definition of HEB nanobridges by UV lithography

To quickly assess whether we can fabricate HEB mixers using the deposited ultrathin NbN films, we are developing a straight fabrication process based on the use of the standard low-cost UV 320nm lithography technique. Because of UV light limits, it is obvious that the optical lithography cannot pattern the superconducting HEB nanobridge whose length should be ideally in 100-200nm range. To overcome the UV photolithography limits, the HEB is viewed as being made up

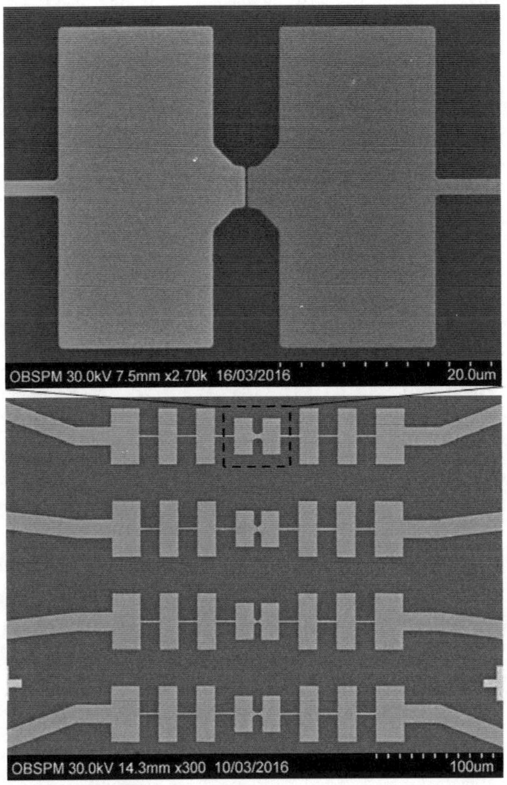

Fig. 6 Electron microscope picture of four HEB devices featuring two Au electrodes defined by UV photolithography. The zoom in picture shows the

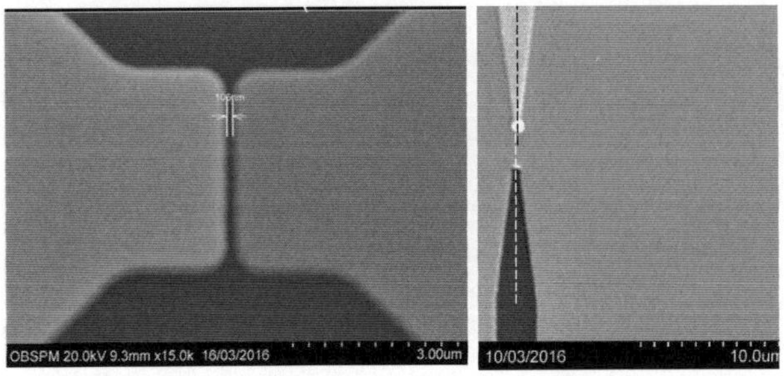

Fig. 7 Electron microscope picture of electrodes separated by a gap of ~100nm patterned by UV lithography. This is achieved when the arrows are misaligned by Δw (right picture).

of two microscale electrodes which can be separately defined in two successive UV lithographic steps. In this case, the issue is no longer the UV light limits but the positioning accuracy of one electrode to the other. This can be addressed by designing proper alignment marks on the mask. For example, we can use a set of arrows whose points must align to ensure an accurate alignment of electrodes with the desired gap between electrodes (i.e., length of the nanobridge) as shown in Figs. 4 and 5. As electrodes must be both horizontally and vertically well aligned, we define arrows along horizontal and vertical axes. In Fig. 4, we summarize the fabrication process sequence using the standard 320nm UV lithography. This is done thanks to the widely used MGB 4 aligner mask of SUSS Micro Tech. After the deposition of the ultrathin NbN layer on 3-inch Si substrate,

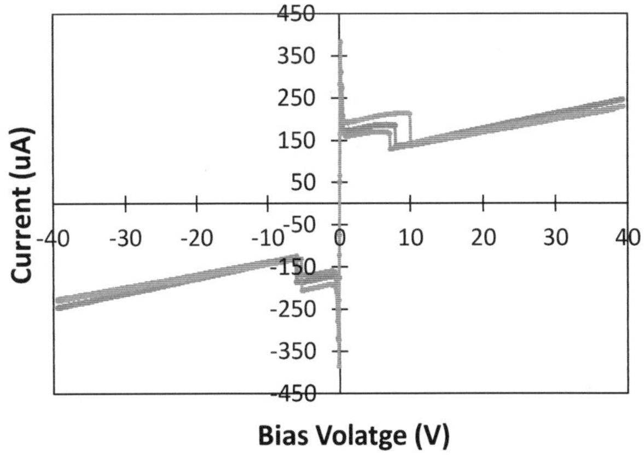

Fig. 8. First I-V curves of HEBs fabricated by UV lithography process using NbN layer with $R_{\square}\approx 550\Omega/\square$ and nanobridges of $L\times w\approx 0.25\times 2$, 0.3×2 and 0.4×2 μm.

we first define one of electrodes as well as the first set of arrows by UV lithography using SPR700 positive photoresist. The deposition followed up by the lift-off of Au 150 nm-thick allowing the realisation of the first Au electrode as shown in Figs.4b and 5a. Using again the SPR700 photoresist, the second step consists in pattering the second electrode which must be precisely and closely aligned to the first Au electrode. This is done when the points of the second set of arrows align with those of the first set as shown in Fig. 5. The width of the nanobridge is defined using 2μm-width rectangular photoresist or insulator (SiO) layer which must be well aligned in the center of electrodes. Finally, the uncoated NbN layer is removed by reactive ion etching. Thus, this process requires a limited number of steps. Fig. 6 shows the electron picture of four HEB devices. On the zoom-in picture, the gap between electrodes is around 250nm while the targeted one is 200nm. This difference is not due to the alignment process but to the used mask which features a deviation of patterns in XY plane. It is possible to recover this by adjusting the position of arrows as illustrated in Fig. 7 (right picture).

To assess the capability of this technique to achieve a smaller HEB nanobridge length, we could further reduce the gap between the electrodes by shifting and misaligning the arrows as shown in Fig. 7. This shows a gap of ~100nm obtained when the arrow marks are shiftted by around $\Delta w\approx 150$nm.

Fig. 8 shows first I-V curves of HEB featuring nanobridges of $L\times w\approx 0.25\times 2$, 0.3×2 and 0.4×2 μm.

CONCLUSION

NbN films of 3-5 nm thick deposited on Si substrate exhibiting critical temperatures of 9-10K have been achieved using a new sputtering machine at Paris Observatory. These films are deposited on 3-inch Si substrates heated up to 700°C. However, we expect to achieve higher T_c with a temperature substrate of 1000°C. In order to fabricate HEB mixers, nanobridges of typically 100-400 nm length have been patterned using the standard low-cost 320 nm UV lithography. The UV light limits were overcome by pattering the HEB electrodes in two separated UV lithographic steps. HEB devices with lengths up to ~200nm were achieved.

ACKNOWLEDGMENT

This work has been carried out at Centrale Technologique de l'Observatoire de Paris (CTOP)-GEPI and LERMA and is partly supported by Région Ile-de-France.

REFERENCES

[1] F.Boussaha, J.Kawamura, J.Stern, A.Skalare,and V.White,"Alownoise2.7 THz waveguide based superconductingmixer," IEEETrans. THzSci.Technol., vol. 2, no. 3, pp. 284–289, May 2012.

[2] R. Lefèvre et al., "Terahertz NbN hot electron bolometer fabrication process with a reduced number of steps", 23rd International Symposium on Space Terahertz Technology, Japan, 2012.

[3] S. Miki, Y. Uzawa, A Kawakami, and Zhen Wang, "Fabrication and Bandwidth Measurements of NbN Hot-Electron Bolometers," Electronics and Communications in Japan, Part 2, Vol. 85, No. 7, 2002.

A New Two-way Power Divider/Combiner Based on Magic Tin W-Band

Hong Tang*, You Lei Pu, Ran Yan, and Yong Luo

School of Physical Electronics, University of Electronic Science and Technology of China, Chengdu, Sichuan610054, China,
* Contact: 906578759@qq.com

In this paper, a W-band two-way power divider/combiner usinga new matching structurebased on magic T is proposed. This divider/combiner has properties of low loss, equal power splitting, compact structure , good return loss, and good heat dissipation. The 3D EM simulation results show that , from 102GHz to 108GHz, the return loss is better than 22dB , the insertion loss is less than 0.08dB , and the isolation between the output ports is less than 23dB. Thedivider/combiner can be used in power combing systems to obtain solid-state high power with a high efficiency in W-band.

Electron gun design for a 170 GHz megawatt-level corrugated coaxial gyrotron

Kun Dong[1], Wei Jiang[1], Yong Luo[1]

[1] *School of Physical Electronics, University of Electronic Science and Technology of China, Chengdu, Sichuan610054, China,*

* Contact: dkun1990@126.com

Abstract— **This paper presents the design of a triode type magnetron injection gun (MIG) for a 170 GHz megawatt-level corrugated coaxial gyrotron. The genetic algorithm (GA) is introduced to optimize the beam quality. According to the design acquirements, the predicted transverse velocity spread is 3.03% with a transverse-to-axial velocity ratio of 1.3. The preliminary design procedure is accomplished by an in-house developed code. A multi-objective genetic algorithm (MOGA) code GUNOP written by MATLAB is used to perform the optimization. 2-D electron trajectory code EGUN and 3-D CST particle studio (CST-PS) code are employed to do the calculation and simulation. The results agree well with each other. The sensitivity analysis has also been carried out to estimate the practical operation stability.**

INTRODUCTION

Gyrotrons are capable of generating hundreds of kilowatts of electromagnetic (EM) power in the millimeter and sub-millimeter wave regime [1]. By adopting a longitudinal corrugated tapered insert inside the cavity, the coaxial gyrotron can effectively suppress the mode competition and eliminate the restrictions of voltage depression and limiting current [2].

As a crucial part of the gyrotron, MIG provides the hollow electron beams to interact with the EM wave. Triode type MIG has a (modulating anode (M-anode) and an accelerating anode (A-anode)). Fig. 1 gives the schematic view of triode type MIGs. By tuning the M-anode voltage in triode type MIG, one can readily acquire the desired velocity ratio (defined as $\alpha=v_t/v_z$, v_t and v_z are transverse and axial beam velocity components, respectively). This paper presents the design of a triode type MIG for a 170 GHz megawatt-level corrugated coaxial gyrotron. The design procedure is given in detail. A GA based code GUNOP is introduced to optimize the beam parameters. A 2-D beam trajectory code EGUN is adopted to do the simulation and optimization [3]. 3-D software CST-PS is employed to verify the results.

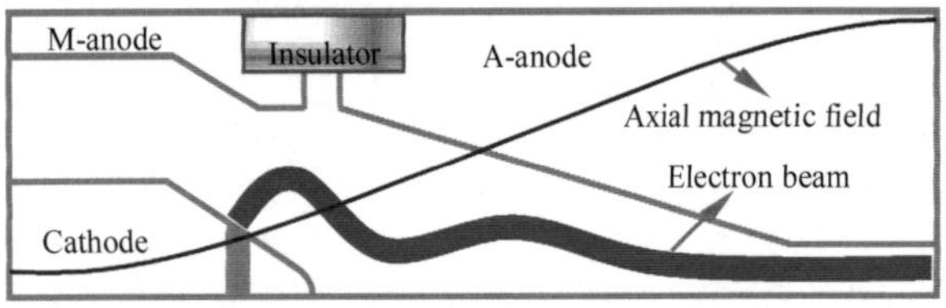

Fig. 1 Schematic views of triode type MIGs.

DESIGN PROCEDURE of the MIG

Taken from [3], the specifications of this coaxial gyrotron are summarized in Table I. The $TE_{31,12}$ mode, which lies in a relative sparse spectrum, is chosen as the operating mode to weaken the mode competition. The electron beam is launched at the first radial maxima of the transverse electric field, corresponding to a radius of 9.48 mm. Considering the electronic efficiency of 48.4% and output power of 1.716 MW, the electron beam is expected to give more than 3 MW power. The operating current is 48 A and voltage is 73.5 kV. Furthermore, a moderate velocity ratio of 1.3 with a transverse velocity spread of ≤ 5% is the design target.

TABLE I
SPECIFICATIONS of the COAXIAL GYROTRON

Operating mode	$TE_{31,12}$
Beam voltage (V_a)	73.5 kV
Beam current (I_b)	48 A
Output power (P_{out})	1.716 MW
Efficiency (*eff*)	48.4%
Magnetic field (B_0)	6.64 T
Beam radius (r_{g0})	9.48 mm
Electron velocity ratio (α)	1.3
Velocity spread ($\Delta\beta_t$)	≤ 5%

An in-house code is developed to determine the initial parameters of the MIG. The code mainly adopts a synthesis approach of MIG design which makes use of the analytical trade-off equations derived by Baird and Lawson [4].

DESIGN OPTIMIZATION of the MIG

Simulated results show that the previous beam quality is poor, so we must perform the deeper optimization. Manual calculations are time-consuming, and sometimes, the results may not satisfy our demands. To perform the optimizations efficiently and automatically, GA is employed. GA is an optimizing method which is based on the biology evolutionary theory driven by natural selection [5]. Based on the concept of GA, a multi-objective GA code written by MATLAB is accomplished. The population size is set as 100. Table II lists the final gun dimensions and optimized beam parameters. The final beam transverse velocity spread is 3.03% when the velocity ratio is kept at 1.29. To verify the EGUN results, CST-PS code is introduced to simulate the MIG in three-dimensions. The CST used magnetic field data is exported from EGUN and other parameters are guaranteed the same with EGUN. Fig. 2 shows the radial beam position at the MIG exit. It is revealed that the average beam radius is about 9.5 mm with a small position spread. Further calculations show that the transverse velocity spread is approximately 3.67% with a velocity spread of 1.32. The results obtained separately by EGUN and CST-PS are in good agreement.

TABLE II
Gun DIMENSIONS and OPTIMIZED BEAM PARAMETERS

Mean emitter radius (r_c)	42.4 mm
Magnetic compression ratio (f_m)	24
Emission current density (J_c)	6 A/cm^2
A-anode voltage (V_a)	73.5 kV
M-anode voltage (V_m)	51.1 kV
Magnetic field at interaction region (B_0)	6.64 T
Magnetic field at cathode (B_c)	0.278 T
Beam guiding center radius (r_{g0})	9.47 mm
Velocity ratio (α)	1.29
Transverse velocity spread ($\Delta\beta_t$)	3.03%

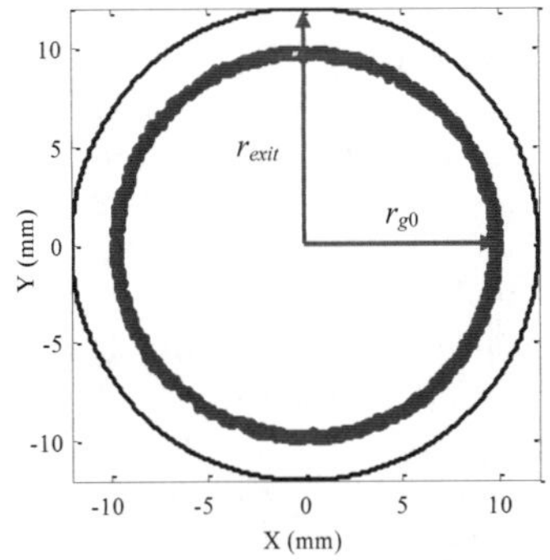

Fig. 2 Radial beam position at MIG exit calculated by CST-PS (r_{exit} is the MIG exit radius)

SENSITIVITY STUDY

The real conditions are different from the nominal in the practical operation of gyrotrons [6]. It is necessary to perform the sensitivity study. Fig. 3 plots the effect of the variation of cathode magnetic field B_c and M-anode voltage V_m on the electron beam quality parameters. As illustrated in Fig. 3(a), the beam velocity ratio α is sensitive to the cathode magnetic field. When B_c varies from 0.275 T to 0.281 T, α grows significantly from 1.16 to 1.4. Transverse velocity spread $\Delta\beta_t$ also increases, but never exceeds 4%. Fig. 3(b) shows the parametric dependence of beam quality on the M-anode voltage α is almost linearly increased with the increase of V_m as shown in Fig. 3(b).

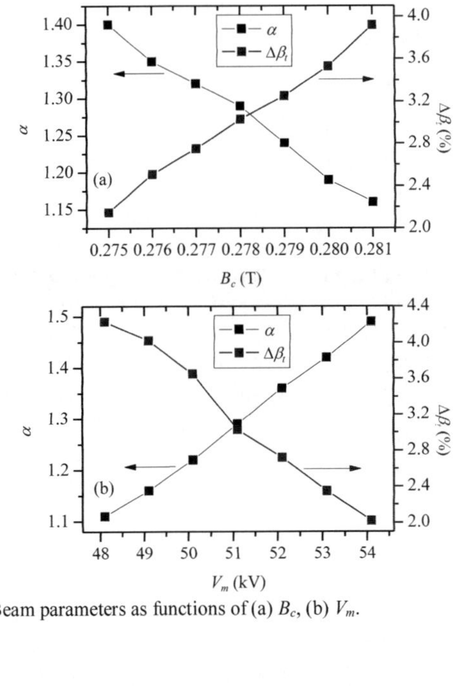

Fig. 3 Beam parameters as functions of (a) B_c, (b) V_m.

CONCLUSION

This paper is aimed at presenting an optimal design of a triode type MIG for a 170 GHz megawatt-level corrugated coaxial gyrotron by introducing the genetic algorithm method. A high-quality electron beam with a transverse velocity spread of 3.03% and velocity ratio of 1.29 is obtained. The design results acquired by EGUN are validated by CST-PS. The results agree with each other well. Sensitivity analysis has also been performed to demonstrate the gun reliability in real operation.

REFERENCES

[1] M. Thumm, "Novel application of millimeter and submillimeter wave gyro-devices," Int. J. Infr. Millimeter Waves, vol. 22, no. 3, pp. 377–385, Mar. 2001.

[2] Mimi Qin, Yong Luo, Kuo Yang, et al, Numerical study and simulation of a 170 GHz megawatt-level corrugated coaxial-gyrotron. Vacuum, vol. 109, no.42, pp. 34-42, Jun. 2014.

[3] W. B. Hermannsfeldt, "EGUN—An electron optics and gun design program," Stanford Linear Accelerator Center, Stanford, CA, USA, Rep. SLAC-226, 1979.

[4] W. Lawson, "Magnetron injection gun scaling," IEEE Trans. Plasma Sci., vol. 16, pp. 290–295, Apr. 1988.

[5] Maulik U and Bandyopadhyay S, "Genetic algorithm-based clustering technique," Pattern Recognition, vol. 33, no. 9, pp. 1455–1465, Sep. 2000.

[6] K. Dong, Y. Luo, R. Yan, and S. Wang, "Magnetron injection gun design for a Q-band 300 kW 30 A gyrotron traveling wave tube," Phys. Lett. A, vol. 379, no. 36, pp. 2078–2082, Sep. 2015.

Design of Q-band Broadband Rectangular Waveguide TE_{10} Mode to Circular Waveguide TE_{01} Mode Converter

Shuai Zong*, Yong Xu, and Guo Liu
School of Physical Electronics, University of Electronic Science and Technology of China, Chengdu, Sichuan610054, China,
* Contact: paradise_zs112@126.com, xuyong01@uestc.edu.cn

Abstract-In this paper, a novel design of rectangular TE_{10} mode to circular TE_{01} mode waveguide converter with compact structure is presented. To begin with, the theoretical analysis is provided. Based on the principle of waveguide mode converter, a creative project with crisscross structure is obtained. As is shown in the project, this kind of waveguide mode converter totally consists of three segments. The first segment performs the transformation from rectangular TE_{10} mode to rectangular TE_{20} mode, and the second accomplishes the conversion from rectangular TE_{20} mode to TE_{22} mode of the crisscross waveguide, while the last section achieves the switching from TE_{22} mode to circular TE_{01} mode. According to the project suggested above, the 3D model of the converter is established in the powerful commercial software named by HFSS where the simulation and optimization of the waveguide converter is completed. Remarkable results are obtained via abundant simulation and optimization of the critical geometric dimension of the mode converter. The conversion efficiency of the converter over 33GHz to 50 GHz is considerably perfect approaching an average level at 99.9%, while the return loss is generally below -25dB.

INTRODUCTION

In recent years, domestically and abroad, plenty of magnificent progresses are achieved in the design and experiment of the gyrotrons and gyro-amplifiers (including gyro-TWTs and gyro-klystrons that generally operate at TE_{01} mode [1]. During transmitting in the circular waveguide, the TE_{01} mode has pretty small attenuation, which is of great significance when TE_{01} mode is applied as the transmitting mode for long distance transmission and as the operating mode of high-Q resonance cavity.

In the design of gyrotrons and gyro-amplifiers operating at TE_{01} mode, it is considerably necessary to perform the cold test experiment to ensure the property of the interaction circuit system of these devices, including testing the resonant frequency, Q-value of the resonance cavity [2], the transmission and reflection characteristics of the input couplers and the output windows [3]. As far as we know, the primary operation mode of dominant microwave sources in the laboratories is rectangular TE_{10} mode. Thus, it is of remarkable value to design more perfect converters to complete the transformation from rectangular TE_{10} mode to circular TE_{01} mode [4].

From the study of the conventional research of the circular waveguide TE_{01} mode converters, a fact is obtained that there are mainly three forms of the TE_{01} mode converter, namely fan-shaped converter, turning magnetic surface excitation converter and the sidewall coupling converter [5],[6]. The fan-shaped mode converter is produced by electrotyping technology, resulting the high cost and the complexity of manufactory [7], [8]. What's worse, the conversion efficiency is absolutely low and there are spurious modes during the transformation. Considering the different direction of the input and output of the magnetic surface excitation mode converter, it is not convenient for the system to be assembled. The arc-shaped structure of the second kind mode converter is difficult to be produced and its bandwidth is narrow [9-11]. While sidewall coupling mode converter achieves high conversion efficiency and broad bandwidth and has compact structure, due to the Y-type power divider network, the product technology demand is severe.

In order to meet the demand of cold test experiment in Q-Band, the principle and design scheme of broad bandwidth mode converter from rectangular TE_{10} mode to circular TE_{01} mode with compact structure is presented in the paper [12]. Based on the design scheme, 3D model is built in the popular commercial electromagnetic simulation software HFSS. In the paper, the first part is the theoretical analysis and design scheme of the TE_{01} mode converter, the second is the simulation results and the relative analysis and the third is the conclusion of the work.

DESIGN AND ANALYSIS OF THE TE_{10}-TE_{01} WAVEGUIDE MODE CONVERTER

A. The principle of the mode converter

To ensure that the rectangular waveguide TE_{10} mode is efficiently coupled into the circular waveguide TE_{01} mode and the waveguide wavelength is kept unchanged along the whole length of the device in the operating band, the physical structure of the converter needs to satisfy certain conditions that the structure of the converter must have ideal symmetry and the transition process between different parts should be as level and smooth as possible. Only by this method, can the perfect result be acquired that the purity of the circular TE_{01} mode is ensured and the conversion loss becomes low. In order to produce the converter successfully and easily, there is some necessary compromise between theoretical design and actual fabrication.

B. The design scheme of the TE_{10} -TE_{01} mode converter

The whole structure of the mode converter mainly consists of three sections. The structure of the first part is linearly gradient from rectangular waveguide to T-shaped waveguide, and then to another rectangular waveguide, realizing the conversion from rectangular TE_{10} mode to rectangular TE_{20} mode. The second one adopts the structure of a rectangular waveguide linearly gradient to the crisscross waveguide, completing the transformation from rectangular TE_{20} mode to the crisscross waveguide TE_{22} mode. The last one is formed by crisscross waveguide linearly gradient to circular waveguide, achieving the change from crisscross TE_{22} mode to circular waveguide TE_{01} mode. The 3D model of the mode converter based on crisscross structure is shown in figure 1. Through the transformation of the three segments, the conversion from rectangular TE_{10} mode to circular TE_{01} mode is achieved. And the mode conversion sequence is presented in figure 2.

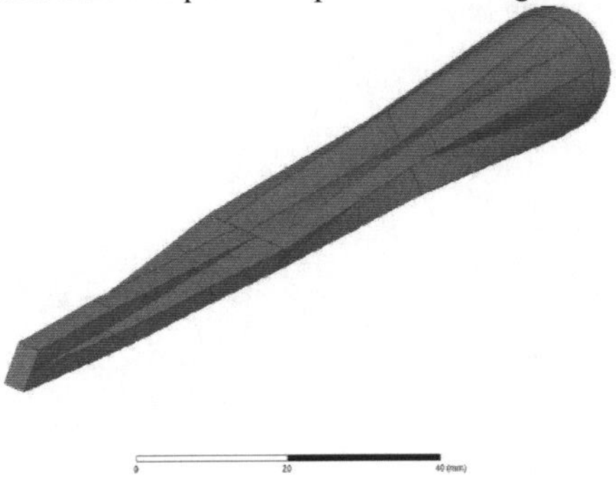

Fig. 1 The 3D model of TE_{10}-TE_{01} mode converter

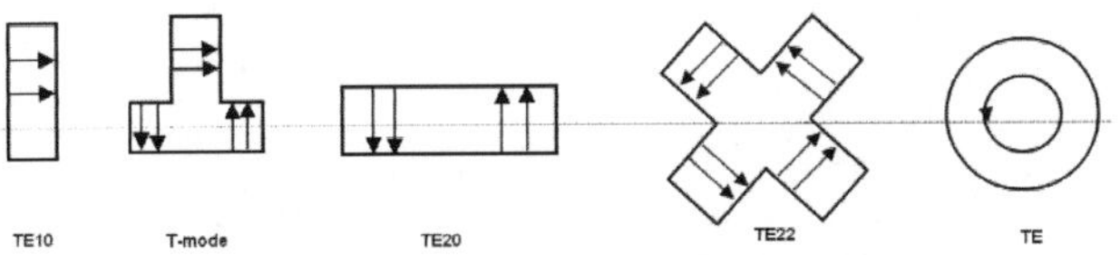

Fig. 2 The mode changing sequence of TE_{10}-TE_{01} mode converter

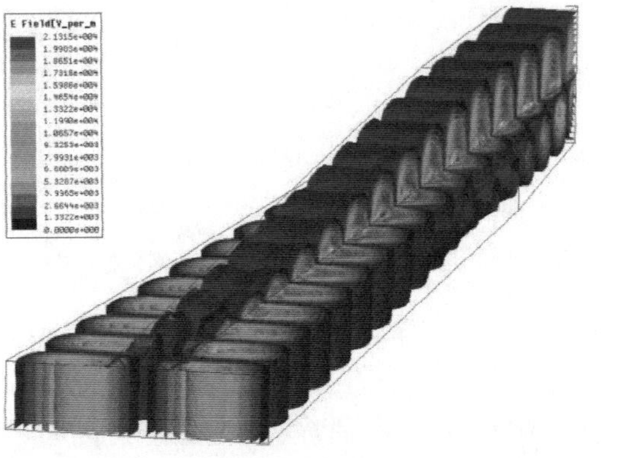

Fig. 3 The electric field distribution of the output port of the first part

THE SIMULATION RESULTS AND THE ANALYSIS OF THE TE_{10}-TE_{01} MODE CONVERTER

The simulation is performed in HFSS by building the 3D model in the simulation software. According to the technical index, it is significantly important to minimize the length of the converter on the condition that the conversion efficiency is kept perfect in the operation bandwidth. Every length of the three parts of the converter is built as an optimization variable that can be scanned to determine the most optimized length of each segment. As is shown in the following figures, better optimized results are presented.

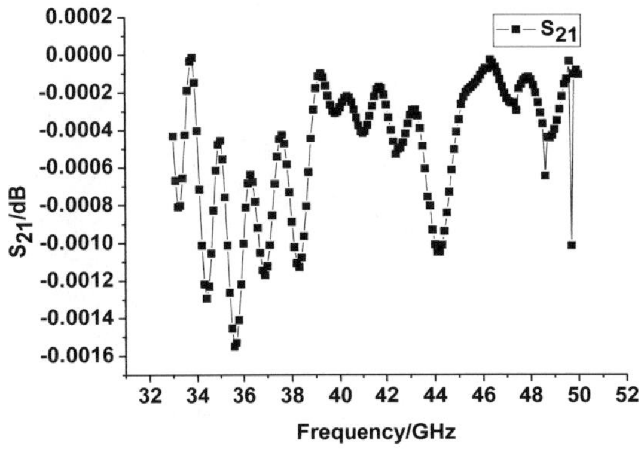

(a)

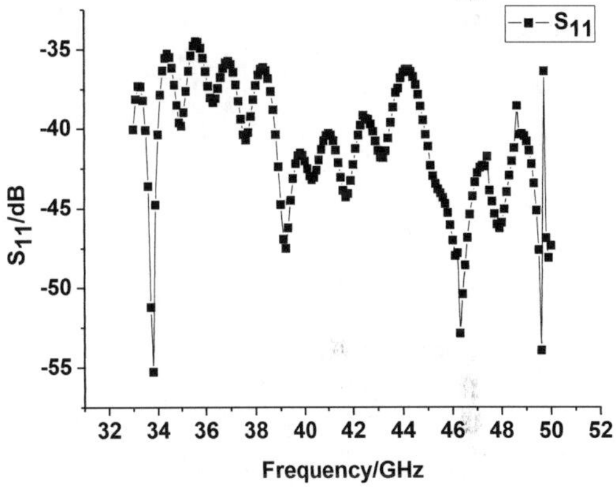

(b)

Fig. 4 The transmission/reflection parameter of the first part
(a) The transmission parameter (b) The reflection parameter

In figure 3, the electric distribution of the output port of the first part is given, from which a conclusion can be acquired that the input TE_{10} mode is efficiently converted into TE_{20} mode.

As is shown in figure 4, the transmission and the reflection characteristic is presented. Without the loss in the conversion considered, the S_{11} is below -35dB in the whole Q-band, identifying the energy of TE_{10} mode reflected is less than one thousandth and the transmission energy is above 99.9%. Therefore, TE_{10} mode is considerably converted into TE_{20} mode.

Figure 5 is the electric field distribution of the output port of the second section of the converter. As the figure 5 shows, after transmitting in the second section, TE_{20} mode is converted to crisscross TE_{22} mode with high purity.

What's is shown in Figure 6 is the S_{21} parameter and S_{11} parameter of the second section. The figure 6 presents that S_{11} parameter of TE_{20} mode is less than -35dB and S_{21} is better than -0.0012dB in the whole Q-band, demonstrating that the transmitting energy is nearly 99%. Therefore, most of the energy of rectangular waveguide TE_{20} mode is successfully converted into the crisscross waveguide TE_{22} mode.

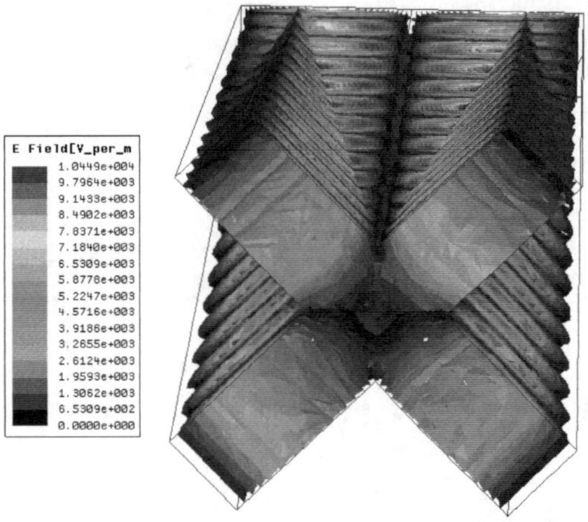

Fig. 5 The electric field distribution of the output port of the second part

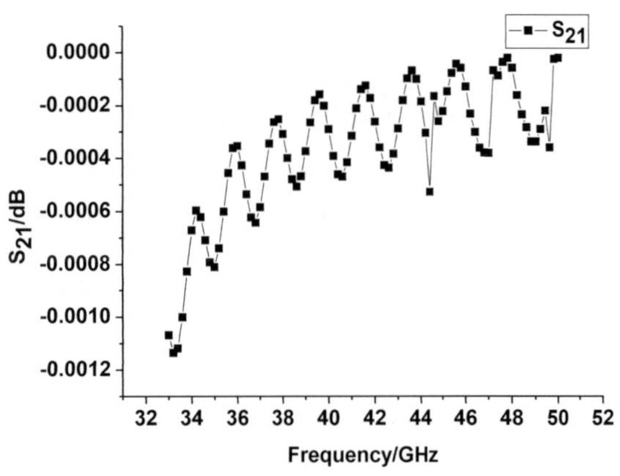

(a)

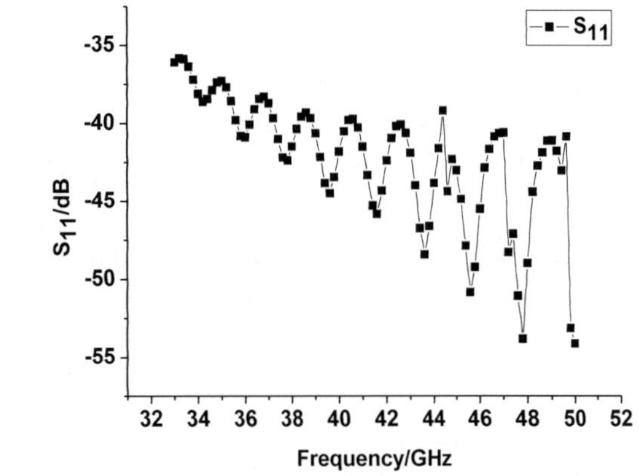

(b)

Fig. 6 The transmission/reflection parameter of the second part
(a) The transmission parameter (b) The reflection parameter

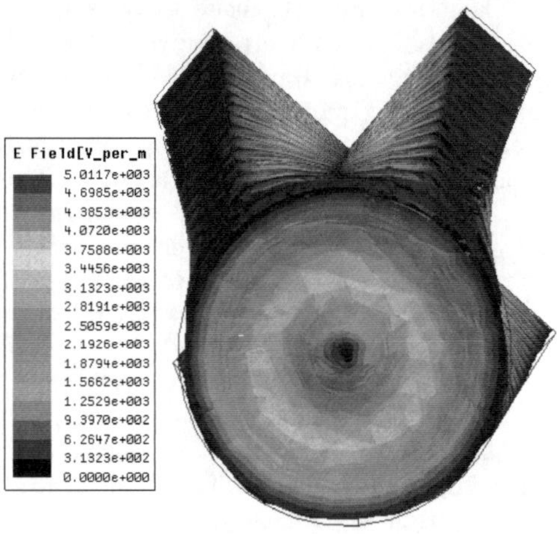

Fig. 7 The electric field distribution of the output port of the third part

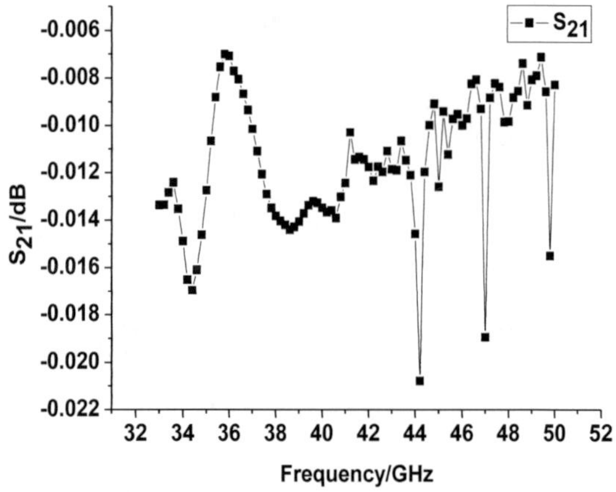

(a)

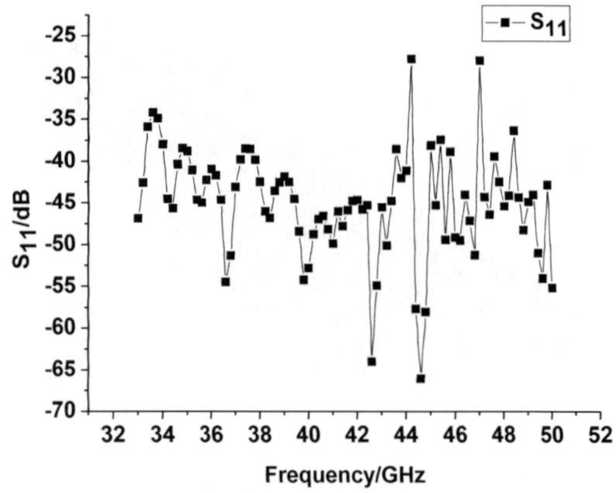

(b)

Fig. 8 The transmission/reflection parameter of the third part
(a) The transmission parameter (b) The reflection parameter

Figure 7 is electric field distribution of the output port of the third section, showing that crisscross waveguide TE_{22} mode is transformed into circular waveguide TE_{01} mode with high purity. As is shown in figure 8, the S_{21} parameter from TE_{22} mode to TE_{01} mode is above -0.018dB, while the S_{11} parameter of the TE_{22} mode is generally below -35dB, identifying that TE_{22} mode is efficiently converted into TE_{01} mode.

When the optimization of each of the three sections is completed, the whole TE_{10}-TE_{01} mode converter is assembled. As the figure 9 shows, the final field distribution of the output of the mode converter is vividly presented, indicating that the TE_{10} mode is efficiently converted to TE_{01} mode. And as the figure 10 presents, both the S_{11} parameter of the TE_{10} mode and the S_{21} parameter from TE_{10} mode to TE_{01} mode are perfect enough to meet the demand of the TE_{10}-TE_{01} mode converter.

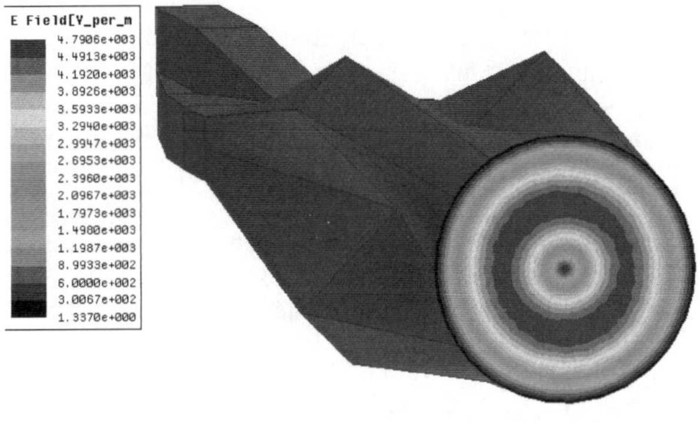

Fig. 9 The electric field distribution of the output port of the whole converter

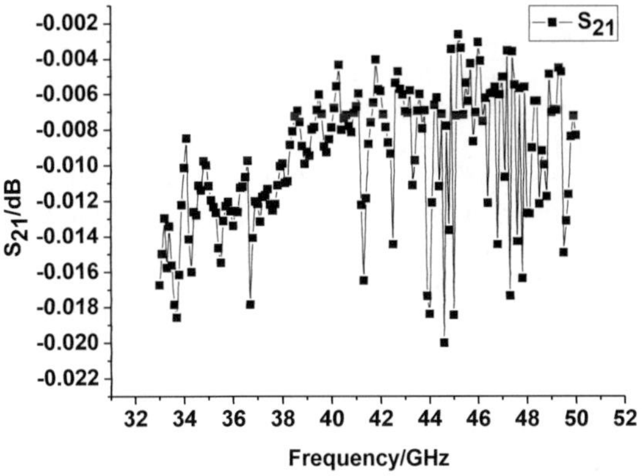

(a)

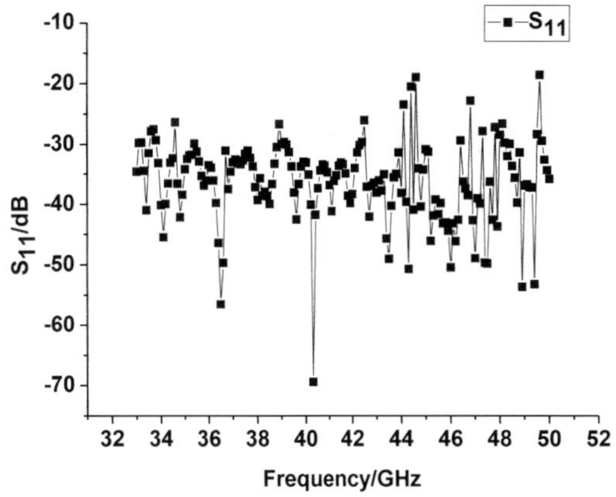

(b)

Fig. 10 The transmission/reflection parameter of the whole converter
(a) The transmission parameter (b) The reflection parameter

CONCLUSIONS

The design of a broad-band TE_{10}-TE_{01} mode converter in Q-band is presented in the paper. Through plenty of simulation and optimization of the converter model by FEM (finite element method) in HFSS, a good result is obtained. As the simulation and optimization results shows, the conversion efficiency is more than 99% while the return loss is below -25dB. At the same time, the work presented in the paper can also supply some suggestion in the research of other mode converters.

REFERENCES

Chu K R. Overview of research on the gyrotron traveling-wave amplifier [J]. IEEE Transactions on Plasma Science, 2002, 30(3):903-908.

XU Yong, LUO Yong, WANG Jian-Xun. Design and experiment of high frequency circuit: fundamental wave gyroklystron amplifier with 34 GHz TE01 mode [J]. J. Infrared Millim. Wave, 2013, 32(6):531-537.

Xu Yong, Luo Yong, Li HongFu, Xiong CaiDong, Wang JianXun. Design of a New Type Broad-band Output Window for High Power Gyro-TWT. J. Infrared Millim.Waves.2013; 32(NO.2):6.

Chang C-F Ya T-H. High Performance of Circular TE01 Mode Converter. IEEE Transactions on Microwave Theory and Technologies. 2005; 53(12):5.

Du RenBo, Luo Yong, Niu XinJian. Design of High-power Millimeter Wave Corrugated Circular Waveguide Mode Converter. High Power Laser and Particle Beams. 2008; 20(1):4.

Ding YanFeng,Liu QingXiang, Zhang JianQiong. Design and Experiment Study of Overmode Bend Circular Waveguide. High Power Laser and Particle Beams. 2011; 23(8):6.

Zhu Min, Luo JiRun, Luan YuanTao, Guo Wei, Yuan GuangJiang. Bandwidth Property of the Converter with Sector Taper from a Rectangular Guide TE10 Mode to a Coaxial Guide TE01 Mode. Journal of Electronics & Information Technology. 2007; 29(6):3.

Wang WenXiang. Microwave Engineering Technology. Beijing: National Defense Industry Press; 2009 4.

Fu Hua, Li Hao, Jiang LiHui, Wu ZeWei, Li TianMing. Circular TE01 Mode Transducer at V-band. High Power Laser and Particle Beams. 2014; 26 (6):5.

Niu XinJian, Li HongFu, Xie ZhongLian. Analysis of High-power Millimeter Wave Circular Waveguide TM01-TE11 Mode Converter. High Power Laser and Particle Beams. 2002; 14(5):4.

Wang Qiang, Zhou HaiJing, Yang Chun, Li Biao, Ye Jian. Bandwidth Study of TE01-TE11 Mode Converter. High Power Laser and Particle Beams. 2013; 25(3):4.

Xu Yong, Luo Yong , Wang JianXun , Yan Ran, Pu YouLei, Wang Hui, Wang Li, Li HongFu, Xiong Caidong. Design of Ka-band TE10-TE01 Broad Band Mode Converter. Chinese Journal of Vacuum Science and Technology. 2013; 33(4):6.

A Novel Wideband Antipodal Fin-line Waveguide-to-Microstrip Transition Structure for Ka-band Applications

Bo Fang*, Youlei Pu, and Yong Luo
School of Physical Electronics, University of Electronic Science and Technology of China, Chengdu, Sichuan610054, China,
* Contact: 18200516559@163.com, phone +18200516559

Abstract— **In this paper, a wideband low loss antipodal fin-line waveguide-to-microstrip transition structure for operation in the Ka-Band (28 ~ 40 GHz) has been designed .The design uses a novel fin-line transition structure to eliminate the resonance caused by the traditional resonant cavity and thus broadens the bandwidth. The transition structure is realized by clamping the printed circuit board (PCB) between two halves of the metal body. The compact transition design could decrease the effect of manufacture and assembly error. All parameters of the designed structure are optimized by using a high-frequency structure simulator (HFSS) .The design was verified by experimental results at Ka-band, which has a coincidence with the results of simulation. Test results show that insertion loss fluctuates between 0.09 dB and 1.0 dB covering 28 ~ 40 GHz. In the frequency range of 29.1 ~ 32.8 GHz and 36 ~ 39.4 GHz, the insertion loss fluctuates in the range of 0.09 dB to 0.6 dB .The return loss fluctuates between 10 dB and 30 dB covering 28 ~ 40 GHz.**

INTRODUCTION

With the millimeter-wave technology broadly used in short-distance wireless communications and radar systems, the use of low-cost and high reliable Monolithic Microwave Integrated Circuit (MMIC) is becoming increasingly widespread. In the millimeter-wave receiving system, microstrip line is used as the connection between MMICs. However, current millimeter-wave test system mostly use a rectangular waveguide interface which requires the system to find a low cost, low loss, easy to manufacture, wide bandwidth rectangular waveguide to microstrip ferry. Currently, the transition structure commonly used between waveguide and microstrip is: stepped ridge waveguide transition [1], antipodal fin-line waveguide-to-microstrip transition [2-4], microstrip probe coupling transition [5-6]. These transition structures have wide bandwidth, low insertion loss characteristic. Wherein, the stepped ridge waveguide transition has the characteristics of complex processing. Microstrip probe coupling transition due to the outlet direction perpendicular the circuit, it does not meet many systems' requirements. With regard to antipodal fin-line waveguide-to-microstrip transition, it provides low insertion and reflection loss levels in comparison with the aperture coupled transitions and also are easier in realization comparing with the probe transitions. Those transitions were first demonstrated in [7] for the 18 ~ 26GHz frequency band. Later in [8-9] their modifications were presented for frequencies up to 48 GHz.

This paper presents a design of a wideband waveguide-to-microstrip transition operating in the Ka-Band (28 ~ 40 GHz) with a low insertion loss. The transition is formed on the PCB with realized antipodal fin-line inserted into a standard WR28 (26.5 ~ 40GHz) rectangular waveguide structure. The PCB is disposed along the E-plane of the waveguide. The PCB in the proposed transition is based on high frequency technology RO5880 (with dielectric permittivity $\varepsilon = 2.2$ and loss tangent $\tan\delta = 0.0009$). The transition characteristics are optimized with 3D electromagnetic simulation (HFSS) and confirmed by experimental verification.

WAVEGUIDE-TO-MICROSTRIP TRANSITION STRUCTURE

In this paper, all parameters of the designed structure are optimized by using a high-frequency structure simulator (HFSS). Fig. 1 shows the structure of antipodal fin-line waveguide-to-microstrip transition. As we can see, the transition structure is realized by clamping the printed circuit board (PCB) between two halves of the metal body. The tapered antipodal fin-line smoothly transforms the incident TE10 waveguide mode to a quasi-microstrip mode [10] which is propagating in the area of overlapping antipodal fins of the transition. E field in the transition area concentrates and rotates by 90 degrees until waveguide mode transforms to the microstrip mode. Smooth transformation of the field mode allows the transition to operate in a wide frequency band.

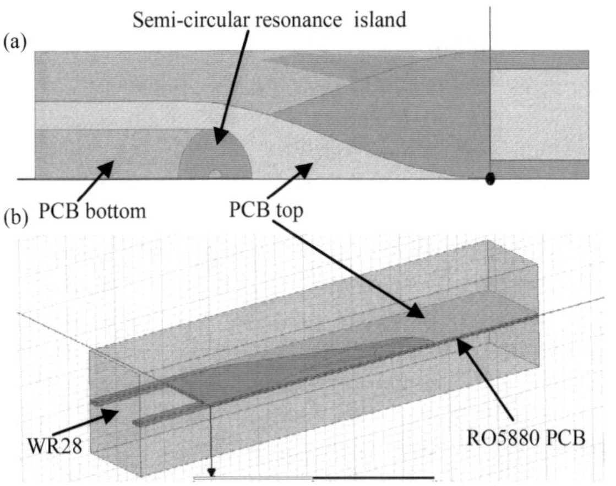

Fig. 1 A typical antipodal fin-line waveguide-to-microstrip transition structure

In order to obtain good impedance matching between waveguide and microstrip, smooth gradient segment between the top and bottom of PCB is necessary. Generally, parabolic curve, exponential curve and cosine squared curve are widely used in fin-line transition. In consideration of easy processing, this paper will use cosine squared curve.

In tradition, to prevent electromagnetic field resonate with the cavity wall, a semi-circular resonance island is introduced and it achieves the elimination of resonance. In this paper, a novel triangle resonance island will be used, this structure eliminates resonance more easily and is easier to be fabricated.

ELECTROMAGNETIC SIMULATION RESULTS

Fig. 2 shows the back to back model which achieves antipodal fin-line waveguide-to-microstrip transition. Extending to both sides of the portion of the substrate is to make the substrate assemble in the waveguide cavity more stably. The metal via holes extends the waveguide walls inside the PCB structure along the fin length. That allows increasing stability of the transition and improving its reflection characteristics by eliminating the slot mode excitation between two halves of the metal body.

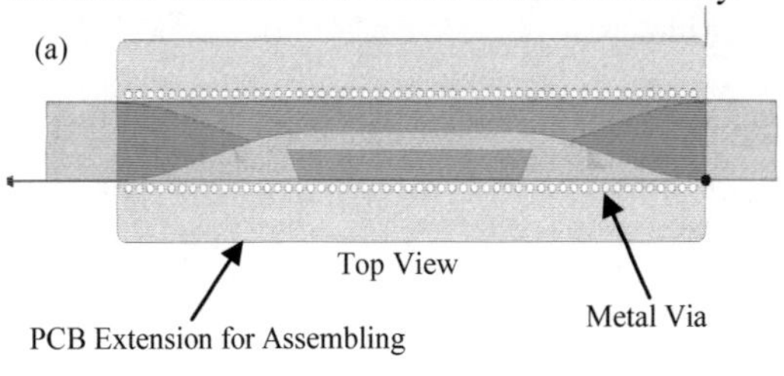

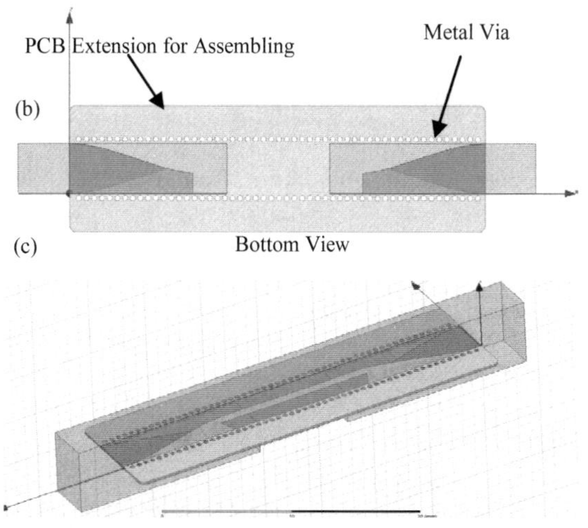

Fig. 2 Back to back model in HFSS simulation

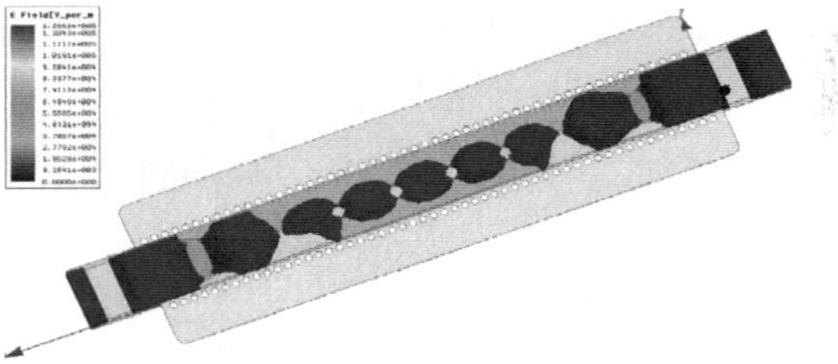

Fig. 3 E field simulation result

From Fig. 3, we can see that electromagnetic energy achieve good transition from waveguide to microstrip. The tapered antipodal fin-line smoothly transforms the incident TE_{10} waveguide mode to a quasi-microstrip mode.

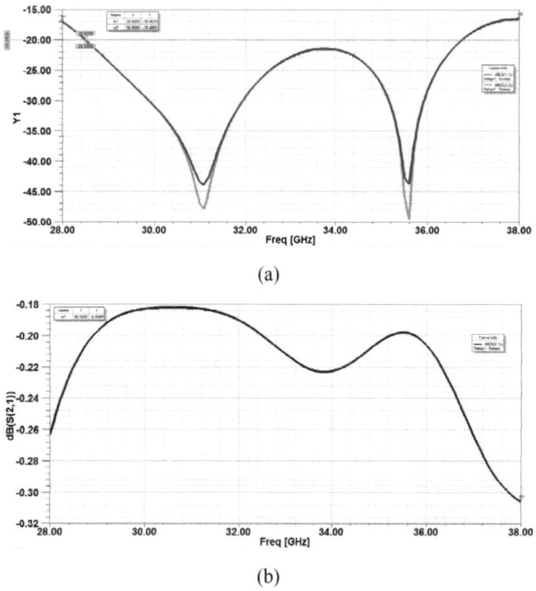

Fig. 4 Simulated S_{11}, S_{22} (a) and S_{21} (b) of the designed waveguide to microstrip transition

By means of using HFSS software' optimization function, we achieved some good results which proved the rationality of this design. Finally, results are shown in Fig. 4. Simulation results show that insertion loss fluctuates between 0.18 dB and 0.3 dB covering 28 ~ 40 GHz. The return loss fluctuates between 16.98 dB and 45 dB covering 28 ~ 40 GHz.

EXPERIMENTAL RESULTS

In order to further verify the reasonableness of the design, we use SolidWorks software design the mechanical processing map about antipodal fin-line waveguide-to-microstrip transition structure as Fig. 5 shows. Physical process is shown in Fig. 6.

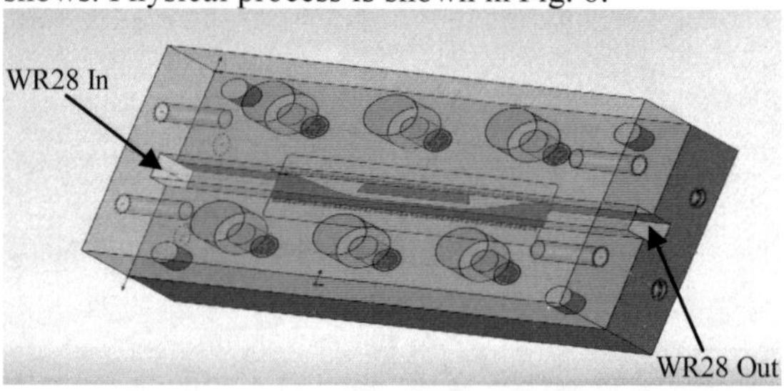

Fig. 5 3-D model of design diagram

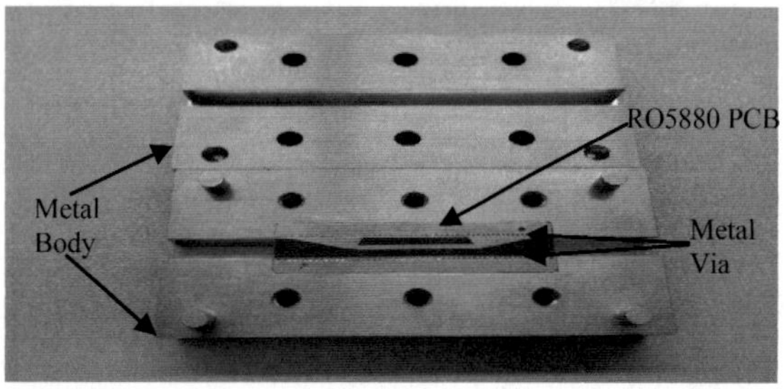

Fig. 6 Fabricated PCB and metal body of back-to-back antipodal fin-line waveguide-to-microstrip assembled transition

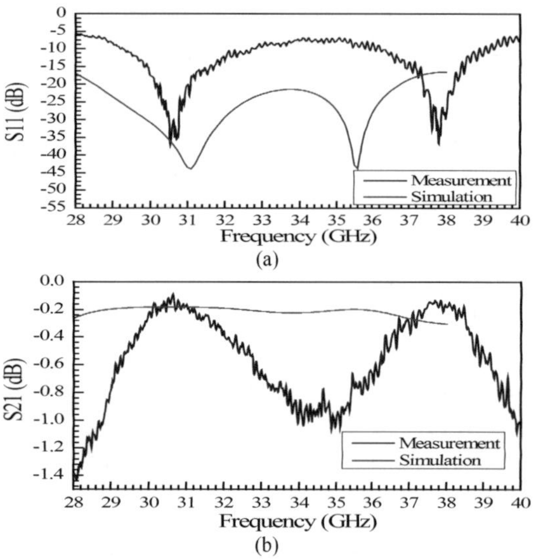

Fig. 7 Comparison of experimental results and simulation results

The insertion loss of single waveguide-to-fin-line-to-microstrip structure (S_{21}) is about half of the test result. Physical test results are shown in Fig. 7. The test results show that the maximum value of insertion loss is 1.0 dB while at the minimum of 0.09dB covering 28 GHz ~ 40 GHz. In the frequency range of 29.1 ~ 32.8 GHz and 36 ~ 39.4 GHz, the insertion loss fluctuates in the range of 0.09 dB to 0.6 dB. The return loss（S_{11}，S_{22}） fluctuates between 10 dB and 30 dB covering 28 ~ 40 GHz.

By comparing the experimental and simulation results, the experimental results and simulation results exists some deviation, but the trends about experimental results are much the same as simulation results. The fabricating error, assembling error, physical dimensions and conductor loss are the main reason for this difference. Therefore, this design can be considered as a success.

CONCLUSION

A novel Ka-Band waveguide-to-microstrip antipodal fin-line transition was designed in this paper. The transition was fabricated to verify the design. The results of simulation and experimental show the transition has a broadband, low insertion loss and low return loss performance. It is helpful for millimeter band devices and circuits. Simultaneously, due to these characteristics, the transition structure also has a certain potential applications at higher frequencies.

REFERENCES

Yao Hui-Wen, Abdelmonem Amr, Liang Ji-Fuh. A full wave analysis of microstrip-to-waveguide transition. IEEE MTT-S, 1994, 1: 213-216

J H C van Heuven. A New Integrated Waveguide-Microstrip Transition. IEEE Trans. MTT, 1976, 24 (3): 144—147

Dydyk M, Moore B D. Shielded microstrip aids V-band receiver designs. Microwaves, 1982, 21 (3): 77—82

Ponchak George E, Downey Alan N. A new model for broad-band waveguide-to-microstrip Transition Design. Microwave Journal, 1988, 5:333 — 343

Villegas Frank J, Stones D Ian. Hung H Alfred. A novel waveguide-to-microstrip transition for millimeter2wave module applications. IEEE MTT, 1999, 47(1): 48—55

Yoke2Choy Leong, Sander Weinreb. Full band waveguide-to-microstrip probe transitions. IEEE MTT-S, 1999, 4: 1435 — 1438

J. H. C. van Heuven, "A New Integrated Waveguide-Microstrip Transition," IEEE Trans. on Microwave Theory Tech., vol. 24, is. 3, pp.144-147, March 1976.

A. K. Sharma, "Tunable Waveguide-to-Microstrip Transition for Millimeter-Wave Applications," Proc. IEEE MTT-S int. Microwave Symp. Dig., vol. 1, pp. 353-356, 1987.

T. Djerafi, A. Ghiotto, and K. Wu, "Antipodal fin-line Waveguide to Substrate Integrated Waveguide Transition," Proc. IEEE MTT-S int. Microwave Symp. Dig., pp. 1-3, 2012.

J. B. Knorr and P.M. Shayda, "Millimeter-wave fin-line characteristics," IEEE Transactions on Microwave Theory and Techniques ,vol. 28, no. 7, pp. 737-43,July 1980.

Design of a Novel Nonlinear Curve Coupling Waveguide Coupler for Sheet Beam Travelling Wave Tube

Yang Liya*, Wang Jianxun, and Liu Guo

School of Physical Electronics, University of Electronic Science and Technology of China, Chengdu, Sichuan610054, China,
* Contact: yly1990@126.com

A new type of input/output coupler for sheet beam travelling wave tube (TWT) is designed in this paper. The coupler adopts a single coupling waveguide with Chebyshev nonlinear curve profile distributed between two paralleled rectangular waveguides in H-plane, to compact dimensions, ease fabrications, possess a high power capacity and be more convenient for periodic cusped magnet-quadruple magnet (PCM-QM) focusing system. Based on the analytical investigation of the electronic field contour plot, a certain sloping angle of the coupling waveguide is obtained to correct the phase shift and extend the operating frequency bandwidth and suppress the reflection to avoid generating oscillation in slow wave structure (SWS). Simulation results based on the numericalcalculation and genetic algorithm built-in Ansoft High Frequency Structure Simulation (HFSS), imply that the coupler achieves a broadband bandwidth (S11<-20 dB) of 23 GHz and 10 GHz in W-band, with the coupling coefficient above -0.2 dB.

Design of a Ka-band HE_{11} Mode Corrugated Horn for the Faraday Rotator

Fang Li*, Guo Liu, Li Wang and Yong Luo

School of Physical Electronics, University of Electronic Science and Technology of China, Chengdu, Sichuan, 610054, China,

* Contact: 201421040332@std.uestc.edu.cn

Abstract—**Faraday rotator is an important components used as isolator in transmission line of millimeter wave radar system. To verify the property of faraday rotator for the application, a microwave measurement is required, which needs a HE_{11} motivation mode. In this paper, a short Ka-band HE_{11} mode corrugated horn for faraday rotator was designed and simulated. This corrugated horn converts a TE_{11} mode into HE_{11} mode in Ka-band with a reflection under -28 dB and a HE_{11} mode conversion efficiency of above 99%. In addition, the mode conversion contents closes to desired HE_{11} mode over a bandwidth of 4 GHz. The length of this corrugated horn is 44mm and the input radius is 6mm, and has produced a well-formed lineal polarized HE_{11} mode. In addition to exhibiting a high converter and broad bandwidth, this corrugated horn is short and simple for microwave measurement.**

INTRODUCTION

A gyro-TWT (gyrotron traveling wave tube) [1] is strong attractive coherent radiation source in millimeter wavelength rage for its perfect characteristics of high peak power, high gain, and high efficiency, which are suited to many applications such as high resolution radar, communication system, military electronic countermeasure systems and space research. However, Gyro-TWT as a radiation source may be destroyed by the reflection caused by load mismatching or environment changing, which needs to be isolated effectively. Faraday rotator [2] as nonreciprocal component is widely used to make isolators and circulators in the transmission line to solve this problem with its property of making the linearly polarized plane wave in propagation has a non-reciprocal rotation. In contrast to waveguide isolator, faraday rotator has many advantages such low insertion loss, high isolation and high power handing capabilities. Since 1980s, many faraday rotators based on ferrite were proposed including transmission-type and reflection-type. Transmission-type obtains broader bandwidths while reflection-type can get lower insertion loss and higher power handing capability with reduced thickness of ferrite plate.

In order to characterize the rotation of faraday rotator before high power testing, a low power microwave measurement should be performed, where HE_{11} excitation mode is essential, which possesses perfect linear polarization with very low side-lobe and cross-polarization, and 98% of whose power concentrated on the main lobe. In microwave measurement system, the HE_{11} mode can be generated from TE_{10} via the mode conversion sequence: TE_{10}-TE_{11}-HE_{11}. A lot of TE_{11}-HE_{11} mode converter have been analyzed and developed for several decades. Some employ circumferentially corrugated waveguide [3], the HE_{11} mode purity was 99%, the cross-polarization and reflection were below -29dB and -50dB. Some employ smooth-walled horn which can be relatively easier to fabricate with approximates properties of main beam efficiency, cross-polarization response and beam symmetry. Others employ profiled corrugated horn with the sine squared, exponential profile or dual profile, which had shown to meet electrical requirements typical of radio astronomy application with more compact structure. And the dual profile corrugated horn will be studied in this paper for microwave measurement of faraday rotator.

In this paper, measurement and corrugated horn design are given in Sec.2, modeling and simulation results are given in Sec.3. A brief conclusion about this paper is given in Sec.4.

DESIGN OF MEASUREMENT SYSTEM AND CORRUGATED HORN FOR FARADAY ROTATOR

A. Microwave Measurement System Designing

Fig.1 showed microwave measurement system of faraday rotator. The output signal from vector network analyzer (VNA) through a sequence of mode converter becomes a HE_{11} mode and radiated. The polarization filter on each side of the faraday rotator are set at 45° to the axis, and it passes a wave with its polarization perpendicular to the wires and reflect the parallel polarization. The vertical axis of polarization is first rotated 45° clockwise as it passes through the faraday rotator. After reflection, it undergoes an additional 45° rotation by the nonreciprocal property of the faraday rotator and is then deflected by the wire-grid polarization filter. Thus transmission path is port 1 to port 2 or port 2 to port 1 with the reflection been isolated [4, 5].

As in the microwave measurement system, two ports of VNA is not a standard waveguide, a coaxial-waveguide is required to convert the coaxial to standard waveguide, which works on TE_{10} mode. And the HE_{11} mode can be generated from TE_{10} via the mode conversion sequence: TE_{10}–TE_{11}–HE_{11}. Using the TE_{11} mode as polarized intermediate mode as it has the advantage that all converters can be made without bends. TE_{10}-TE_{11} mode converter is realized by a square-to-circle waveguide which is coaxial. The length of TE_{10}-TE_{11} mode converter is several waveguide wavelength and can get a wide bandwidth. The input size of TE_{10}-TE_{11} mode converter is a standard rectangle waveguide (W22) at Ka-band. The output radius of TE_{10}-TE_{11} mode converter should agree with the input radius of TE_{11}-HE_{11} corrugated horn, so that the different components can be connected without space and eliminate the reflection.

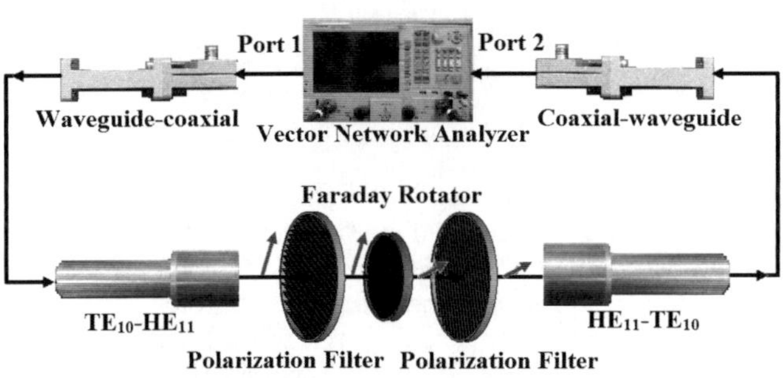

Fig. 1 Design of the microwave measurement system for faraday rotator

B. Corrugated Horn for Faraday Rotator

A corrugated horn with excellent performance in our application possesses low reflection, desired mode contents, and high conversion efficiency in a wide frequency. In addition, it should be compact, easily fabricated and convenient for microwave measurement. In our design, the TE_{11}-HE_{11} corrugated horn is achieved by a double-profiled circular waveguide, which was circumferentially corrugated. Conversely, the HE_{11} mode can be converted to TE_{11} mode by use this spline-profile corrugated horn in the opposite direction, which can be used as receiving components. In some application, a pure sin squared profile, compact horns with overall dimensions about two-thirds those of the original linear horn. Double-profiled corrugated horn can make the size more compact, and the amount of excitation of hybrid modes depends on the position of the point connecting. In this article, two different profiles (shown in eq.1) was connected. At the connecting point, the derivative of these two profiles are 0. In this way, the two profile can realize transition smoothly without break, and it can eliminate the wave reflection because of the mismatch. This connecting point L_1 in this article is $L/2$, and the total length of corrugated horn is 44mm. The slot depth is gradually changed from a half of wavelength to a quarter of wavelength to make the impedance matched. The profile $R(z)$ (shown in Fig. 2 (a)) of the circular waveguide are as follows:

$$R(z)=\begin{cases} R_{in}\sqrt{1+\left(\dfrac{z}{1.3\cdot k_0\cdot R_{in}^2}\right)^2} & 0\le Z\le L_1 \\ 2R_a-R_{in}\sqrt{1+\left(\dfrac{L-z}{1.3\cdot k_0\cdot R_{in}^2}\right)^2} & L_1\le Z\le L \end{cases} \quad (1)$$

Where k is the free space wave number, L is the total length of the horn, R_{in} is the input radius, R_a is the first profile's output radius.

Besides, the depth d, period p and duty ratio (r) of corrugated are very important parameters for the corrugated horn. In this paper, the depth varied from a half of wavelength to a quarter of wavelength, which satisfied the law shown in eq. 2. The period is chosen as $\lambda_0/3$ and the duty cycle equals to half of period. The output radius of corrugated horn is 8mm and operates at the frequency of 34 GHz. The structure along the axis shows in Fig. 2 (b).

$$d(z)=\frac{\lambda}{2}-\left(\frac{\lambda}{4}\right)^N \quad (2)$$

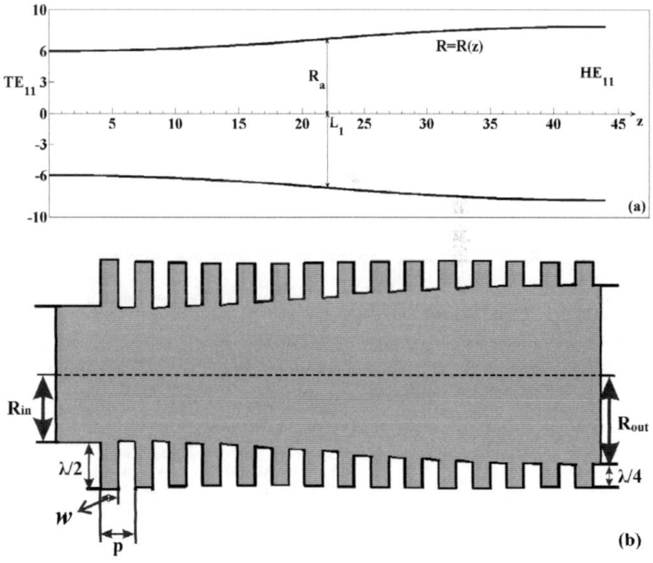

Fig. 2 Geometry and profile of the corrugated horn. (a) Inner diameter R(z) of the waveguide (b)The cross-section of corrugated horn along the axis.

SIMULATION

The structure parameters were set, optimized and verified by the CST MWS (CST Microwave Studio). It was found in the simulation that this double-profile corrugated horn was able to meet requirements in mode conversion efficiency, desired mode contents, and a lineal polarization, which was important for microwave measurement. Simulated mode contents, desired mode contents and conversion efficiency of corrugated horn is shown in Fig .4. The normalized powers of TE_{11} and TM_{11} mode are close to the desired consist ratio 85%

TE_{11} and 15% TM_{11} [7], and the total conversion efficiency are above 99% over a bandwidth from 32-36 GHz, while reflection under -28dB over a bandwidth from 30-38 GHz (shown in Fig. 3). Fig. 5 shows the output field distribution of the corrugated horn compared to standardized HE_{11} mode. The simulation were carried out for 32, 34 and 36 GHz. The normalized power shows an axisymmetric pattern and the center of the output containing about 98% of the radiated power, which were close to distribution of standardized HE_{11} mode. It can be demonstrated that a well-formed HE_{11} mode over a bandwidth from 32-36 GHz with low reflection, high conversion efficiency, desired mode contents and field distribution pattern.

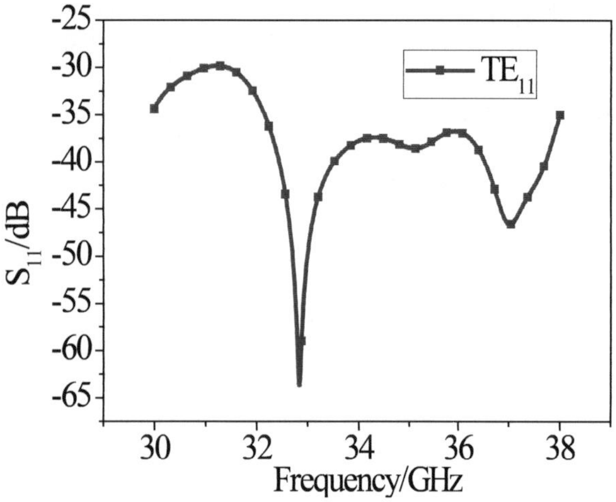

Fig. 3 Reflection coefficient of corrugated horn.

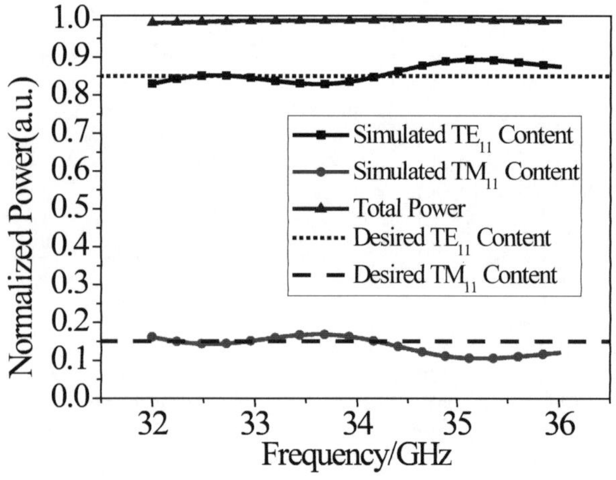

Fig. 4 Simulated vs. desired mode contents and conversion efficiency of corrugated horn.

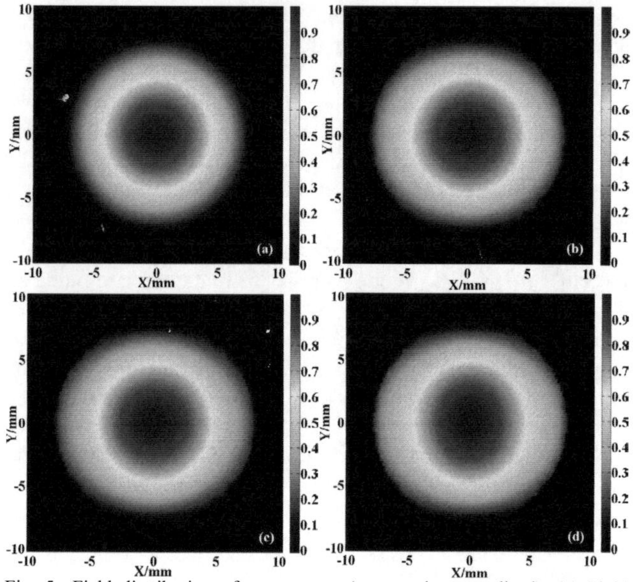

Fig. 5 Field distribution of output port (contour in normalized). (a) Field distribution of standard HE_{11} mode; (b)field distribution of corrugated horn output at 32 GHz; (c)field distribution of corrugated horn output at 34 GHz; (d)field distribution of corrugated horn output at 36 GHz

CONCLUSIONS

A HE_{11} mode corrugated horn for the performance microwave measurement of a Ka-band faraday rotator had been designed and simulated. Through optimization and simulation, it was demonstrated that the HE_{11} mode corrugated horn can meet requirements of a well-formed HE_{11} mode with high mode conversion efficiency, low side-lobe and cross-polarization. It is expected to fabricated this HE_{11} mode corrugated horn and testing its transmission and reflection.

REFERENCES

Denisov G. G., Bratman V. L., Cross A. W., et al.: 'Gyrotron traveling wave amplifier with a helical interaction waveguide', *Physical review letters*, 1998, 81, (25), pp. 5680-5683

William D. Fitzgerald: 'A 35-GHz Beam Waveguide System for the Millimeter-Wave Radar', The Lincoln Laboratory Journal, 1992, 5, pp. 245-272

Thumm M.: 'High power mode conversion for linearly polarized HE_{11} hybrid mode output', *International Journal of Electronics*, 1986, 61, (6), pp. 1135-1153

Dionne G. F., Weiss J., Allen G.: 'Nonreciprocal magneto-optics for millimeter waves', *IEEE Transactions on Magnetics*, 1988, 24, (6), pp. 2817-2819

Hunter R., Robertson D., Goy P., et al.: 'Design of High-Performance Millimeter Wave and Sub-Millimeter Wave Quasi-Optical Isolators and Circulators', *IEEE Transactions on Microwave Theory & Techniques*, 2007, 55, (5), pp. 890 - 898

Liu H, Lou Z, Wang H, et al.: 'Precise Design of Millimeter Wave Quasi-Optical Faraday Rotators', *Journal of Infrared Millimeter & Terahertz Waves*, 2009, 30, (4), pp. 401-409

Sakaguchi M., Idei H., Kubo S., et al.: 'Quasi-Optical High Purity HE_{11}-Mode Exciter for Oversized Corrugated Waveguide Transmission', *Plasma & Fusion Research*, 2012, 7, (2405037), pp. 1-4

T.-L. Zhang, Z.-H. Yan, F.-F. Fan, et al.: 'Design of a Ku-Band Compact Corrugated Horn with High Gaussian Beam Efficiency', *Journal of Electromagnetic Waves & Applications*, 2011, 25, (1), pp. 123-129

Mun Seok Choe, Kwang Hoon Kim, EunMi Choi: 'A comprehensive analysis of a TE_{11} to HE_{11} mode converter for an oversized F-band corrugated waveguide', *Journal of Electromagnetic Waves & Applications*, 2013, 27, (17), pp. 2221-2238

This page intentionally left blank.

High Current Density Impregnated Scandate Cathode for Terahertz Vacuum Devices

Yafen Shang*, Qiang Zheng, Hao Fu

School of Physical Electronics, University of Electronic Science and Technology of China, Chengdu, Sichuan610054, China,
* Contact: a1264186011@163.com

To meet the demand of terahertz devices, we develop a new type scandium aluminate impregnant. Scandium aluminate was synthesized *via* solid phase mixing and sintered in air atmosphere. Powder X-ray diffraction (XRD) measurement and Rietveld refinement revealed that scandium aluminate impregnants is $Ba_5CaAl_4O_{12}$Phase with Ba_2ScAlO_5 secondary phase.The Current-Voltage characteristics of this impregnated scandate cathode was measured in the DC mode and the maximum current density is 17.6 A/cm^2 occurring at 1100ºC for the applied voltage from 0 to 540 V. It illustrates that the scandate aluminate impregnant is promisingfor applications in THZ electronic devices.

Research on Gyrotron Traveling Wave Amplifier with Lossy Dielectric-Load Waveguide

NaLiu*
School of Physical Electronics, University of Electronic Science and Technology of China, Chengdu, Sichuan610054, China,
* Contact: 635733312@qq.com

Abstract—Gyrotron traveling wave tubes (gyro-TWTs) are ofconsiderable interest for high-power millimeter and submillimeter radiation sources. Lossy dielectric waveguide to improve the stability of the cyclotron traveling wave tube amplifiers and other properties have a positive effect. The combined appropriate selection of the lossy waveguide thickness, permittivity, voltage, the applied magnetic field and the velocity ratio can effectively give attention to bandwidth and instability to ensure the stable operation of the gyrotron traveling wave amplifier. It is revealed that due to the lossy property of the dielectric, the energy in the dielectric slots is absorbed effectively and the high order Bloch harmonics induced by the periodicityof the structure are suppressed, which changes the discrete spectrum under lossless condition into a continuous one .

INTRODUCTION

As a high power broadband millimeter wave sourceGyrotron amplifier is the inevitable choice for the next generation high power millimeter wave imaging radar transmitter. Gyro-TWT with high average power capacity, high efficiency, high gain and wide bandwidth and other characteristics, can be applied in high resolution radar and high capacity intensive communication systems, but an actual gyro-TWT interaction system is highly susceptible to potential absolute instabilities, which bring up oscillations and spread out in the entire interaction system. The National Tsing Hua University reported that an ultra-high gain gyro-TWT amplifier employing distributed wall losses produced 93 kW peak power, with 70 dB saturated gain, 26.5% efficiency and a -3 dB bandwidth of 8.6% (Chu, Chen, Hung, Chang, and Barnett1998; Chu et al. 1999).Recently, a new type of distributed loss scheme has been employed by NRL for high average power applications. A Ka-band TE11 mode gyro-TWT loaded with high thermal conductivity ceramic elements produced 78 kW power, 60 dB saturated gain and 19% efficiency with a 3 dB bandwidth of 17.1%.This paper aims to reach an interaction structure with distributed loss for gyro-TWT, which demonstrates that loading lossy dielectric is excellent to stabilize the spurious oscillations. Furthermore, the designs of magnetron injection gun, interaction circuits, and input and output structures are also achieved to satisfy the requirements.

STRUCTURE AND ANALYSIS OF DISPERSION

In our gyro-TWT research, we employed a long loaded section of constant radius which consists of lossy ceramic rings spaced with metal rings to provide controlled loading of the fundamental TE01 mode. The scheme of periodical dielectric loaded waveguide and normalized radical E-field are described in Fig.1 and Fig.2.

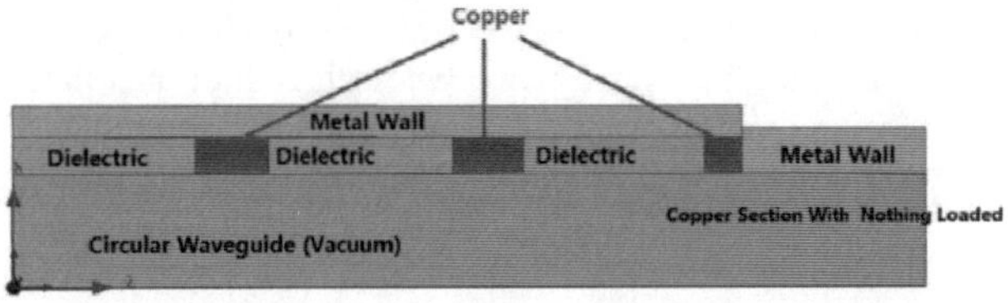

Fig. 1. Schematic view of the simulation model for gyro- TWTinteraction structure.

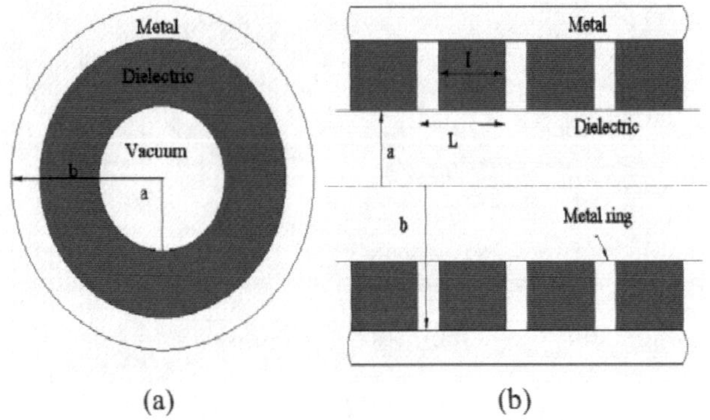

Fig. 2. (a) Transverse structure of the dielectric-loaded metal cylindrical waveguide. (b) Periodical dielectric loaded waveguide for gyro-TWT.

The interaction circuits are structured with a lossy section (AlN-SiC) followed by a copper section. The nonlinear highest power portion of the amplification occurs in the short conducting wall section at the end of the interaction region.The dispersion diagram of TE_{01}gyro-TWT interaction circuits with unloaded structure is shown in Fig. 3.

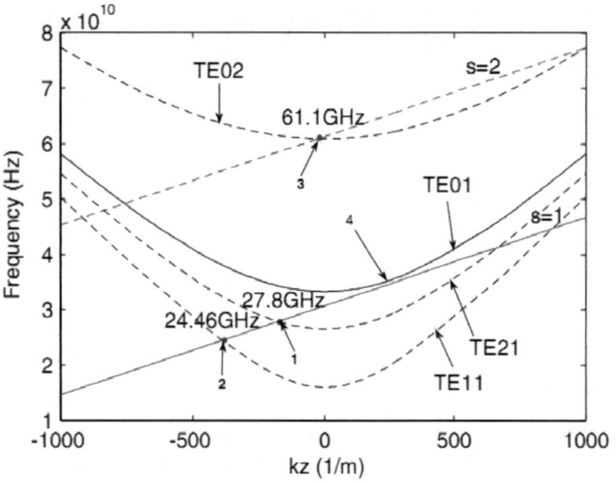

Fig. 3. Dispersion diagram of TE_{01} gyro-TWT

Small-signal analysis is the foundation of gyro-TWT beam-wave interaction analysis andgives a clear physical interpretation to the amplification and self-induced oscillation ofgyro-TWT. Taking advantage of the small-signal analysis, start current and start length oprimary modes in smooth and dielectric-loaded waveguide arecalculated. According to actual gyro-TWT parameters such as operating band and outputpower et al. and to the status of our high power Gyrotron hot test Lab, essential parametersof distributed-loss loading circuit of Ka-band gyro-TWT including beam voltage, current,velocity ratio, waveguide and guiding center radius are determined.

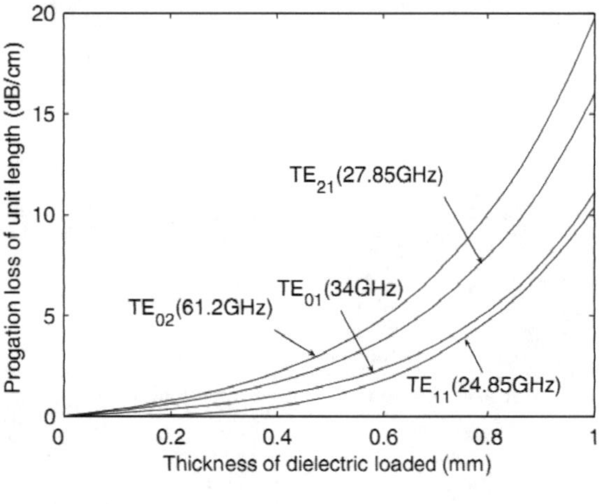

Fig. 4 Propagation loss of unit length vs. thickness of loss layer d=0.07 cm (relative permittivity ξ″=11−6j, waveguide radius rw=0.56cm).

Whenbeam current I_b=10 A, even the most susceptible oscillation mode TE_{02} whose start lengthin lossless circuit is 4 cm, longer than designed lossless circuit length 3.5 cm. Hence, thedielectric loading scheme is capable of suppressing oscillation of both operation andparasitic modes (Fig. 4).

CONCLUSION

The theoretical predictions ofa Ka-band TE01 gyro-TWT has been presented. Distributedloaded lossy dielectric rings are introduced to suppress theunwanted modes TE_{11},TE_{21}, and TE_{02}. The performanceresulted by loss wall plays an important role for the gyro-TWT oscillation stability. Dielectric loading results in reducing of operating mode gain of unit length. Adjustment of interaction section structure is necessary.

ACKNOWLEDGMENT

The authors would like to thank Dr. D. K. i, Dr. Y. R,and Prof. L. Y. for many useful discussions aboutan interaction structure with distributed loss fora Ka-band gyro-TWT and calculation of gyro-TWT.

REFERENCES

H. H. Song et al., "Theory and experiment of a 94 GHz gyrotrontraveling-wave amplifier," Phys. Plasmas, vol. 11, no. 5, p. 2935,May 2004.

C. H. Du and P. K. Liu, "Linear full-wave-interaction analysis of agyrotron-traveling-wave-tube amplifier based on a lossy dielectric-lined circuit," IEEETrans.Plasma Sci., vol. 38, no. 6, pp. 1219–1226, Jun. 2010.

R. Yan, Y. Luo, G. Liu, and Y. Pu, "Design and experiment of a Q-bandgyro-TWT loaded with lossy dielectric," IEEE Trans. Electron Devices,vol. 59, no. 12, pp. 3612–3617, Dec. 2012.

K. T. Nguyen, J. P. Calame, D. E. Pershing, B. G. Danly, M. Garven, B. Levush, and T. M. Antonsen,"Design of Ka-band gyro-TWT for radar applications," IEEE Trans. Plasma. Sci., vol. 48,pp. 108–11,Jan.2001

D. E. Pershing et al.,"ATE11 Ka-band gyro-TWT amplifier withhigh-average power compatible distributed loss," IEEE Trans. PlasmaSci., vol. 32, no. 3, pp. 947–956, Jun. 2004.

B. T. Liu, J. J. Feng, E. F. Wang, Z. L. Li, X. Zeng, L. J. Qian, andH. Wang, "Design and experimental study of a Ka-band gyro-TWT withperiodic dielectric loaded circuits," IEEE Trans. Plasma Sci., vol. 39,no. 8, pp. 1665–1672, Aug. 2011.

J. P. Calame, M. Garven, B. G. Danly, B. Levush, and K. T. Nguyen,"Gyrotron-traveling wave-tube circuits based on lossy ceramics," IEEETrans. Electron Devices., vol. 49, no. 8, pp. 1469–1477, Aug. 2002.

H. Wang, H. Li, Y. Luo, and R. Yan, "Theoretical and experimentalinvestigation of a Ka-band gyro-TWT with lossy interaction structure,"J. Infr. Millim. Terahertz Waves, vol. 32, no. 2, pp. 172–185, Feb. 2011.

D. E. Pershing, J. P. Calame, K. T. Nguyen, B. G. Danly, and B. Levush, "A distributed loss technique for high average power gyro-TWT's," Proceedings of the 2nd International Vacuum Electronics Conference, Noordwijk, The Netherlands, p.145,2001.

Y. S. Yeh, C. L. Hung, C. W. Su, T. S. Wu, Y. Y. Shin, and Y. T. Lo, "W-band second-harmonic gyrotron traveling wave famplwith distributed-loss and severed structures," Int. J. Infrared Millim. Waves, vol. 25, no. 1,pp. 29–42, Jan. 2004.

Chu, K.R., Lau, Y.Y., and Barnett, L.R. (1981), 'Theory of a Wideband Distributed GyrotronTraveling Wave Amplifier', IEEE Transactions on Electron Devices, 28(1), 866–871.

C. H. Du and P. K. Liu, "A lossy dielectric-ring loaded waveguidewith suppressed periodicity for gyro-TWTs applications," IEEE Trans.Electron Devices, vol. 56, no. 10, pp. 2335–2342, Oct. 2009.

Garven, M., Calame, J.P., Danly, B.G., Nguyen, K.T., Levush, B., Wood, F.N., and Pershing, D.E.(2002), 'Experimental Studies of a Gyro-TWT Amplifier with a Lossy Ceramic InteractionRegion', IEEE Transactions on Plasma Science, 30, 885–893.

Measurements of Dielectric Properties near 100GHz Using a Reflection-Type Hemispherical Open Resonator

Hao Li*, Q.S.Zhang, Ying liu, Liusha Yang
University of Electronic and Science Technology of China, Chengdu, Sichuan610054, China
*Contact: lihao_uestc@126.com, phone +15908175447

Abstract—**As low loss dielectric materials play an important role in application for microwave devices. To obtain their dielectric properties, Based on both theoretical analysis and numerical simulation by a 3-D finite element electromagnetic code, HFSS, two reflection-type hemispherical open resonators are designed to excite TEM0030 mode at 94.7GHz and TEM0055 mode at 100.5GHz.In contrast with two ports measurements, the system with only one coupling hole was directly connected to a W-band vector network analyser (VNA) provide a simple method. Calculating through the VNA measured port reflection coefficient (S11) resonant curve can get dielectric properties. The automated measurement system has addressed several key technologies of how to determine precision value of cavity length and how to choose correct solution from a lot of solutions. The certified measurement system after a series of checking is used to measure sapphire. Lots of measurement results show that the standard deviation of measurement error is less than 0.154% in permittivity and 20.42% in loss tangent. Meanwhile, some experimental summaries on the open resonator technique are provided. Software that controls the measurement system is developed and it improves the testing efficiency greatly.**

INTRODUCTION

Low loss dielectric materials are the key of millimeter-wave vacuum electronic devices (VEDs) components like input-output window, helix support rods, cavity loss-buttons, Therefor, precisely measurements of dielectric materials become important especially in high frequency-band. Lots of measurement techniques have been developed such as the closed cavity method [1], the open resonator technique [2], the free space method [3] and so on. Owing to great advantages of high Q value, good single mode performance and high measurement accuracy, the open resonator technique has been proved to be the most powerful tool in measuring loss tangent and dielectric permittivity of low loss dielectric material in millimeter wave, submillimeter wave and THz wave regime.

Generally, the open resonators can be divided into two types. One of them is the confocal type which often is built with two symmetric concave mirrors and two ports. The other one is the hemispherical type, which is composed of a concave mirror, a plane mirror and one port. Due to the traits of two ports, dielectric properties are measured through simulation of the transmission coefficient (S21).Unfortunately, the coefficient always would be too small to precisely obtain because of its high sensitive to input signal even with the change in noise level would make the coefficient great changes. In addition, vector network analyzer (VNA), scalar network analyzer, spectrum analyzer, and power meter are used to construct the measurement system. The complexity of confocal open resonator system brings lots of system errors .Thus, Compared to the complex system [4]-[6], we choose the simple one- hemispherical open resonator system which just consists of a reflection-type hemispherical open resonator ,a VNA and the sample. To a large extent decreasing of the numbers of measurement instruments will reduce influence of system errors.

In this paper, the design method of reflection-type open resonator is presented. And the measurement system is constructed well. The measurement system is introduced in the next sections briefly. In section II a design methodology of the measurement system is described. In section III, construction of the system, measurement procedure and results are discussed .In section IV, the work is concluded for this paper.

DESIGN OF REFLECTION-TYPE HEMISPHERE RESONATOR

A reflection-type hemispherical open resonator, as shown in Fig.1,which consists of a hemispherical concave mirror, a plane mirror and a coupling aperture. Through connecting coupling aperture with a standard rectangular waveguide, the electromagnetic energy of TE_{01} mode in the input waveguide is coupled to the coupling hole to excite operation modes (Gauss beam modes $TEM_{p,l,q}$).Since the fundamental modes ($TEM_{0,0,q}$) have smaller Gauss beam radius, smaller Gauss beam divergence and more energy distribution than high-order modes, $TEM_{0,0,q}$ is used for our measurement system. The coupling aperture is placed exactly the center of the concave mirror so that $TEM_{p,l,q}$ are easily obtained.

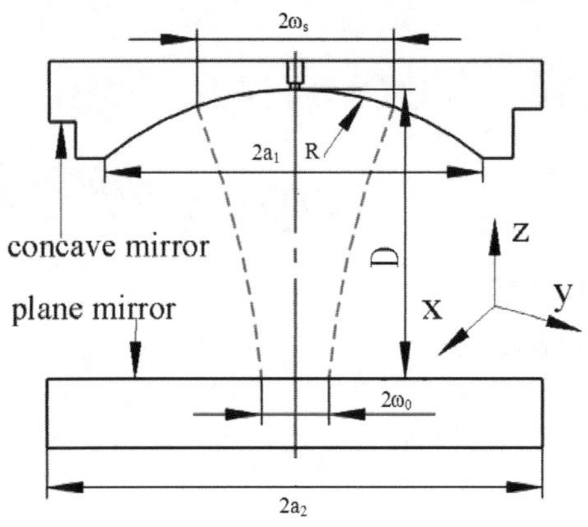

Fig 1 Schematic of open refleciton-type resonantor

A. Design theory

The scalar theories of openresonator are well established in many literatures such as [7-8].As the fundamental modes, the $TEM_{0,0,q}$ resonant modes are excited in the open resonator and the resonator frequencies of the empty resonator are obtained from

$$f_0 = c/(2D)\left[q+1+\frac{1}{\pi}\arctan\sqrt{D/(R-D)}-\frac{1}{\pi}\arctan(\frac{1}{kR})\right]^{-1} \quad (1)$$

Where R is the curvature radius of the spherical mirror, D is the cavity length, q represents the axial modal orders corresponding with $TEM_{0,0,q}$ mode , c is the velocity of the light and k is the free space wavenumber. The Gauss beam radius is expressed by

$$w(z) = w_0\sqrt{(1+(2z/kw^2{}_0)^2)} \quad (2)$$

$$w_0 = \sqrt{\lambda\sqrt{D(R-D)}/\pi} \quad (3)$$

Where λ is the wavelength. w_0 is the Gauss beam radius at the plane mirror It should be noted that , in order to enhance accuracy of the measurement results , w_0 is much less than the sample radius. The unloaded quality is given by

$$Q_0 = 1/[1/(\frac{D}{2\delta_c}\frac{1}{(1-1/(k\sqrt{D(2R-D)}))})+ \quad (4)$$
$$1/(\frac{2\pi D}{\lambda\alpha_d}\frac{1}{(1-(2p+l+1)/(k\sqrt{D(2R-D)}))})]$$

where δ_c means the penetration depth of the silver surface, k represents wave number of free space. Diffraction loss is expressed by $\alpha_d = \exp(-2D_s/w_d{}^2)$ and D_s is the diameter of the spherical mirror, w_d is the Gauss beam radius at the concave mirror.

When a sample is placed on the plane mirror, the resonator frequency will shift to lower frequency $f_s$due to the dielectric permittivity and the bandwidths will become boarder due to the dielectric loss tangent. The permittivity is obtained through the following equations [9-12]

$$\frac{1}{n}\tan(nkt-\phi_t) = -\tan(kd-\phi_d) \quad (5)$$

Where $\phi_t = \arctan\left(\frac{t}{nz_0}\right) - \arctan\left(\frac{t}{nkR_1(t)}\right)$,

$$\phi_d = \arctan\left(\frac{d'}{z_0}\right) - \arctan\left(\frac{1}{kR}\right) - \arctan\left(\frac{t}{nz_0}\right) - \arctan\left(\frac{t}{kR_2(t)}\right)$$

$R_1(t) = t + \frac{n^2 z_0{}^2}{t}$, $R_2(t) = \frac{R_1(t)}{n}$, $k = \frac{2\pi f_s}{c}$, $z_0 = \sqrt{d'(R-d')}$,

$d' = d + \frac{t}{n}$, $d = D - t$, t is the thickness of the sample, and $n = \sqrt{\varepsilon'}$ is the refraction index of the sample. In addition, the loss tangent can be expressed by

$$\tan\delta = \frac{1}{Q_s}\frac{2nk(t\Delta+d)}{2nkt\Delta - \Delta[\sin 2(nkt-\phi_d)]} \quad (6)$$

In which

$$\frac{1}{Q_s} = \frac{1}{Q_{LS}} - \frac{1}{Q_{00}}\frac{D(\Delta+1)}{2(d+t\Delta)} \text{ and}$$

$$\Delta = \frac{n^2}{n^2\cos^2(nkt-\phi_t)+\sin^2(nkt-\phi_t)}$$

Q_{LS} denotes the loaded quality factor of the resonator containing the sample, and Q_{00} means the Qfactor of the empty resonator.

B. Goal of the design

Based on the theories presented above, a reflection-type hemisphere open resonator is designed. In order to maintain the repeatability, accuracy, reliability and stability of our measurement system, a few factors are took into consideration. First, Gauss beam radius should be as small as possible to improve the resolution; Secondly, loss must be low and Q_{00}should be as large as possible; thirdly,electromagnetic field in the resonator should be as pure as possible and the whole system could be easy to installation and debugging. In our system, a design goal is made: 85GHz<f_s<105GHz,Q_0 >9×10^4,ω_0<4.5mm.

C. Design procedure

A Matlab optimized code has been written to search the basic parameters according to equation (1)-(4).Based on the code, several groups values of R, D, q are obtained ,which satisfy the design goal above. Some other factors should be obeyed during design process: First, the mode-spacing (MI) between operation mode and parasitic mode should be as large as possible. Second, Spherical surface should fit the Gauss beam phase front should be as well as possible. To decrease the loss of fundamental mode, the diameter of mirror should be as large as possible. However, decreasing the mirror can suppress the loss of parasitic mode. Thus, tradeoff should be made to decide the diameter of the sphericalmirror ($2a_1$) and the one of the plane mirror ($2a_2$). After considering these factors, table1 gives optimized parameters of measurement system. Fig.2 gives the theoretical electrical field distribution of the working mode and parasitic mode. As we can see from the figure, the fundamental mode has a smaller waist radius and more centralized field distribution. From eq(4),three kinds of quality coefficient are showed in table2.Q_d, Q_r, $Q_o$respectively, correspond to the diffraction loss L_d , mirror surface resistance loss L_r ,and the total loss of the resonator

L_t.Fig.3 is showed that the designed spherical curve has a good agreement with the operation mode.

Table 1 Structure parameters and electrical parameters for hemispherical resonator

R(mm)	*D*(mm)	a_1(mm)	a_2 (mm)	*q*	$f_{0\_Th.}$ (GHz)	$Q_{0\_Th.}$ (10^5)	w_0(mm)	*MI* *(MHz)*
55	49.8	70	90	30	94.5	11.9	4.03	610
90	84.4	70	90	55	101.1	20.3	4.20	559

Table 2 Three kin*ds* of quality coefficient

Mode(q)	50	51	52	53	54	55
f[GHz]	91.307	93.012	94.858	96.634	98.409	10.019
$Q_d\times10^4$	2.23	2.51	2.81	3.14	3.52	3.95
$Q_r\times10^5$	9.86	8.89	8.04	7.29	6.62	6.03
$Q_0\times10^5$	2.00	2.02	2.04	2.05	2.07	2.09

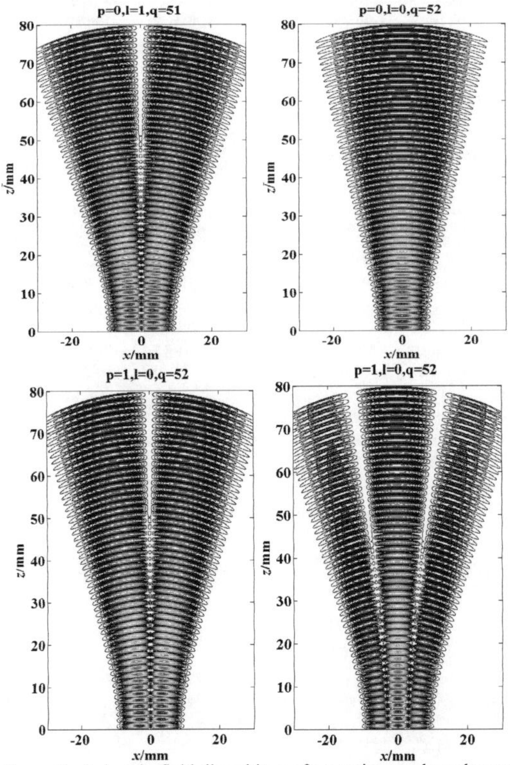

Fig 2 Theoretical electric field distrubiton of operation mode and parasitic modes

Few of references have mentioned about the influence of the coupling hole in open resonator system. Just in [13,14], the whole open resonator system with the hole can be analyzed using the equivalent circuit. In this paper, with the usage of 3-D analysis tool ANSYS HFSS, the coupling aperture diameter(*Φc*) and depth (*dc*) are analyzed. The simulated electronic field distribution is shown in fig.4. The number of axial field peaks is 31 for TEM_{0030} and 56 for TEM_{0055}, which agree well with theoretical analysis.Compare theoretical analysis about simulation analysis, there is a relativity large difference in upload quality Q_0, Mainly because of the neglect of coupling loss in the theoretical method .Fig.5denotes the reflectioncoefficientS_{11}. According to the simulated performance parameters and considering the actual fabrication level and cost, trade-off is made to decide the final dimensions of the coupling hole: h_c=0.2mm; Φ_c=0.9mm.

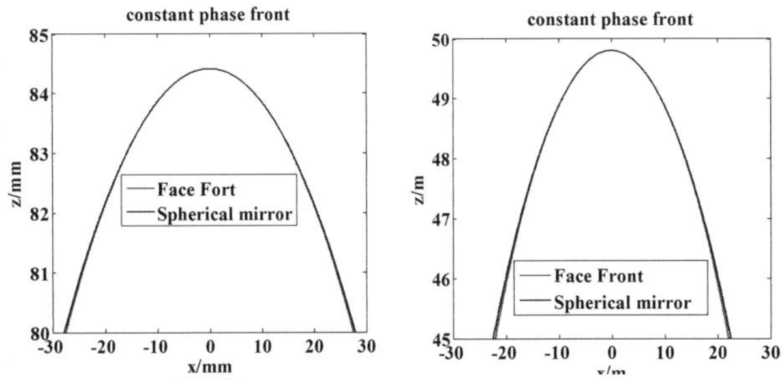

(a) (b)

Fig 3 Face front of the operation mode(a)$TEM_{0,0,55}$,spherical mirror radius(*R*)84.414mm (b)$TEM_{0,0,30}$,spherical mirror radius*R*=55mm.

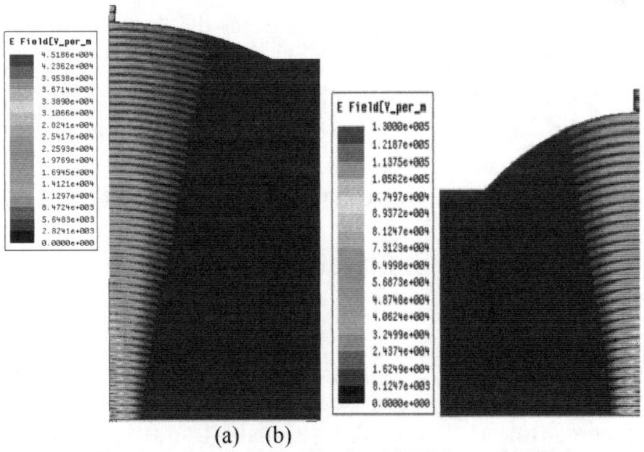

(a) (b)

Fig 4 Simulated electric field distribution (a)$TEM_{0,0,55}$,(b)$TEM_{0,0,30}$.

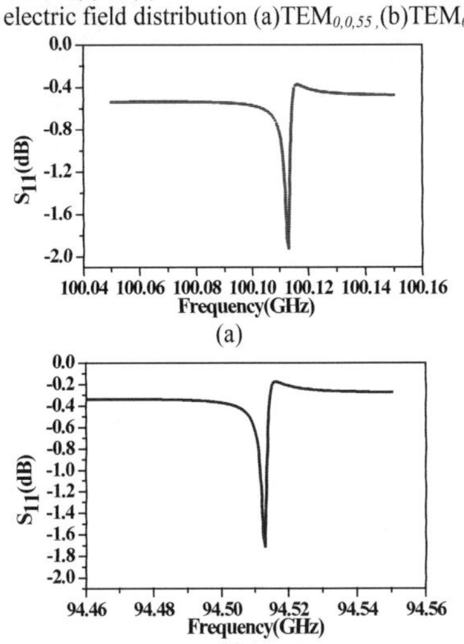

(a)

(b)

Fig 5 Resonant curve of HFSS simulation (a)$TEM_{0,0,55}$,(b)$TEM_{0,0,30}$.

D. System construction and measurement

In several references, lots of materials such as copper, bronze, and aluminum, are used as body mirrors in measurement system. In this work, body mirrors are both made of K9 optical glass coated silver film. Compared to metal mirrors, it is more convenient to decrease surface roughness to avoid the Gaussian Beam scattering out of the resonator. In addition, curvature radius of the spherical mirror can be better processed.

Several steps have been made to obtain the final mirror.

Step 1 ： The rectangle waveguide hole and coupling aperture are punched using ultrasonic wave punching machine at the center of mirrors. A kind of locating device is used to keep the center of rectangle hole, coupling aperture and spherical mirror be coaxial.

Step 2： Mirrors are polished with optical method and chemical silver .The thickness of the silver film is about five times of the skin depth at resonant frequency, which can prevent the microwave escape form the mirror surface.

Step3: Indium is used to connect the glass waveguide aperture and the external metal rectangle waveguide located on the upper surface of the spherical mirror.

Agilent N5247A vector network serves as I/O signal separation device in this work. The measurement system is connected to VNA with a spectrum extension module through E-face bend waveguide.

To improve the measurement accuracy, the sample should be well prepared by keep the following tips. First of all, the radius of the sample should be large enough to prevent the influence of diffraction of at specimen edge. Secondly, the sample should be flat enough and without spur at its edge. At last, thickness over entire surface of the sample should be uniform.

The measurement procedure of the system is presented as follows:

Step 1: TRL method is used for calibration of VNA with the help of WR-10 waveguide kit.

Step 2:We identify operation mode ($TEM_{0,0,q}$) of the measurement system. Search for resonant frequency using a piece of paper with low-loss material on it .When the paper is put on the center of the plane mirror, the peak of operation mode would disappear or be extremely weakened. Then substitute resonant frequency (f_{00}) of operation mode into eq.(1) ,we can determine the accurate cavity length (D) . After obtaining the whole mechanical parameters, empty resonant quality (Q_0) also can be calculated with the help of eq.(4)

Step 3: Put the sample on the center of the plane mirror. The resonant frequency (f_{os}) would shift and use the same method in step2 to find operation mode. Then record fos and calculate the unload quality (Q_{OS}) with sample.

Step 4: obtain permittivity and loss tangent of the sample through eq.(4) and eq.(6).

E. Measurement results

The electrical parameters and measured results of permittivity and loss-tangent are shown in Table 3. Quality factor value and resonant frequency of the resonator without and with sample were all recorded or calculated through the resonant curves. The permittivity of the sample can be solved numerically using above equation.Compared to record measurement results [15], the standard deviation of measurement error is less than 0.154% in permittivity and 20.42% in loss tangent.

Table 3 Comparison of measured results with other record results

Parameter	This work	IAP
ε_r/f(GHz)	9.40/92.1206	9.4/140
$\tan\delta\times10^4$/f(GHz)	1.8/92.1206	1.7/90

F. Conclusion

Two reflection-type open resonant measurement system have been designed in this paper. One of them is constructed well and the other one is in our plan. Although measurement results have a good agreements with the record results, a smaller Gauss beam radius of the open resonator is also needed to measurement the distribution of the dielectric Properties of the low loss materials.

REFERENCES

B. B.Yang , S. L. Katz, K. J. Willis,etal,A high-Q Terahertz resonator for the measurement of electronic properties of conductors and low-loss dielectrics. IEEE Trans. Terahertz Science and Technology. 2 (4), 449-459, 2012

G.X.Shu, Y.Luo,Q.S.Zhang, Millimeter Wave Measurement of the Low-Loss Dielectricin Vacuum Electronic Devices with Reflection-TypeHemispherical Open Resonator. J Infrared MilliTerahz Waves , 2015 36:556 – 568

J. J. Choi, Dielectric measurements of CVD diamonds at millimeter wavelength using a Fabry-Perot open resonator. Int.JInfrated and Millimeter waves. 26(10). 1427-1436, 2005

Y. F. Gui, W. B. Dou, K. Yin, and P .G .Su, Open resonator system for automatic and precise dielectric measurement at millimeter wavelengths. Int.JInfrated and Millimeter waves. 29, 782-791, 2008

T. M. Hirvonen, P. Vainikainen, A. Lozowski, and A. V. Raisanen, Measurement of dielectrics at 100GHz with an open resonator connected to a network analyzer. IEEE Trans.Instrum. Meas. 45(4), 780–786 , 1996

P. Coquet, T. Matsui, and M. Kiyokawa, Measurements of low loss dielectric materials in the 60 GHz band using a high-Q gaussian beam open resonator, IMTC'94, pp.1265–1268, May 10–12, 1994.

C. R. Jones, J. Dutta, G. F Yu and Y. C. Gao, Measurement of Dielectric Properties for Low-Loss Materials at Millimeter Wavelengths.J Infrared MilliTerahz Waves.32,838-847, 2011

F. R. Cui, Z. X. Luo, F. Ji, B. J. Hu, and S. I. Lai. Quasi optical resonator for measuring surface resistance and its distribution of high temperature superconductor film. Int.JInfrated and Millimeter waves.20(6), 1037-1045 , 1999

M. N. Afsar, H. Y. Ding, A novel open-resonator system for precise measurement of permittivity and loss-tangent. IEEE Tran.Instrum. Meas. 50(2), 402-405, 2001

M. N. Afsar, H. Y. Ding, K. Tourshan, A new 60GHz open resonator technique for precision permittivity and loss tangent measurement. IEEE Trans. Instrum. Meas. 48(2), 626–630, 1999.

R. G. Jones. Precise dielectric measurements at 35GHz using an open microwave resonator. Proc.IEE. 123(4), 285-290, 1976

J. Xia, C. H Liang, The calculation for the Q factor of an electromagnetic open resonator. Journal of Electronics. 17(1), 103-107, 1995

R. K. Mongia and R. K. Arora, Equivalent circuit parameters of an aperture coupled open resonator cavity. IEEE Trans. Microwave Theory Tech. 41, 1245-1250, 1993

Guofen Yu, JyostnaM.Dutta, Charles R.Jones, The Influence of a Coupling Film on Ultra-Low-LossDielectric Measurement Using an Open Resonator ,J Infrared MilliTerahz Waves (2011) 32:935 – 942,2011

R. Heidinger, "Dielectric measurements on sapphire for electron cyclotron wave systems," J. of Nucl.Mat. 212–215, 1101–1106, 1994

A Novel Design of Waveguide-Coax Millimeter-wave Equalizer

Liusha Yang*, Yong Xu, and Yong Luo
School of Physical Electronics, University of Electronic Science and Technology of China, Chengdu 610054, China.
*Contact: 18782952283@126.com, phone +18782952283

Abstract—**Based on theories of transmission line and resonance, a novel waveguide-coax equalizer for millimeter wave is presented through studying the features of various equalizers. The structure utilizes stepped waveguide as the main transmission line, coaxial cavity loaded with absorbing material as the resonant unit. The influences, which the coaxial cavity length, the radius and the inserted depth of the probe, the characteristic of absorbing material have, on the attenuation amplitude and resonant frequency are analysed. With the HFSS simulation software, the attenuation amplitude and the operating frequency are adjusted by changing the factors mentioned above, the stepped waveguide height and the location of the absorbing material. According to the gyro-traveling wave tube output power curve, the required equalization value is obtained. The designed equalizer preferably achieves the goal in 33-37GHz.**

INTRODUCTION

Gyro-TWT (Gyro-traveling wave tube) is a high-output-power device in millimeter wave. It generally reaches the saturated output power in the required frequency band under the same excitation condition. However, the output power usually varies with the working state in the utilization, which affects the performance. In order to get a stable output power, an equalization network is necessary. According to the best excitation curve of the gyro-TWT, the aim line should be in accordance with the output power line when the equalizer is at input port of the tube. If the equalizer is at output port, the aim line should be mutually complementary with the output power line.

There are three types of equalizer, microstrip line, waveguide and coaxial structure. Microstrip line is characterized with the small size, light weight and easy fabrication, but the power capacity is low and it is unsuitable in the high frequency. The theory of microstrip line equalizer design is quite abundant [1][2].Waveguide equalizer is usually used in high frequency with electromagnetic wave absorber in its cavity resonators. However, it is so hard to decide the position and amount of the wave absorber that the reflection coefficient and the max attenuation are difficult to control. Coaxial equalizer is used widely in high frequency and high power due to the big power capacity and easy tune [2-5]. It contains several sub-structures and obtains corresponding attenuation through adjusting the probe length, which undoubtedly changes the boundary condition of the main transmission line and increases the reflection coefficient.

The paper suggests a novel design of waveguide-coax equalizer, which combines the advantages of waveguide and coaxial structure to improve the transmission characteristic. The equalizer adopts the stepped waveguide as the main transmission line to get matched. The coaxial resonator is connected with the broadside of the waveguide. The inner conductor is used as a probe to couple the energy. The other end of the probe is loaded with tapering wave absorber to change the attenuation. With the help of HFSS(high frequency structure simulator), the equalizer presents an equalization value of 9.1dB in 33-37GHz, the minimum attenuation at 34.3GHz is -11.5dB, the maximum attenuation at 37GHz is -2.4dB and the reflection coefficient is under -7.3dB, which satisfies the design goal preferably.

FORMULATION AND SIMULATION

A. Theory

A single branch of equalizer structure is shown as Fig. 1. The rectangle waveguide is the main transmission line. In order to transmit the microwave in 33GHz-37GHz and avoid the competition of high order mode, the waveguide adopts the standard BJ320 waveguide. The coupling probe inserts the waveguide in center of the broadside so that the TE_{10} mode can be excited efficiently. The wave absorber is a dented cone which can increase the relative absorbing area and decrease the reflection.

Since the structure of waveguide-coax equalizer is complex, it is hard to find an analytic solution to decidethedistribution of electromagnetic field and the change of energy. Thus, we use the transmission line theory and resonance theory to study the structure.The coupling probe is equivalent to capacitance and the absorber is equivalent to resistance,so the coaxial resonator can be equivalent to a resonant circuit.

Fig. 2 shows the equivalent series resonant circuit. Based on the resonance theory [6-8], we build the transcendental equation [9] about the resonant frequency, capacity value and the cavity length:

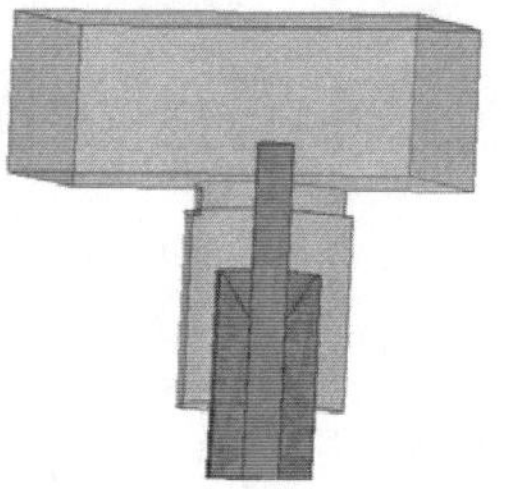

Figure 1.A single branch of the equalizer structure.

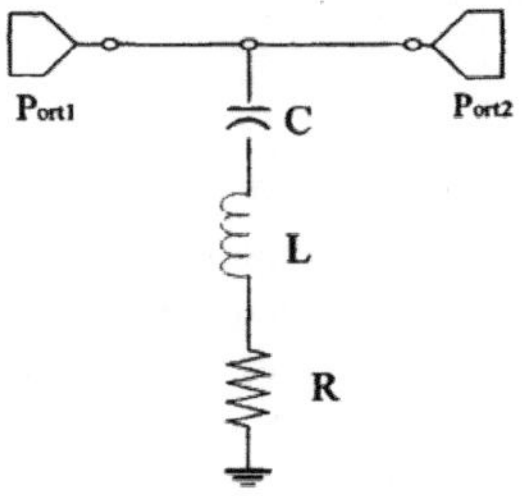

Figure 2.The equivalent series resonant circuit

$$S_{21}(\omega)=\frac{2(1-\omega^2LC+j\omega RC)}{2(1-\omega^2LC+j\omega RC)+j\omega RC} \quad (1)$$

It can be seen that the resonant frequency can be changed by the capacitance and the inductance. Besides, the resistance can regulate the attenuation.

The relation between attenuation and Q factor is:

$$Q_L=\omega_0\frac{W}{P_l} \quad (2)$$

$$BW=\frac{1}{Q} \quad (3)$$

With the attenuation increases, the Q factor decreases. So it sacrifices the Q factor to improve the flatness.

Since the single resonance branch has the low equalization value and narrow frequency band, it can't satisfy the demand. If we put a series of resonance branches together, we may get an appropriate equalization curve through choosing the right resonance frequency and Q value. Besides, considering the mismatch introduced by the branches, stepped waveguide is adopted. According to the transmission line theory, the input impedance of the point in the transmission line is:

$$Z_{in}=Z_c\frac{Z_L+jZ_c\tan\beta l}{Z_c+jZ_L\tan\beta l} \quad (4)$$

When the length of transmission line is $\lambda_g/4$ (λ_g is guide wavelength), the equation is simplified as:

$$Z_{in}=\frac{Z_c^2}{Z_L} \quad (5)$$

In this way, the two parts of the transmission line get matched. But, the single segment can only be effective in a specific frequency. In order to expand the operating frequency band, cascading $\lambda_g/4$ impedance transformers is utilized.

B. Simulation

The main indexes of equalizer include resonant frequency and attenuation. The radius and length of the probe, the coaxial cavity length andthe resistance value are the primary factors of the equalizer performance. In the following part, the effect of these factors is analyzed using HFSS software.Fig 3 shows the varying curve about attenuation versus the radius of probe. It is obvious that the larger the probe radius is, the higher the resonance frequency is and the attenuation changes little. Fig 4 suggests that the deeper the inserted probe is, the larger the attenuation is and the lower theresonancefrequency is. This phenomenon can be explained that the coupling capacitance coefficientbetween the probe and the waveguide grows when the probe inserts deeper, so that the resonance frequency gets down.

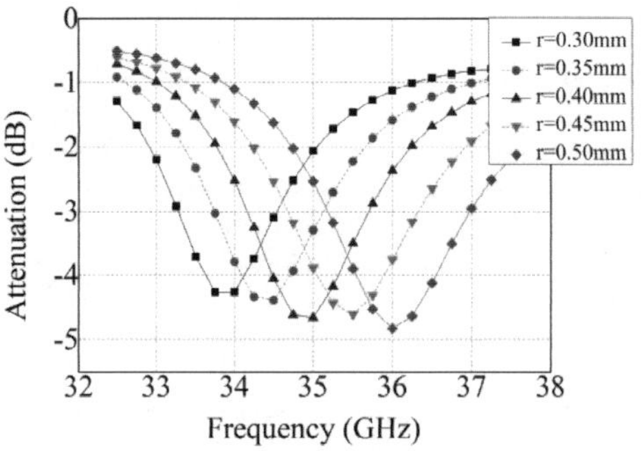

Figure 3.Attenuationvary with the probe radius

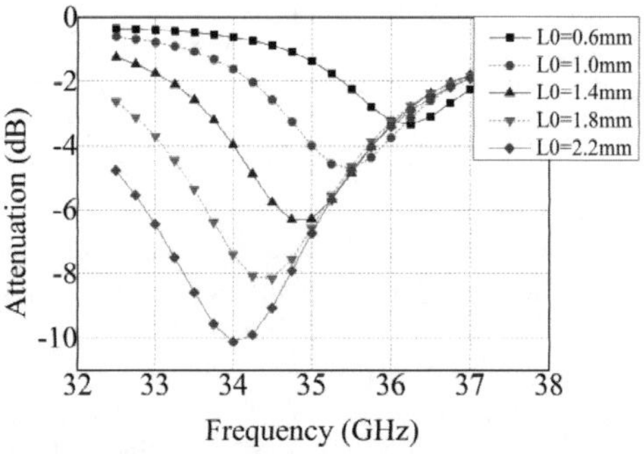

Figure 4.Attenuationvary with input length of the probe

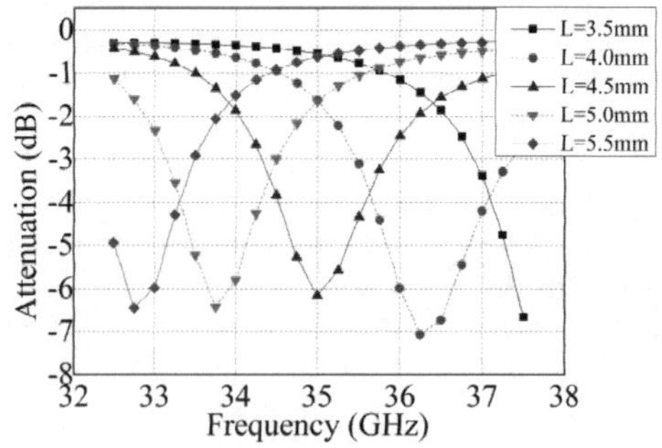

Figure 5.Attenuationvary with the coaxial length

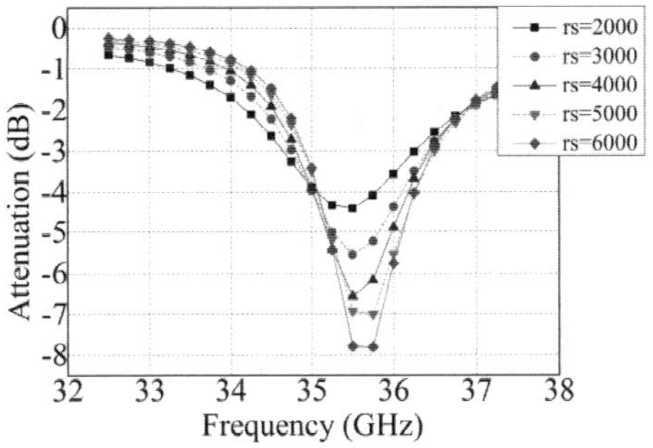

Figure 6.Attenuationvary with the resistance value

Fig 5 indicates that in a resonant period, the resonant frequency becomes lower witha longer cavity lengthand the attenuation has small changes. From Fig 6, we can see that the attenuation increases when the resistance increases.

In conclusion, the adjustment of attenuation can be changed by the inserted probe length and resistance.The resonance frequency can be changed by the probe radius, the inserted probe length, and the coaxial cavitylength.

Generally, there are 2-6 resonance branches in a cascade structure. Considering the big equalization value and small volume, we adopt the four resonance branches to construct the equalizer, as shown in Fig 7. To match neighboring branches and reduce reflection, each stepped waveguide length is $3\lambda_g / 4$. The simulation results are shown as Fig 8 and Fig 9. From Fig 8, we can see that the simulation line is quite approximate to the aim line. The equalization value isabout 9.1 dB in 33-37GHz. And the low inserted attenuationis about -11.5dB at 34.3GHz, the high inserted attenuation is about -2.4dB at 37GHz.

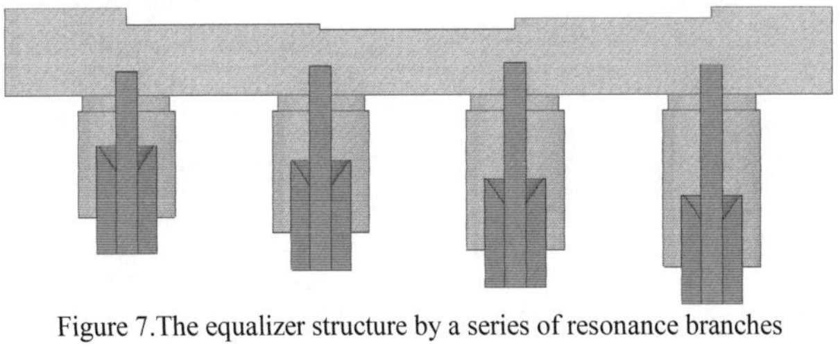

Figure 7.The equalizer structure by a series of resonance branches

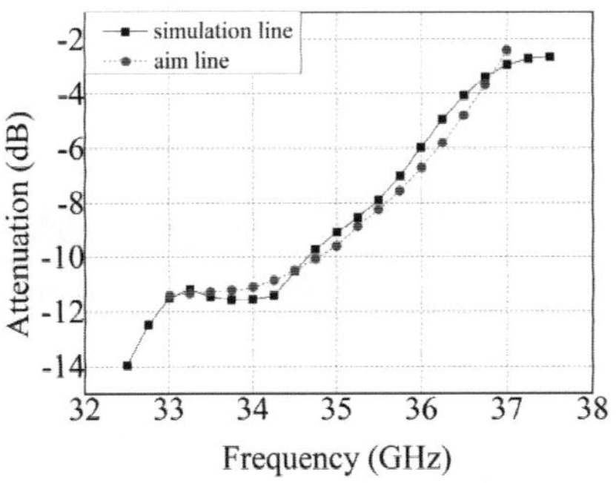

Figure 8.The comparison between simulation line and aim line

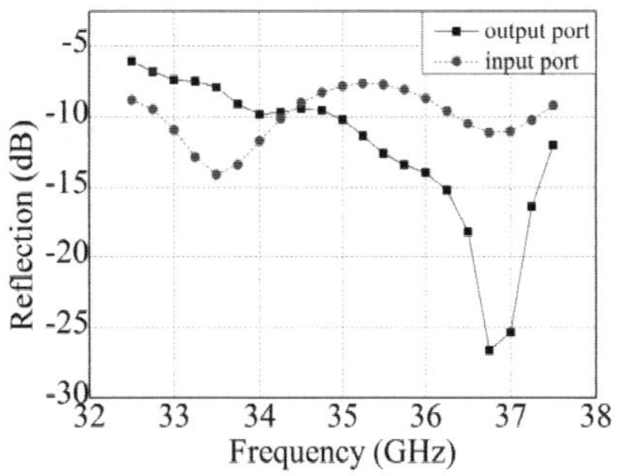

Figure 9.The reflection of input port and output port

Conclusion

In this paper, we research a novel design of waveguide-coax millimeter wave equalizer. The structure has the advantage of large power capacity, high operating frequency band, convenient tune and good transmission performance. Through simulation, an equalizer operating at 33-37GHz is presented and meets the design requirement well. This design will promote the study of the millimeter wave equalizer in the future.

References

Zhao Ying, Zhou Dong-fang, Niuzhong-xua, Zhang De-wei, "ANewly Designed Broadband Micro strip equalizer", Microwave Conference Proccedings, 2005. APMC 2005.

Jiao Li, Bo Yan, Yong Zhang., "Design of a microwave power equalizer." IEEE International Conference, Microwave Technology &Computational Electromagnetics (ICMTCE), pp. 145-147, May 2011.

Liu Ya-nan, He Qing-guo, Hao Jin-zhong, Wang Kang-han, "Design and Implementation of 6-18 GHz Super-Broadband Microstrip Equalizer", Semiconductor Technology, 35(1):23-26, 2010.

Tai-fu Zhou, Yong Zhang, Riu-min Xu. "Design of Microwave and Millimeter Wave Gain Equalizer."2011 China-Korea-Japan Electronics and Communication Conference. pp. 26-28, 2011.

Tai-fu Zhou, Yong Zhang, Rui-min Xu. "Research on the Millimeter Wave Gain Equalizer." Microwave Technology & Computational Electromagnetics(ICMTCE). 2011 IEEE International Comference, pp. 180-182. May 2011.

Zhang Yi, Zhang Feng, NiuZhong-xia, et al. "A more accurate analysis and design for the millimeter wave equalizer." Proceeding of 2005 Asia-Pacific Microwave Conference. Suzhou: IEEE, 2005.

Yang Ming-shan, Zhang De-wei, NiuZhong-xia, "Relationship between the Depth of Probe and Resonant Frequency of Microwave Amplitude Equalizer." Journal of Zhengzhou University (Natural Science Edition), 37(4),60-62 ,2005.

Hong Jia-zhen, Xu Hu-min, BaoZheng-guang, NiuZhong-xia, "Design and Simulation of a Magnitude Equalizer." Journal of Information Engineering University.44-46, 2005.

Yong Zhang," Research on the Millimeter Wave Power Equalizer.' IEEE 2007 International Symposium on Microwave, Antenna, Propagation, and EMC Technologies for Wireless Communications, pp. 446-449, Aug. 2007.

A TE13 Mode Input Converter for 0.1THz High Order Mode Gyrotron Travelling Wave Amplifiers

Yan Wang*, Guo Liu, Guoxiang Shu, Li Wang and Yong Luo

School of Physical Electronics, University of Electronic Science and Technology of China, Chengdu, Sichuan610054, China

* Contact: wangyanuestc@163.com

A techniqueto launch a circular TE_{13} mode to interact with the helical electron beam of a 0.1THz gyrotron travelling wave amplifier is proposed and verified by simulation in this paper.The converter consists of a Y-type power divider, a cutoff waveguide, an output cylinder waveguide, grooves and convex strips to suppress the unwanted modes.The high order TE_{13} mode is excited by a broadband Y-type power divider with the aid of a cylindrical waveguidesystem. Usinggrooves and convex strips loaded at the lateral planes of the output cylindrical waveguide,the electric fieldsof the potential competing TE_{32} and TE_{71} modes are suppressed to allow the transmission of the dominant TE_{13} mode. The converter performance with and without grooves and convex stripsare studied and compared, and excellent results have been achieved. The converter without grooves and convex strips has an average transmission ~-3 dB to TE_{13} mode, and the conversion to the TE_{32} and TE_{71}modesare respectively at -8dB and -10 dBlevel. After introduced grooves and convex strips, the simulationpredictsthat the average transmission is ~-1.8 dB with a 3 dB bandwidth of 7.3 GHz (96.3-103.6 GHz) and port reflection is less than-15 dB. The conversion to the TE_{32} and TE_{71}modesare respectively under -15dB and -24dB in the operating frequency band. It shows the loading grooves and convex stripswork well to suppress the spurious modes and improve the conversion efficiencyof the TE_{13} mode.

Optical Testing of the CAmbridge Emission Line Surveyor (CAMELS)

Lingzhen Zeng[1*], Raymond Blundell[1], Dorota Glowacka[2], David J. Goldie[2], Paul Grimes[1], Eloy de Lera Acedo[2], Scott Paine[1], Christopher N. Thomas[2],Edward C. Tong[1], Stafford Withington[2]

1 Smithsonian Astrophysical Observatory, 60 Garden Street, Cambridge, MA 02138, USA

2 Cavendish Laboratory, JJ Thomson Avenue, Cambridge, CB3 0HE, UK

*Contact: lingzhen.zeng@cfa.harvard.edu

Abstract— **The motivation for this work is to develop submillimeter wave and far-infrared imaging technology in which each detector in a focal plane is intrinsically capable of yielding detailed spectroscopic information.A first step towards this is the development of the CAMELS instrument, which will eventually be used to survey nearby galaxies at 3 mm with a spectral resolution, R = δλ/λ, of about 3000.The CAMELS instrument is based on the Microwave Kinetic Inductance Detector (MKID), operating at 100 mK, combined with an integrated filter bank design to provide 512 spectral channels between 103.0 and 114.7 GHz.In this work, we present the ongoing optical measurements of the CAMELS detectors, including dark tests and planned line source and gas cell tests.**

INTRODUCTION

On-chip spectrometers for the millimeter and submillimeter waveband have generated increasinginterest in the past few years. In these devices, each detector pixel consists of a filter bank feeding multiple detector devices that are multiplexed onto a single readout line. Several projects, such as DESHIMA [1], SuperSpec [2], X-Spec[3],andWSpec[4] have been proposed for the submillimeter and far-IR wavebands. CAMELS is an instrument operating ata wavelength of about 3 mm.The scientific motivation of the CAMELS project is to map the $^{12}CO(1\text{-}0)$ and $^{13}CO(1\text{-}0)$ line emission from local galaxies with redshift in the range of 0.05-0.13 (^{12}CO) and 0.003-0.961 (^{13}CO), providing information about the evolution of the molecular gas content in galaxies throughout the cosmic epochs It will also serve as a pathfinder instrument for investigating the use of integrated filterbank spectrometers for astronomical observations.

Fig. 1 is a system diagram of the CAMELS instrument. The top panel shows the layout of the optics, cryostat and readout electronicsThebottom panel is a schematic representation of the CAMELS detector chip. It consists of a pair of horn-coupled microstrip feedlines, carrying two orthogonal polarization signals. The feedlines are coupled to an integrated bank of narrow-band superconducting resonator filters that provide spectral selectivity. The power admitted by each spectral selection filter will be detected by microwave kinetic inductance detectors (MKIDs) through a millimetre-wave coupler. Finally, all the MKIDs will be readout via a microstrip line using frequency domain multiplexing. The top right panel of Fig.1 shows the frequency multiplexing readout system of the instrument. The details of this readout scheme were presented in [5]. Fig. 1 shows a single-pixel system, however the compact design of the spectrometer will allow packing a powerful multi-pixel spectrometer system on a focal plane.

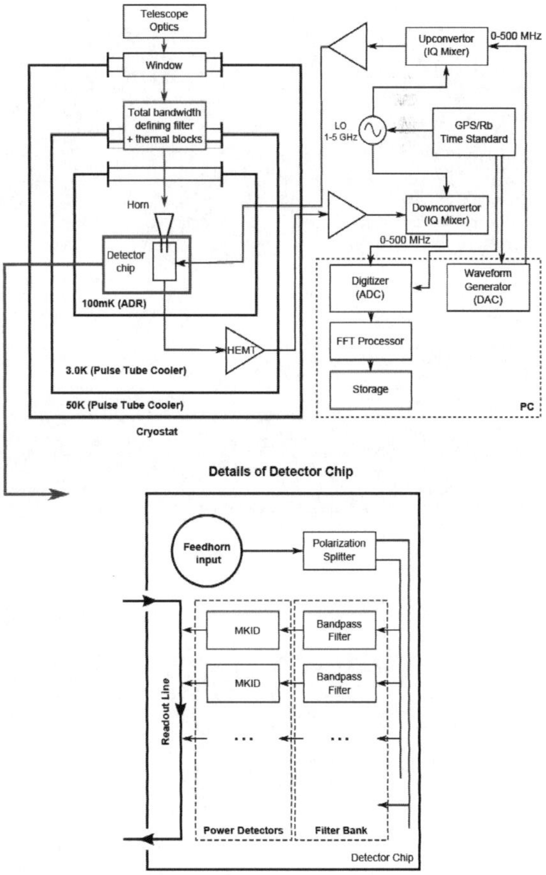

Fig. 1 System block diagram of the CAMELS instrument.

OPTICAL TESTS

Fig. 2 shows photos of a prototype CAMELS chip that is being tested to prove the technologies required for a full chip.The top panel shows the sample holder and the detector chip. The bottom panel shows the details of the detector chip, including the optical coupler (top right) and signal filter (bottom right).Key to the operation of the CAMELS devices is the use of β-Ta (Tc ~ 700mK) as a sensing material, which allows operation at millimeter-wavelengths.This is incorporated into the termination of quarter-wavelength NbN resonator for readout, similar to [6]. However, a novel optical coupler design must be used as both the milllimeter-wave and resonator line are microstrip. We use a scheme in which the millimetre-wave line overlaps the end of the resonator line, such that the β-Ta strip line of the resonator forms the ground plane of the signal line [5]. The test chips are to designed to verify this coupling scheme and allow investigation of loss and noise mechanisms for optimization of future designs.In addition, the signal is coupled into the chip via a planar antenna in the prototype devices, rather than a waveguide probe as planned for the final devices. The antenna is a centre-fed single slot design, which is illuminated from the underside of the detector chip through a window in the holder.

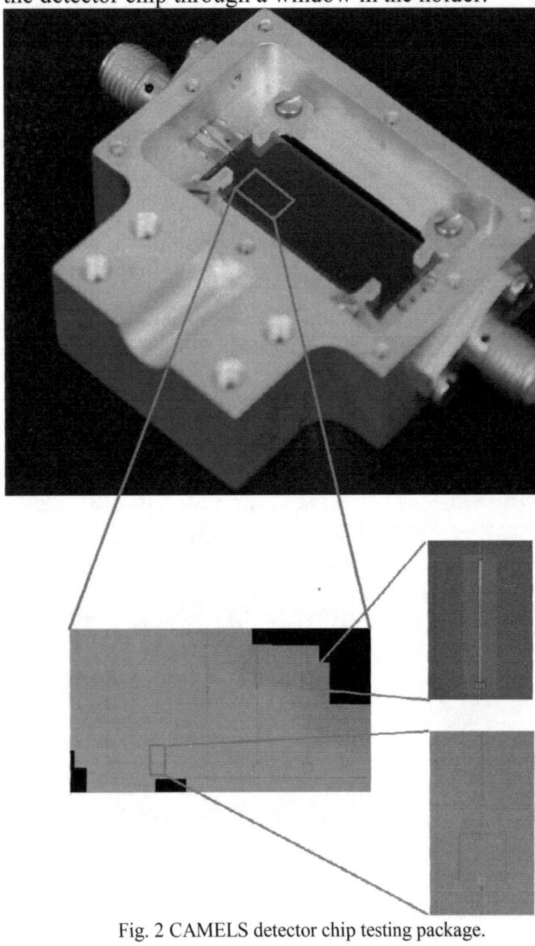

Fig. 2 CAMELS detector chip testing package.

A. Dark tests

In the dark tests, the detectors were sealed inside a metal jig to isolate them from the radiation of the surrounding environment. We performed frequency sweeps and bath temperature sweeps to measure the detector response using a vector network analyzer and a single-channel prototype of the broadband readout electronics which will eventually be used to read out an array of detectors multiplexed onto a single RF line. We fitted the resonant frequencies and quality factors of the resonators as a function of temperature. We have also studied the sources of noise in the detector and have determined the maximum readout power at which the onset of nonlinear behavior is observed.

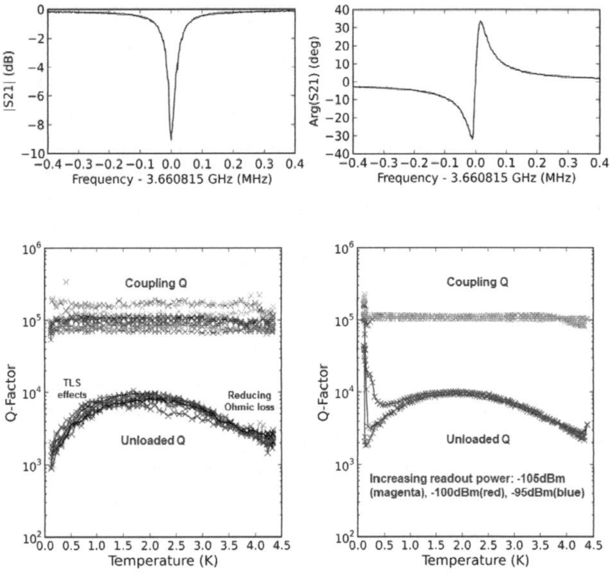

Fig. 3 CAMELS MKID resonator measurement results.

Fig.3 shows the measurement results of the CAMELS resonators on chip. Data are shown for a via-shorted NbN microstrip resonator without the millimetre-wave coupler. The top plots are the measured S_{21}magnitude and phase of a resonator centred at 3.660815 GHz. The bottom left panel shows the measured coupling quality factor Q_c and internal quality factor Q_i (unloaded Q) of different resonators at a low readout power. The variation in Q_c results from the varying resonant frequency for different resonators. The internal coupling factors drop at low temperature due to the two-level system effects in the dielectric and at high temperature due to the ohmic loss in the superconductor. The bottom right panel shows Q_c and Q_i of a resonator at different readout power level and at different temperature. At the operating temperature of100mK, we measured thepeak $Q_i$to be around 10^5.

B. Line source measurement

To measure the frequency-response of the integrated filters, we have developed a narrow-band cryogenic source that can be continuously tuned over the range 100-115 GHz. This line source is realized using a Pacific Millimeter Products harmonic mixer (FM model) mounted on the 4K plate as a cryogenic multiplier. We pump the multiplier with a <30 GHz signal produced by an external room-temperature synthesizer and coupled into the cryostat through a coaxial cable. The multiplied signal is then free-space coupled to the detectors on the 100mK stage, avoiding the need for a window or waveguide plumbing to low temperature. It also allows fast switching of the source power, which is useful for response time measurements and for distinguishing between optical response and thermal response from source loading, taking advantage of the different timescales of these processes. Initially, the power output of the line source wascalibrated in a 4K cryostat by transmission measurements to an external room-temperature detector (see Fig. 4), but we are also investigating in-situ monitoring at cryogenic temperatures via a directional coupler and cryogenic power detector. A good knowledge of the power output as a function of frequency is necessary to separate variations in detector response due to the filters from those due to variations in source.

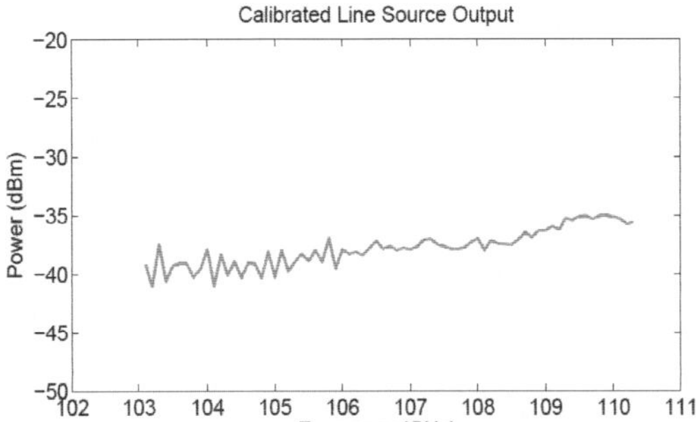

Fig. 4Line source output that is calibrated from 103 GHz to 110 GHz.

C. Gas cell measurement

In order to demonstrate the telescope-readiness of the CAMELS instrument, we intend to carry out measurements of spectral emission lines in a gas cell mounted outside the instrument cryostat. A liquid nitrogen cooled 77K load placed behind the cell will provide a low brightness temperature background to the gas emission lines in the ambient temperature cell, realistically simulating a typical astronomical observation (particularly in terms of coupling to the telescope). The use of a naturally abundant mix of carbonyl sulphide (OCS) isotopologues in the gas cell will provide five spectral linesspread across the CAMELS band (see Fig. 5 for details).These will provide a wide range of line temperatures,with pressuretuneable line width to cover a few CAMELS channels.

Fig. 6 shows aCAD model of the CAMELS gas cell measurement set up. The gas cell is coupled to the detectors by an optical path consisting of a room temperature parabolic off-axis mirror; vacuum window of the cryostat; IR blocking filters mounted on 50K and 4K shields; a secondary parabolic mirror mounted on the 4K plate and a conical feedhorn mounted on the detector package at the 100mK Adiabatic Demagnetization Refrigerator (ADR) stage.

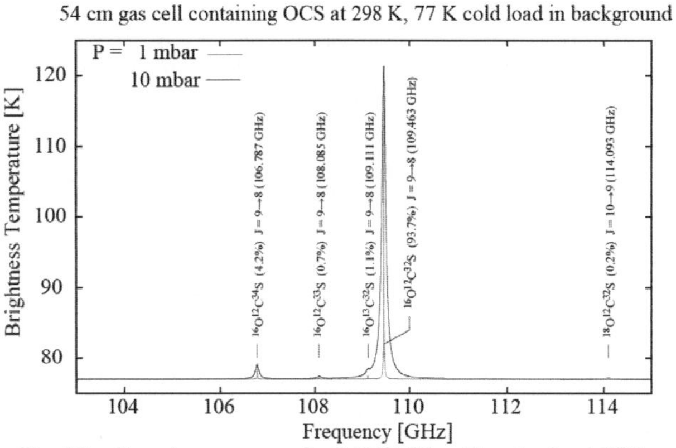

Fig. 5Five lines show up across the CAMELS band in asimulated OCS gas cell spectrumproduced with the*am* radiative transfer code [7].

CONCLUSIONS

Wehave presented the ongoing optical measurements of the CAMELS MKID detectors, including dark tests and planned measurement using a calibrated line source and an OCS gas cell.

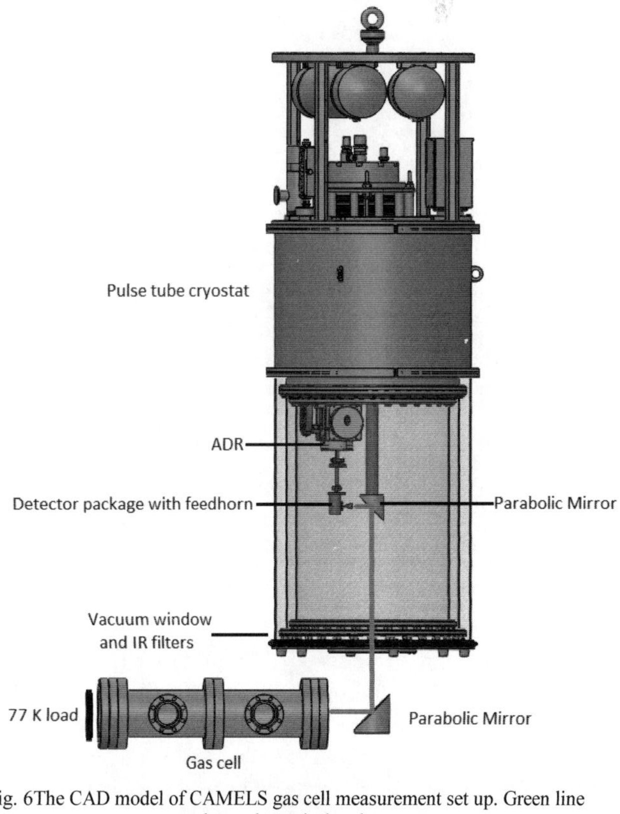

Fig. 6The CAD model of CAMELS gas cell measurement set up. Green line shows the optical path.

REFERENCES

[1] A. Endo, J.J.A. Baselmans, P.P. van de Werf and others, "Development of DESHIMA: a redshift machine based on superconducting on-chip filterbank", *in Proc. SPIE 8452, paper 84520X*, 2012

[2] E.Shirokoff, P.S. Barry, C.M. Bradford and others, "Design and Performance of SuperSpec: An On-Chip, KID-Based, mm-Wavelength Spectrometer." *Journal of Low Temperature Physics, vol. 176, pp. 657-662,* Sep. 2014.

[3] Scott C. Chapman, Charles M. Bradford, Eric Shirokoff and Steve Hailey-Dunsheath,"X-Spec: A multi-object, trans-millimeter-wave

spectrometer for CCAT", *39th International Conference on Infrared, Millimeter, and Terahertz waves (IRMMW-THz),* 14-19 Sept. 2014

[4] Sean Bryan, James Aguirre, George Che, Simon Doyle, Daniel Flanigan, Christopher Groppi, Bradley Johnson, Glenn Jones, Philip Mauskopf, Heather McCarrick, Alessandro Monfardini, Tony Mroczkowski, "WSPEC: A waveguide filter-bank focal plane array spectrometer for millimeter wave astronomy and cosmology", *Journal of Low Temperature Physics, DOI 10.1007/s10909-015-1396-5,* December 2015

[5] Christopher N. Thomas, Ray Blundell, D. Glowacka, David J. Goldie, Paul Grimes, Eloy de Lera Acedo, Scott Paine, StaffordWithington and Lingzhen Zeng, "Progress on the CAmbridge Emission Line Surveyor (CAMELS)",*26th International Symposium on Space Terahertz Technology , March 2015*

[6] Day, P.K., Leduc, H.G., Goldin, A., Vayonakis, T., Mazin, B.A., Kumar, S., Gao, J. and Zmuidzinas, J., 2006. "Antenna-coupled microwave kinetic inductance detectors".*Nuclear Instruments and Methods in Physics Research Section A: Accelerators, Spectrometers, Detectors and Associated Equipment, 559(2), pp.561-563*

[7] Scott Paine, "The AM atmospheric model", *SMA technical memo #152, version 9.0, (2016),* https://www.cfa.harvard.edu/sma/memos/152.pdf

Design and Simulation of Interaction Structure for 110GHz Second-Harmonic Gyro-TWT

N.Huang*

*University of Electronic Science and Technology of China *, Chengdu, SiChuan610054, China*
* Contact: 294985115@qq.com

In this paper, the application of nonlinear theory in high harmonic gyrotron traveling wave amplifiers (gyro-TWT's) is analysed. And a beam-wave interaction structure of 110GHz second-harmonic gyro-TWT is designed. The simulation indicates that, in ideal case, the interaction structure can produce more than 70kW outer power when 70kV and 10A electron beam is input. Also, employing secondharmonic achieve high-frequency, single-mode and stable output. This paper describes the high stability and wide bandwidth of 110GHz TE_{01} mode gyro-TWT and the technique of design and simulation in particle-in-cell(PIC) simulation software-MAGIC. The interaction structure of gyro-TWT employs ceramic loading, and indicates the effective of dielectric-loaded for suppressing the spurious oscillations and improve stability.

A 15Gps high speed OOK receiver based on a 0.34THz Zero-bias Schottky diode detector

Yaoling Tian[1,2*], Jun Jiang[1,2], Qingfeng Liu[1,2], Changxing Lin[1,2], Kun Huang[1,2], Xianjin Deng[1,2]

[1] *Institute of Electronic Engineering. China Academy of Engineering Physics, Mianyang Sichuan 621900, China*

[2] *Microsystem and Terahertz Research Center, China Academy of Engineering Physics, Chengdu 610200, China*

*Contact: tianyaoling@126.com, phone 2497242

Abstract— **This paper presents a 0.34THz high speed on-off keying (OOK) receiver composed of a direct detector, the external video amplifier chain and a waveguide horn antenna. To evaluate the high speed performance of the receiver, a 2^{31}-1 pseudorandom binary sequence (PRBS) is transmitted and received by the existing OOK transmitter and the proposed OOK receiver respectively. And results show an a bit error rate(BER) below 10^{-12} at 10Gbps and a bit error rate(BER) of 3.15×10^{-7} at data rate up to 15Gbps at room temperature, which proves that the receiver can well meet the requirement of high data rate OOK communication system.**

I. INTRODUCTION

Demand for ultra-high speed wireless communication has increased rapidly recent years, which accelerates the development of various Terahertz (THz) communication systems. Several systems have been demonstrated recently using the modulation scheme such as 16QAM (quadrature amplitude modulation) and high order phase-keying (PSK) [1]-[3]. But these high order modulation schemes require local oscillators and related circuits, which results in a complex system framework. On the contrary, the noncoherent on-off keying (OOK) modulation scheme, though spectrally inefficient, has a very simple architecture with low power consumption since the receiver does not need LOs (local oscillators). Several OOK receivers, transmitters and transceivers have been demonstrated with good performance [4]-[5]. However, most of these OOK scheme systems are operating at frequencies below 300GHz.

In this paper, a 0.34THz noncoherent OOK receiver is designed based on a zero-bias direct detector followed by the wideband video amplifiers. It is shown that a typical responsivity of 1400V/W has been achieved over the frequency range from 315 to 357GHz, which holds the potential to deal with 20Gbps OOK signals. And the receiver has been tested and can achieve 10Gbps communication with a bit error rate (BER) below 10^{-12}. And the 15Gbps communication is also demonstrated with a bit error rate (BER) of 3.15×10^{-7} at room temperature.

II. ZERO-BIAS DETECTOR

The 0.34THz direct detector is designed in the microstrip topology on a 50um-thick quartz substrate, which is placed inside a split block waveguide cavity. Fig.1 illustrates the waveguide to microstrip transition on the right, an impedance matching network designed according to the employed low barrier Schottky diode. The CMRC low pass filter act as an RF short, which can blocks the RF and extract the video signal as a result.

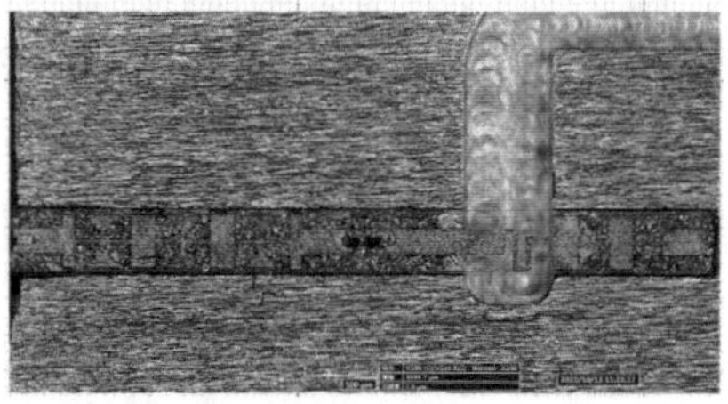

Fig. 1 Photograph of 0.34THz Schottky detector circuit on 50um-thick quartz substrate

Fig.2 (a) presents the measured and simulated responsivity of the proposed OOK detector, with an output impedance of 1MΩ. The measured responsivity into a 1-MΩ load is 910~2210V/W over the frequency range from 315GHz to 357GHz. Meanwhile, the measured output noise voltage is about $7.07nV/Hz^{0.5}$. This results in a measured noise equivalent power (NEP) of $3.2\sim7.8pW/Hz^{0.5}$ at the zero-bias condition. Fig.2 (b) shows the measured noise equivalent power (NEP).

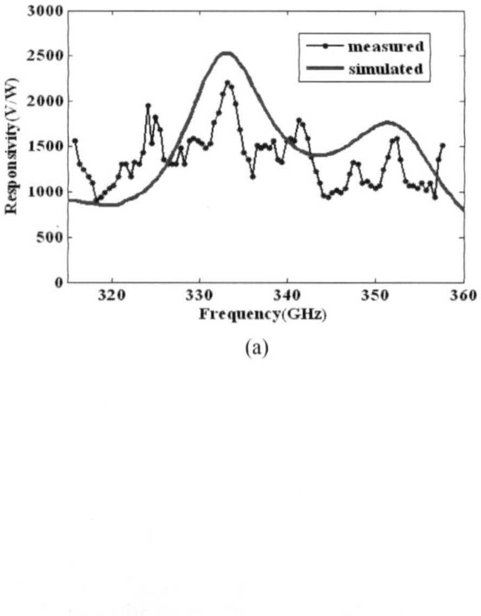

(a)

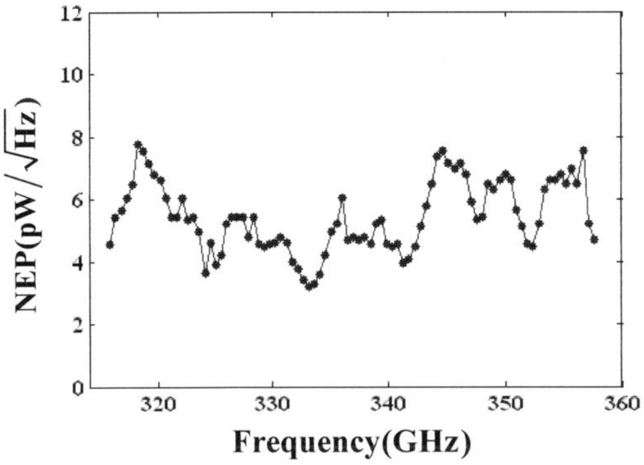

(b)

Fig. 2 (a) measured and simulated detector responsivity versus frequency from 315 to 357GHz and (b) measured noise equivalent power of the detector.

III. 0.34THz OOK Receiver Characterization

A. Detectors and Amplifiers

Due to the absence of the 0.34THz LNA (low noise amplifier), the receiver is a zero-bias detector connected to an external video amplifier chain (Av=26) through an SMA-tee virtually. And the OOK receiver responsivity is simply the video amplifiers' gain multiplied by the detector responsivity into a 50-Ω load. Besides, NEP of the receiver is much larger than the one of the detector, because of the noise brought into by the video amplifier chain.

Fig.3 shows the responsivity and NEP of the receiver versus frequency over the operating frequency band. It can be seen that the receiver has a responsivity of 960~2350V/W, which is similar to the detector with a load impedance of 1MΩ. Besides, the NEP ranges from 37 to 92 $pW/Hz^{0.5}$ over 315~357GHz. And NEP of the receiver is dominated by the wideband video amplifier chain, which can be depressed by the 0.34THz LNA.

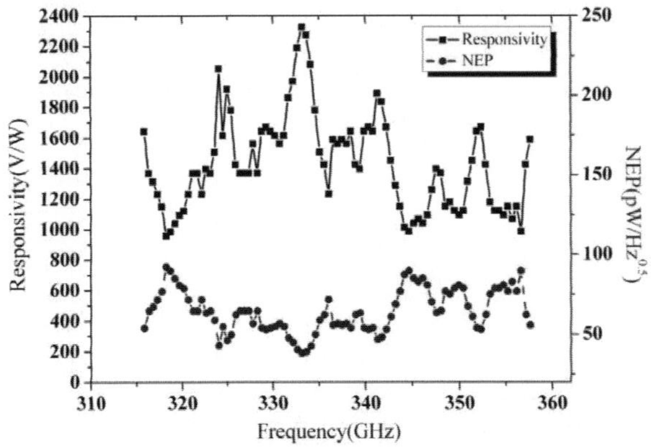

Fig. 3 The measured receiver responsivity and NEP versus frequency from 315 to 357GHz

B. BER Measurements

Fig.4 (a) presents a 0.34THz OOK transmitter system based on the BAC286C BER tester sending 2^{31}-1 PRBS. And a carrier frequency of 340GHz is selected. Moreover, a 0.34THz modulator in [6] is used to achieve the OOK modulation by multiplying the PRBS data with the carrier. In other words, the carrier signal passes through the modulator if the incoming data is 1; else, the modulator has no output. However, because the modulator has a finite isolation, an LO leakage exists at the output, which has a great impact on the performance of the receiver.

The OOK modulated signal is transmitted and received by the WR2.8 waveguide antennas over suitable distance. The video amplifier chain has a NF over 15dB, which is significant in the output noise of the OOK receiver. And for the lack of 0.34THz LNA at the receiver, it is difficult to realize a long distance test. Thus, in order to gain enough signal-to-noise rates (SNR), a carrier frequency of 334GHz is chosen and the output of the modulator is adjusted to -8.5dBm, which reaches the ceil of the multiplier chain. And the transmitter and receiver are set back to back during the test.

Fig.4 (b) presents the BER for different data rates, and a 10Gbps wireless communication link is achieved with a 10^{-12} BER, which is actually determined to be the lower limit because it takes a large amount of time to measure a BER $<10^{-12}$. And 15Gbps and 18Gbps wireless communication links are also achieved with a BER of 3.15×10^{-7} and 2.6×10^{-5} respectively. Moreover, it can be observed in Fig.4 (b) that there is a small distortion in the eye diagram at a data rate of 15Gbps, related to the BER increasing with the data rate.

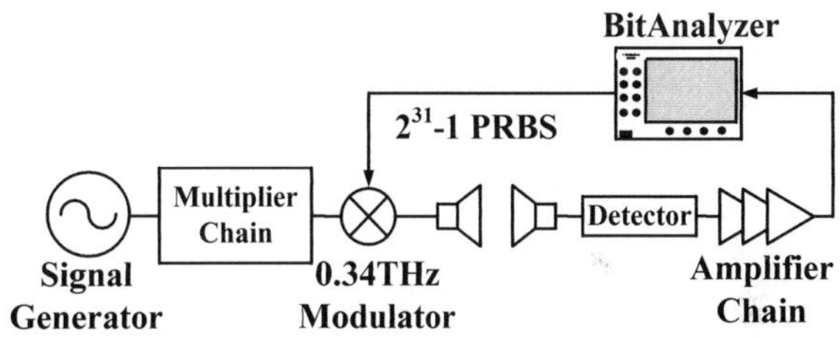

(a)

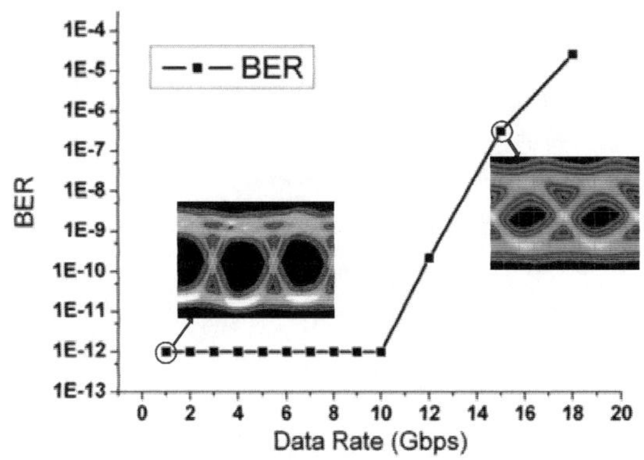

(b)

Fig. 4 (a) BER measurement setup and (b) measured BER versus data rate at a carrier frequency of 334GHz

IV. Conclusion

An OOK receiver at 0.34THz has been developed based on a zero-bias Schottky diode detector with high responsivity and low NEP. And a 10Gbps wireless communication link has been demonstrated with a BER $< 10^{-12}$ at room temperature. Also, operation has been demonstrated at a data rate up to

18Gbps with a 2.6×10^{-5} BER. Finally, performances of the receiver can be improved by reduce the LO leakage from the transmitter as well as developing the 0.34THz LNA, which are the following research.

ACKNOWLEDGMENT

This work is supported by National Key Basic Research Program of China (grant No.2015CB755406). And all of the presented results are accomplished within Microsystem and Terahertz Research Center and the Institute of Electronic Engineering of Chinese Academy of Engineering Physics.

REFERENCES

[1] Bin Lu, Wei Huang, C. Lin， et al. A 16QAM Modulation Based 3Gbps Wireless Communication Demonstration System at 0.34 THz Band. 38th International Conference on infrared, Millimeter, and Terahertz Waves (IRMMW-THz), 2013:1-2

[2] C. Wang, C. Lin, Q. Chen, X. Deng, and J. Zhang, 0.14 THz high speed data communication over 1.5 kilometers[J]. Infrared Millim., Terahertz Waves Dig., 2012:.1 – 2.

[3] Y. Yang, S. Zihir, H. Lin,et al. A 155 GHz 20 Gbit/s QPSK Transceiver in 45nm CMOS[C]. IEEE Radio Frequency Integrated Circuits Symposium, 2014:365-368

[4] Mingquan Bao, Jingjing Chen, et al, 14 Gbps On-Off Keying Modulator and Demodulator for D-band Communication, IEEE International Wireless Symposium, Xian, China, March 2014, 1-4

[5] Tae Jin Chung and Won-Hui Lee, 10-Gbit/s Wireless Communication System at 300 GHz, ETRI Journal, 2013, **35**, (3), 386-396

[6] Xiao-Dong Deng, Yihu Li, Wen Wu, A fundamental OOK modulator for 340 GHz communication using 0.13 μm CMOS technology, Microwave Opt Technol Lett 30 (2015), 438-442.

Improvement in 1.2 Hz Total Power Instability of KVN 129 GHz SIS Mixer Receiver

Jung-Won Lee*, Seung-RaeKim ,andSung-Mo Lee
Korea Astronomy and Space Science Institute, Daejeon, South Korea
*Contact: jwl@kasi.re.kr, phone +82-42-865 2071

Abstract— **We present works for improvement on IF total power instability of Korean VLBI network 129 GHz SIS mixer receiver. To reduce the fractional power instability to 10^{-4}in 1 second integration, which has been set up as the specification ofpower instability in case of a receiver having an SIS mixer. In our 129 GHz SIS mixer receiver, we sometimes suffered from much larger power (~10^{-2}) instability at 1.2 Hz that typically hampered normal pointing and continuum observations in single-dish observations. To improve this instability, vibration-isolation mechanism for the cold head of the receiver was applied but it was proved not influential. Indeed, we found that the spectrum of power instability just follows cold LNA'sl/f noise floor well when the LNA was terminated with 50 ohm loadshowing no trace of the discrete 1.2 Hz peak. After confirming 1.2 Hz instability due mainly to gain change of the mixers itself by temperature fluctuations of the cooling system we measured the instability variation by adjusting mixer bias voltages. Further reduction is planned with mixer temperature varied using a PID temperature controller.**

INTRODUCTION

Possible mechanisms of output power instability that have been so far suspected in the technical literatures can be listed as follows:

- temperature modulation of the SIS mixer and low noise amplifier
- acoustic noise pickup by the LNA and the local oscillator causing direct gain change or variation of LO pumping power to the SIS mixer
- microphonic noise pickupat critical bias wires from motor vibrations of a cryocooler
- SIS mixer bias noise and ineffective suppression of the Josephson effect
- LNA gain fluctuations of the cryogenic low noise amplifier and bias noise

From the beginning of the deployment, 129 GHz SIS mixer receivers on three telescopes of Korean VLBI Network have showed IF total power instability with 1.2 Hz period. Because 1.2 Hz period manifests itself that the origin of the instability has certainly connection with the 4 K cryocooler driven at 60 Hz AC power supply, we can, at least, rule out the last two cases as direct origins of this instability.

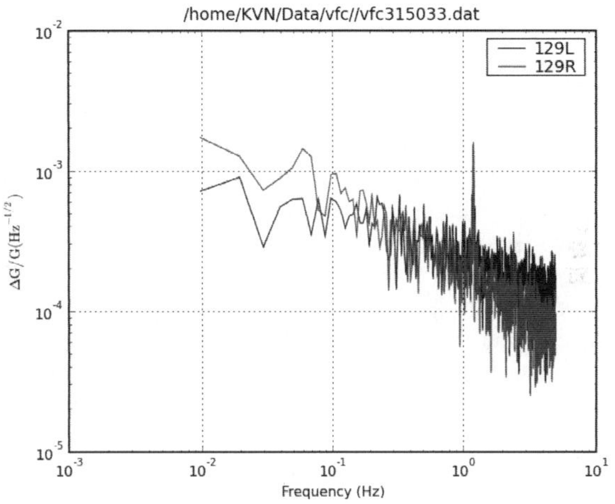

Fig. 1A spectrum of 1.2 Hz fractional power instability typically shown at nominal mixer and LNA bias settings(operation temperature~ 4.7 K)of circular-polarization KVN 129 GHz receiver

Fig. 1 shows a typical spectrum of IF power output normalized to its DC component with a distinct (~10^{-2}) peak at 1.2 Hz as well as 1/f-like noise floor. In general, this kind of instability among receivers are totally uncorrelated in interferometric observations but causes degraded antenna pointing and calibration observationsin single-dish mode. We report how the origin of this instability had been tracedand discuss the direction and implication for future improvement.

VIBRATION ISOLATION

Bias wires of the receiver are wrapped along two G10 supports that separate thermal stages in the receiver cartridge. Wires of constantan span almost one meter and so are intrinsically vulnerable to microphonic noises from vibration. To isolate the vibration from the rotating motor of the coldhead as well as shocks from helium flow, coldhead mount has been changed to one that has shock-absorbing gel mounts and a short cylindrical bellows. Thermal links to the 50 K radiation shield that is secured onto the front lid of the receiver chamber has changed to several flexible OFHC straps. No precise measurement of mechanical accelerations exerted to the thermal stages of the cartridge were taken but it was confirmed that vibration transferred to the receiver chamber was quite reduced. Nonetheless, this vibration isolation was not effective in reducing power instability implying that other mechanisms are under playing.

Fig. 2coldhead mount with shock-absorbing gels and bellows (left), thermal connections of the cold finger inside the receiver chamber (right)

POSSIBLE LNA EFFECTS ON STABILITY

The cryogenic LNAs used in the receiver are CaltechWeinreb group'sCITRYO4-12A using DC power supplies from the original manufacturer. If there would be gain instability due to LNA bias changes, those relevant biases can be servoed. However, in our case, servoing is not readily feasible because the MMIC in the LNA has the drains of all 3 stages connected together.

From communications with Weinreb, gain sensitivity of the LNAto gate voltage change is known to be about 0.3 dB/100 mV i.e. 0.06% / 1mV. Thus for 10^{-2} gain change would need 1.6 mVvoltage change, say, ground difference. Drain voltage sensitivity is designed to be less. Furthermoresince the gate has 11:1 voltage divider in the module 1.6 mV gate voltage would need the gate supply change of 17.6 mV, which is unlikely in our case.

Another possible case may be temperature fluctuation of the LNA. About 1 K of the LNA noise is known due to losses in the input circuit and the MMIC at 15 K. If the instability would originate from this input loss of 1.06, required temperature fluctuation causing 10^{-3} gain change might range as high as 6 K assuming 400 K average input noise. Typical temperature fluctuation at the bare cold finger of 4 K GM cryocooler RDK-415DP is about ± 50 mK.Therefore we need to find other sources in order to explain the measured fractional instability.

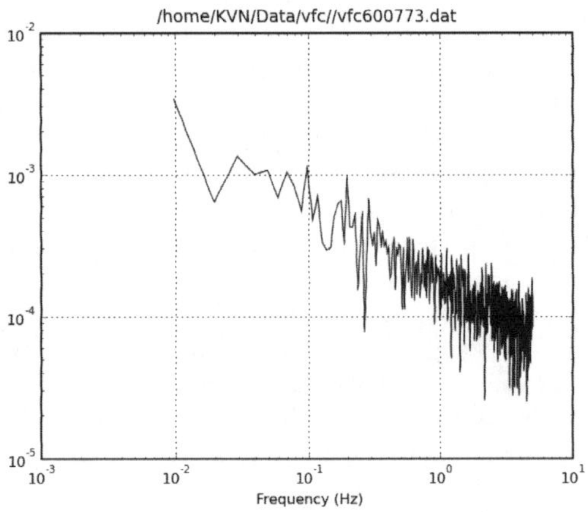

Fig. 3power spectrum of fractional power instabilitymeasured with cold LNA disconnected from the output of the SIS mixers

MIXER GAIN MODULATED WITH TEMPERATURE CHANGE

After some efforts in vain, we measured the spectrum of the output power with only cryogenic LNA terminated with coaxial load disconnecting the SIS mixer from its input. Fig. 3 shows the spectrum in which the 1.2 Hz peak disappears. This finding is implying that the SIS mixer itself causes the power instability. For more detail, correlation between the temperature fluctuation of the mixer block and fluctuation of the output power was investigated. Upper plot in Fig. 4 shows time series data that roughly reflects higher total output power as temperature goes lower and vice versa.

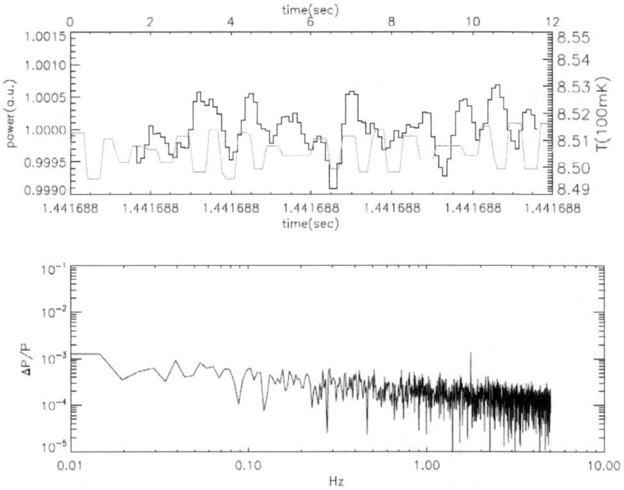

Fig. 4Co-measured power stability(black) and temperature variation of the mixer blocks(green, right axis, in 100 mK unit omitting common 4 K) in the upper plot and power spectrum of the fractional gain(bottom).

Important observation in Fig. 4 is that with only 2 mK p-p temperature variation 10^{-3} power fluctuation results. Fig. 5 shows order-of-magnitude reduction of the peak fractional instability values by adjusting mixer gain with different bias settings than normal.

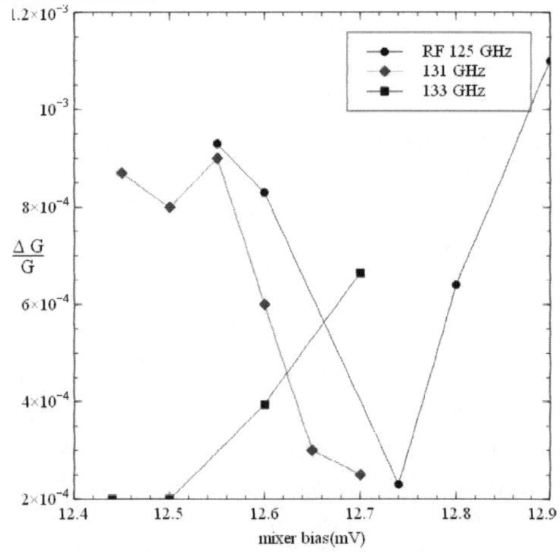

Fig. 5Suppression of 1.2 Hz peak by changing mixer bias voltage to point of lower sensitivity at different LO frequencies (KVN Seoul station)

One previous study[1] reports that given fractional stability of 10^{-4}, typical optimum mixer bias voltage does not guarantee more relaxed temperature modulation, mainly becausemixer's conversion gain varies most sensitively near this bias. But temperature modulation range can be more relaxed at lower-than-optimum bias voltage. The measurements in Fig. 5 show similar trend.

FURTHER REDUCTION OF INSTABILITY

It's not certain that PID controlling will work in stabilizing operation temperature at mixers' location in the receiver cartridge with less than 2 mK p-p resolution. In addition, extensive tests using automated bias sweeping setup is planned during annual maintenance period starting from June.

CONCLUSIONS

We have tracked possible origins of 1.2 Hz total power instability in KVN 129 GHz SIS mixer receiver. Finding the fluctuation due mainly to temperature modulation of the conversion gain of the SIS mixers, we reduced gain change and power instability by lowering bias voltage of the mixers. Test for further reduction is planned using a PID temperature controller.

ACKNOWLEDGMENT

The authors appreciate Sander Weinreb for commenting cryogenic LNA's potential influences on power instability.

REFERENCES

A. Baryshev, R. Hesper, G. Gerlofsma, M. Kroug, &W. Wild, " Influence of Temperature Variations on the Stability of a Submm Wave Receiver," in *Proceedings of the International Symposium on Space Terahertz Technology 2003*, p. 237.

Investigation of the mixing regimes in a superconducting tunnel junction

Anton A. Artanov[1,2,*], KonstantinV. Kalashnikov[1,2],GertdeLange[3], ValeryP. Koshelets[1]
[1]*The Kotel'nikov Institute of Radio Engineering and Electronics, Russian Academy of Science, Moscow, Russia*
[2]*Moscow Institute of Physics and Technology(State University), Dolgoprudny, Russia*
[3]*SRON Netherlands Institute for Space Research, Groningen, the Netherlands*
*Contact: artanov@hitech.cplire.ru

Abstract— **We experimentally investigated different mixing regimes in the tunnel SIS-junction based on three-layer Nb/AlOx/Nb and Nb/AlOx/NbN structures. The SIS mixers were studied in quite unusual modes of operation: in the extremely low frequency range (0.1 – 20 GHz), and as high-harmonic mixers (for the frequencies of about 600 GHz and local oscillator of 20 GHz). The quasiparticle and Josephson mixing regimes have been compared. We demonstrated, that in some applications, such as cryogenic harmonic phase detector, Josephson regime can be more preferable than quasiparticle one due to the possibility to realize larger output signal and better signal-to-noise ratio. This might be caused by partial synchronization of the Josephson current components by powerful local oscillator. Also, we demonstrated the prospects of Josephson mixing regime for up- and down-conversion for the cryogenic multiplexing systems.**

INTRODUCTION

Superconducting mixers based on tunnel junctions are widely used in terahertz receiving systems. In such systems the Josephson effect is usually considered as parasitic, leading to extra noise appearance. That is why critical current is usually suppressed by the external magnetic current. In this work the probability and advantages of Josephson mixing regime in some applications are shown.

CRYOGENIC HARMONIC PHASE DETECTOR

One of the novel applications of the tunnel SIS-junction is cryogenic harmonic phase detector (CHPD) for the broadband phase-locking systems for the cryogenic terahertz generators. Earlier it was demonstrated that functional integration of the harmonic mixer and phase detector in one element allows to significantly increase the synchronization bandwidth of the phase-lock system and improve spectral quality of the radiation [1]. As the spectral quality depends on the amplitude of the CHPD output signal, the optimization of working regimes and mixing regimes investigation is needed. The investigation of the output signal power was conducted in the harmonic mixer regime (HM). Two signals – from flux-flow oscillator(FFO), frequency of about 600 GHz and local oscillator (LO), ~ 20 GHz – were mixed on a tunnel SIS-junction Nb/AlOx/Nb. The intermediate frequency (IF) band was determined by 4-8 GHz amplifiers chain. IF signal was recorded by the spectrum analyzer or power meter. The spectra of the down-converted FFO signal for different mixing regimes are presented in Fig. 1: dotted line is for the fully suppressed by the external magnetic field critical current, dashed line is for the optimal mixing regime without critical current suppression, solid line is for the signal without phase-locking.

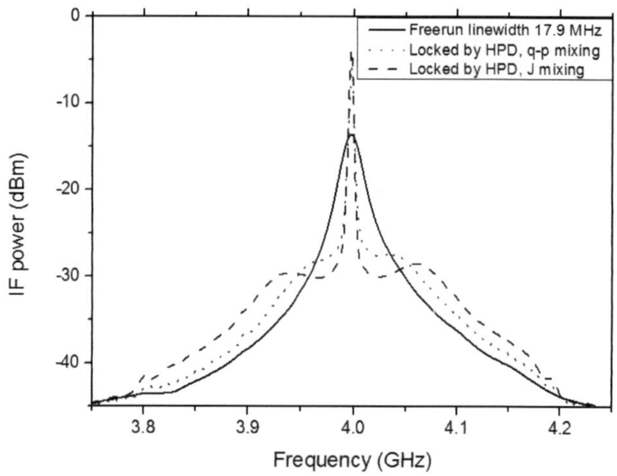

Fig. 1Spectra of FFO radiation at two CHPD mixing regimes:quasiparticle (dotted line) and Josephson (dashed line); and without phase locking (solid)

It is seen from the graphs that using Josephson non-linearity (dashed line) allows to increase the output signal in comparison with only quasiparticle regime (dotted line)due to higher gain; furthermore the signal-to-noise ratio is also better at the first case. Our results demonstrated thatthe CHPD-based phase-lock system is able to phase-lock theFFO with output signal linewidth as wide as 17.9MHz.High effectivity of the Josephson mixing regime in CHPD allows to synchronize up to 83% of the FFO power, while pure quasiparticle regime (atJosephson effect suppression) decreases this value to 70%. So, the Josephson mixing regime is preferable for the effective functioning of the SIS-junction as CHPD..

SIS FREQUENCY CONVERTER

One more promising application of the SIS-based mixers is cryogenic multiplexed readout system for the large arrays of transition edge sensors (TES) [2]. In this system the SIS-mixer works as frequency up- and down-converter for frequencies from hundredsof Megahertz to 1-10 GHz. For the experimental investigation of the SIS-junction as the up-converter we applied to the mixer two signals – 223 MHz and 5 GHz, the IF is 5.223 GHz. The dependencies of the conversion gain on SIS bias voltage at LO power -40 dBm with and without critical current suppression are presented in Fig 2. We also showed the up-conversion of three input signals on a same SIS-device, the IF spectra of the output signal is shown in the Fig. 3.

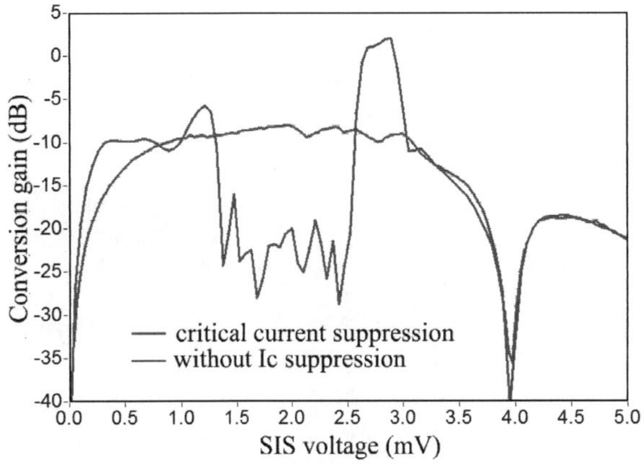

Fig. 2 Dependencies of the conversion gain on SIS bias voltage at LO power -40 dBm with and without critical current suppression.

As seen from the figures 2 and 3, both regimes have their advantages and disadvantages. The Josephson mixing regime provides higher conversion gain;in addition the signal-to-noise ratio at this regime is larger than at quasiparticle one. On the other hand the Josephson mixing has larger noise level, than quasiparticle one. Moreover, it can be realized in comparatively narrow SIS bias voltage and LO power range. Contrariwise, the Josephson regime doesn't require critical current suppression by external magnetic field that would be advantageous for many practical applications.

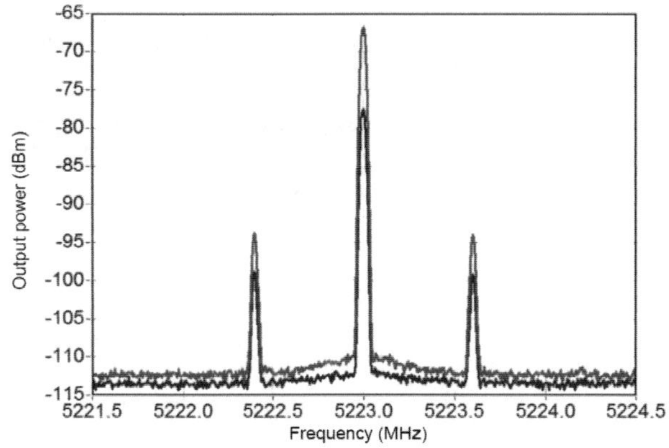

Fig. 3.Up-conversion IF signal spectra (from ~223 MHz to ~5223 MHz)

CONCLUSIONS

So, in this work we have demonstrated that for some practical applications, such as cryogenic harmonic phase detector and cryogenic readout system, the Josephson regime can be more preferable than the quasiparticle one.The Josephson mixing regime of high-harmonic mixer provides 12 dB higher output signal power value, resulting in 4 dB better signal-to-noise ratio (compared to the quasiparticle regime).The Josephson regime allows to achieve larger gain value for low frequency (0.5 – 5 GHz) SIS-mixers.

ACKNOWLEDGMENT

This work was supported by the Ministry of Education and Science of the Russian Federation (No. 14.613.21.0046; ID RFMEFI61315X0046).

REFERENCES

Kalashnikov K.V. et. al.*Harmonic phase detector for phase locking of cryogenic terahertz oscillators*, App. Phys. Lett. 103, 102601 (2013).

de Lange G. *Feasibility of a Frequency-Multiplexed TES Read-Out Using Superconducting Tunnel Junctions*, Journal of Low Temperature Physics. V. 176, 3-4, 408 (2014)

Tucker J.R., Feldman M.J. *Quantum detection at millimeter wavelengths*, Rev. Mod. Phys., v. 4., pp. 1055-1113, 1985

Tien, P. K., Gordon, J. P. *Multiphoton process observed in the interaction of microwave fields with the tunneling between superconductor films*, Physical Review, 129(2), 647, 1963.

Zorin A.B. *Quantum Noise in SIS Mixers*, IEEE Trans. Magn., v. 21., p. 939-942, 1985

Development of a Millimeter Wave Grating Spectrometer for TIME-Pilot

C. T. Li[1]*, C. M. Bradford[2], T. S. Wei[1], Corwin Shiu[3], and A. T. Crites[3]
[1]Institute of Astronomy and Astrophysics, Academia Sinica, Taipei, Taiwan*
[2]Jet Propulsion Laboratory, Pasadena, CA, USA
[3]California Institute of Technology, Pasadena, CA, USA
**Contact: ctli@asiaa.sinica.edu.tw*

***Abstract*—Instantaneous wideband spectral coverage with background-limited sensitivity requires a grating-type spectrometer or filter bank as opposed to a Fourier transform spectroscopy (FTS) or Fabry–Perot. Previously, millimeter wave gratings presented a technological challenge, because conventional echelle grating spectrometers are too large and bulky for cryogenic operation. A unique approach is to use a curved grating in a parallel-plate waveguide to focus and diffract broadband light from a feedhorn to a detector array. This approach markedly reduces the total volume of the spectrometer. The Tomographic Ionized-carbon Mapping Experiment (TIME)-Pilot measures 3-D [CII] fluctuations from 5 < *z* < 9 galaxies by using 32 independent spectrometers. We designed, prototyped, and tested a waveguide grating for TIME-Pilot. Each grating has 190 facets and provides a resolving power in excess of 140 over the full 183–326 GHz range.**

I. Introduction

The Tomographic Ionized-carbon Mapping Experiment (TIME)-Pilot [1] was designed to measure the red-shifted 157.7-µm line of singly ionized carbon [CII] from the Epoch of Reionization (EoR), when the first stars and galaxies formed and ionized the intergalactic medium. For 3-D intensity mapping, TIME-Pilot uses an imaging spectrometer to measure a spatial–spectral data cube, in which the intensity is mapped as a function of the sky position and frequency. The data cube is then analyzed to produce a 3-D power spectrum. Spectral measurements incorporate redshift information that is needed to distinguish faint EoR signals from bright low-red-shift galaxies along the line of sight. [CII] is an energetic emission line in galaxies and a bolometric marker for total star formation activity. [CII] is also well matched to the 1-mm atmospheric windows for z between 5 and 9. As shown in Fig. 1, the instrument is housed in a closed-cycle 4K–1K–300mK cryostat. Thirty-two waveguide grating spectrometers are assembled into two stacks of 16, coupling the same 1-D linear field on the sky through an array of feedhorns illuminated through a polarizing grid. Each grating is similar to that used in Z-Spec [2], but smaller to operate at a lower resolving power. The dispersed light is detected with 2-D arrays of transition edge sensor (TES) bolometers. The spectrometers and detectors are cooled with a dual-stage 250/300-mK refrigerator.

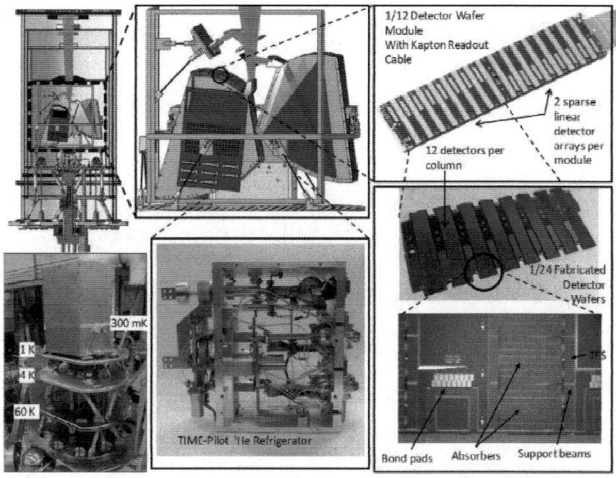

Fig. 1. TIME-Pilot instrument overview.

II. Waveguide Spectrometer Design

The TIME-Pilot spectrometer uses a Waveguide Far-IR Spectrometer (WaFIRS) [3] architecture that employs a curved diffraction grating in a parallel-plate waveguide. Facets on the grating arc both diffract and focus the radiation to locations on a focal curve. Light propagates in the TE_1 mode of parallel-plate waveguide with two degrees of freedom. The grating design begins with a Rowland geometry, as shown in Fig. 2, and then each facet of the grating is positioned such that the total path from the center of the input feed to the facet to the output position changes by exactly one wavelength as the number of facets is incremented at two stigmatic frequencies. Propagation is confined between parallel plates to achieve efficient coupling to the detectors.

For TIME-Pilot, the initial Rowland circle radius is 13.3 cm, and the input position and stigmatic frequency output positions are selected on this circle. The distribution of facets is not centered relative to Rowland's vertex; 35 more lie on the side opposite the input position, because centering the grating does not provide adequate illumination for the upper facets. The blaze angles for the facets vary along the grating arc to accommodate the fact that the input and output angles are varying. They are ranging between 22° and 26°, according to simulations of S polarization blaze efficiencies with the software toolkit PCGrate. The parallel-plate spacing is a compromise between minimizing waveguide propagation loss and avoiding scattering into unwanted waveguide modes. We selected 3 mm for our 183–326-GHz system, which is over-moded by a factor of 4–6. The preliminary grating spectrometer design has 190 facets, with the longest dimension of 31 cm, and provides a resolving power of 140–250. The output arc is approximated by six linear facets so that when the spectrometers are stacked in the two groups of 16, each stack creates six planes, on which the 2-D detector arrays are mounted.

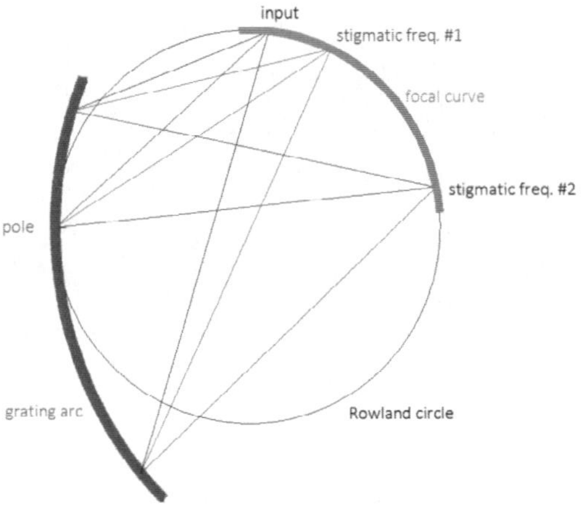

Fig. 2. Rowland geometry.

III. Coupling Structures

The waveguide gratings couple to the incoming radiation through multiple flare-angle (MFA) feedhorns. The MFA feeds are spaced 2.2 f λ apart to balance the desire to couple to the sky with the optimal efficiency per beam, and to pack a large number of horns into the fixed field of view. Light from the feedhorn couples into the grating through a bent split-block waveguide section. The waveguide gratings are single polarization devices, and a polarizing diplexer placed in front of the focal plane feeds two 16-element grating stacks. The waveguide in one stack includes a 90° twist to align the polarization vector of the grating with that of the polarizing grid.

The MFA feeds are smooth-walled, easy-to-machine horns that perform comparably to traditional corrugated feed horns [4]. For TIME-Pilot, we designed a three-section horn that was optimized by varying the positions and magnitudes of these flare angle discontinuities to match the beam widths to an f/3 beam and suppress the sidelobes across the desired band. The simulations were performed with HFSS, a commercial 3-D electromagnetic (EM) simulator, as shown in Fig. 3. The geometry of the optimized horn is given in Table I. The initial waveguide radius R_0 was fixed to 0.56 mm, and the aperture radius R_3 was kept less than 3.9 mm to fit the 8-mm separation between horns. The six parameters of the horn design became variables to be determined. Fig. 4 shows the expected far-field beam patterns for the horn design. Because of the very wide bandwidth, the patterns exhibit better beam circularity and low sidelobes at higher frequencies. The cross polarization remains 30 dB lower across the whole band of interest.

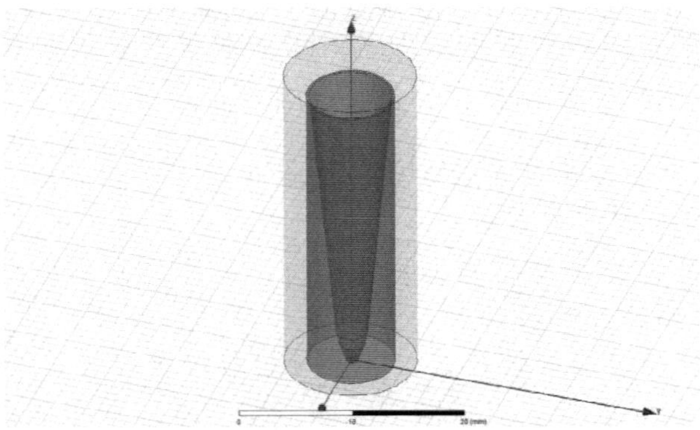

Fig. 3. HFSS model of the three-section MFA horn.

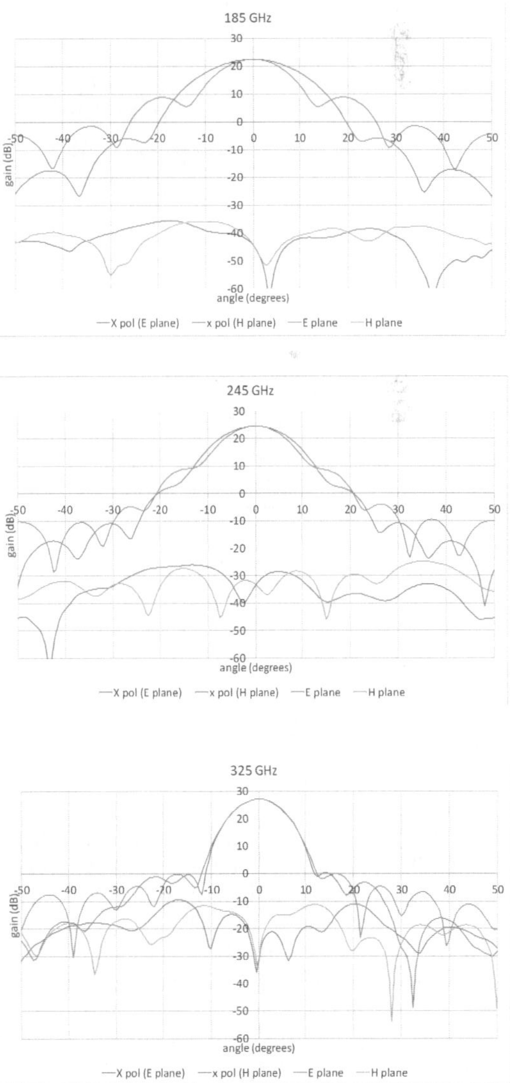

Fig. 4. Beam patterns (E-plane, H-plane, and cross-polarizations) simulated using HFSS at 185, 245, and 325 GHz, respectively.

The MFA feed tapers to a single-mode rectangular waveguide. The rectangular waveguide tapers gradually in height and width and then connects to the spectrometer input horn. Light enters the parallel-plate waveguide through the input horn and illuminates the diffraction grating. Because some portions of the waveguide are over-moded, special care was taken to minimize excitations of the higher order modes. A single-mode section of the waveguide ensures that only one mode propagates, and the higher order modes are cut off within the design bandwidth.

Fig. 5. [Left] 3-D EM simulation model of the MFA feed and waveguide twist, bend, and taper. [Right] Split block design.

TABLE I
GEOMETRICAL PARAMETERS FOR THE THREE-SECTION MFA FEED DESIGN

Parameter	Length (mm)
R_0	0.56
R_1	1.173
R_2	1.511
R_3	3.824
L_1	1.750
L_2	1.236
L_3	25.566

IV. SPECTROMETER TESTING

The prototype spectrometer has two gratings in a ministack and is a simple machined, bolted aluminum assembly, as shown in Fig. 6. The spectrometer requires global tolerances of λ/10 (approximately 100 μm) and low surface roughness on the waveguide plates. The 190 facets on each grating were cut using a wire electrical discharge machine. Operation of the spectrometer requires that the correct spacing be maintained across the entire region to ensure that the same dispersion relation holds throughout.

Preliminary testing of the spectrometer was performed at room temperature. The spectral profile was measured using a sweep-able coherent source and a diode detector in a single-mode waveguide. Several spectra are shown in Fig. 7. The measured profiles represent the convolution of the intrinsic resolving power of the spectrometer with the width of the detector waveguide feed. The spectra show approximately 25%–45% transmission, depending on the width of the output feeds. After deconvolution from the output waveguide width, the measured resolving power is comparable to that is predicted. The warm measurements are affected by some standing waves or multiple reflections because the detector and the source are not well matched. Based on the known sources of loss listed in Table II, the spectra show that the efficiency measurements match expectations within 10%.

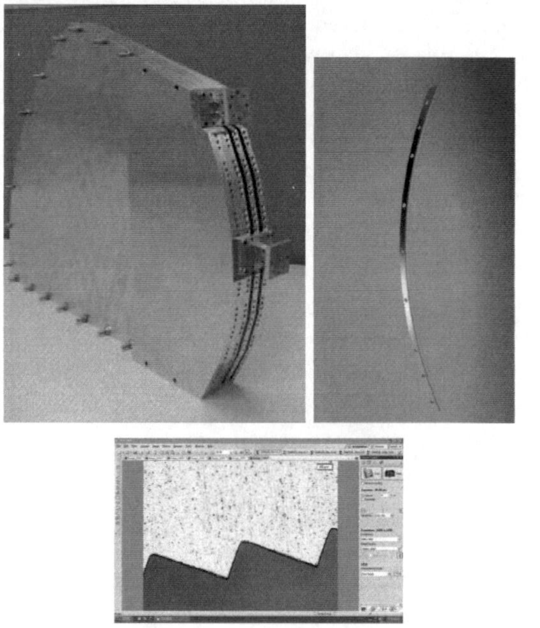

Fig. 6. Two-channel grating spectrometer stack with input and output feeds connected. The 190 facets on the grating were wire-cut. A close-up view of the grating facets is also shown.

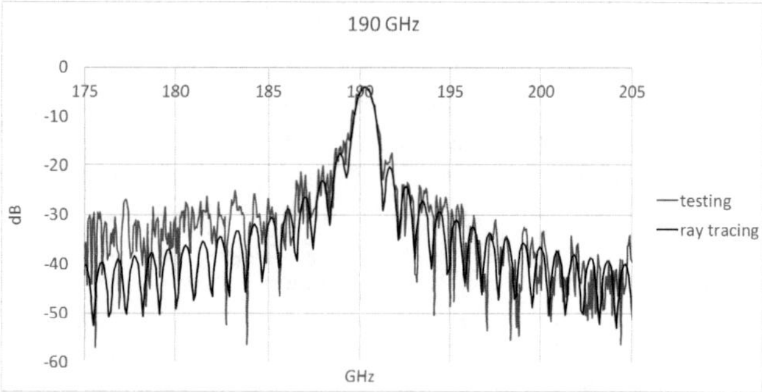

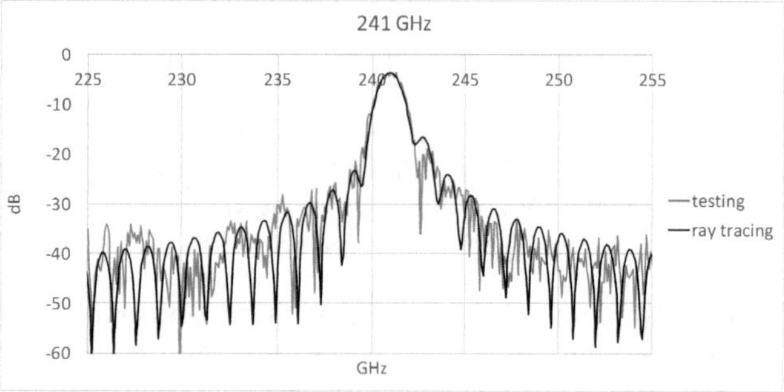

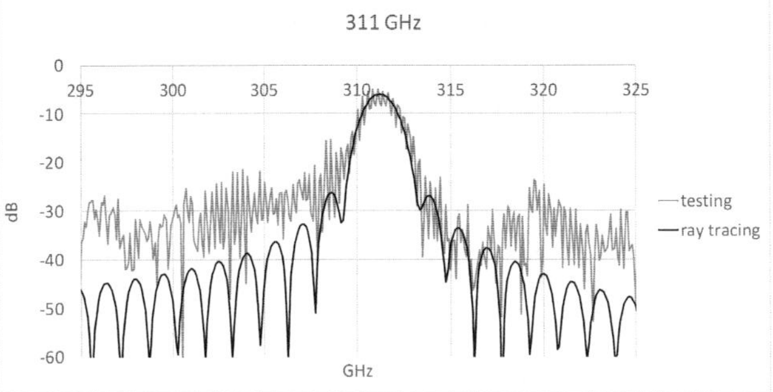

Fig. 7. Spectrometer room temperature test results and responses calculated through ray tracing around 190 GHz, 241 GHz, and 311 GHz, respectively.

TABLE II
SPECTROMETER EFFICIENCY ESTIMATES

Frequency (GHz)	**190.3**	**241**	**311.3**
Grating illumination	0.76	0.85	0.89
Blaze efficiency	0.92	0.97	0.88
WG propagation loss	0.97	0.98	0.99
Output coupling	0.55	0.57	0.33
Product	**0.37**	**0.46**	**0.26**
Measure	0.40	0.45	0.25
Measure/expected	1.08	0.98	0.96

V. CONCLUSION

We present the design and testing of a millimeter-wave grating spectrometer based on the waveguide grating spectrometer architecture WaFIRS for TIME-Pilot. Room temperature testing shows that the efficiency measurements match expectations within 10% with a resolving power above 140. Integration and testing with TES bolometers will follow to verify the performance at cryogenic temperatures.

ACKNOWLEDGMENT

We thank Steve Hailey-Dunsheath and J. C. Cheng for their help in testing the spectrometer.

REFERENCES

A. T. Crites et al, "The TIME-Pilot Intensity Mapping Experiment," Proc. SPIE 9153, 2014.

C. M. Bradford et al, "Z-Spec: a broadband millimeter-wave grating spectrometer – design, construction, and first cryogenic measurements," Proc. SPIE 5498, 2004.

C. M. Bradford et al, "WaFIRS: a waveguide far-IR spectrometer enabling spectroscopy of high-z galaxies in the far-IR and submillimeter," Proc. SPIE 4850, 2003.

J. Leech et al, "Multiple flare-angle horn feeds for sub-mm astronomy and cosmic microwave background experiments," A&A 532, 2011.

TerahertzImaging Progress atCapital Normal University

Guozhong Zhao[1*], Yanchun Shen[1,2], Jia Wang[1], and Jiayi Yu[1]

[1]Department of Physics, Capital Normal University,Beijing Key Lab of Terahertz Spectroscopy and Imaging, Key Lab of Terahertz Optoelectronics, Ministry of Education, Beijing 100048, China
[2]Tang Shan College, Tangshan,Hebei 063000, China
* Contact: guozhong-zhao@126.com

The terahertz imaging at Capital Normal University in Beijing is presented. Our works on Terahertz Imaging include the active and passive imaging. For the active terahertz imaging, the pulse and continue wave terahertz imaging are studied respectively. The active terahertz pulse imaging is based on the terahertz time-domain spectroscopy with the probe-beam-expanded femtosecond pulse laser and an infrared CCD detection. The active terahertz continuous wave imaging is based on a CO_2-laser-pumped terahertz coherent source and a NEC terahertz camera. For the passive terahertz imaging, the low frequency of terahertz radiometers are used to detect the beam-scanned terahertz signal by the point-to-point method. The related components and methods are developed and used for the improvement of the imaging speed and the resolution of images.

The polarization terahertz imaging is studied based on the active continuous wave imaging technology. The higher resolution of terahertz imaging is achieved at 3.1 THz of operating frequency. The polarization imaging provide more information on the measured targets. However, the imaging distance of the high frequency, such as 3.1 THz, of terahertz imaging is limited due to the vapor absorption. The focal plane terahertz imaging is developed to obtain more frequency domain of spectral information. The focal plane imaging can be realized as a quasi-near field imaging so that it can achieve at a higher resolution. But the visual field of focal plane imaging is limited due to the size of electro-optic crystal. The passive terahertz imaging is developed for the longer imaging distance and the larger imaging visual field. The sensitivity of terahertz radiometer is a key factor for the contract and resolution of passive terahertz imaging.

In summary, the active and passive terahertz imaging are investigated at Capital Normal University. The advantage and disadvantage of different terahertz imaging technology can be seen and compared. Of course, they depend on the different requirements of application, too. Some further investigations of terahertz imaging are necessary.

These works are supported by the National Natural Foundation of China, Beijing Natural Foundation of Science, and the National High-Tech Development program.

Development of a 71-116GHz RF module for the EMIR receiver upgrade

A.L. Fontana,D. Maier, Q. Moutote, J. Reverdy, G. Perrin, M. Parioleau and C. Boucher.
Institut de RadioAstronomie Millimétrique, St. Martin d'Hères, France
Contact: fontana@iram.fr, phone +33 76 89 49 17

Abstract— **The Eight MIxers Receiver (EMIR) was installed at IRAM 30-m Pico Veleta Telescope (Andalusia, Spain) in 2009. It is composed of four cryogenic dual polarization side band separating SIS receiver modules covering frequencies from 71 GHz to 365 GHz (and delivering 4-12 GHz IF channels). Outside the EMIR cryostat, some switchable dichroic filters can be inserted in the optical path to allow dual band observations of the same point of the sky.**

In December 2015, the 3 mm band, which covered initially 84-116 GHz, was upgraded to cover the full 3mm atmospheric transmission window which can be observed at Pico Veleta, i.e. the 71-116 GHz band. Excellent performances are obtained with this new 3mm receiver, which covers ~ 50% of bandwidth with noise temperature of almost 30 K in the full band. This work also demonstrates the possibility of covering two ALMA bands simultaneously (ALMA band 2 + band 3 ~ 68-116 GHz) with a single receiver.

INTRODUCTION

The Eight MIxersReceiver (EMIR) has been installed at IRAM 30-m Pico Veleta Telescope in 2009. It has been initially developed to make single band and dual band observations for frequencies covering from ~84GHz to 365GHz. Since it installation, several upgrades have been performed on this receiver to improve it performances. The later one, which has been made at the end of 2015, has consisted in enlarging the frequency coverage of EMIR in the 3mm atmospheric window (shown in Fig. 1), by replacing the 84-116 GHz RF and optical module by a new one allowing to cover the 71-116 GHz band.

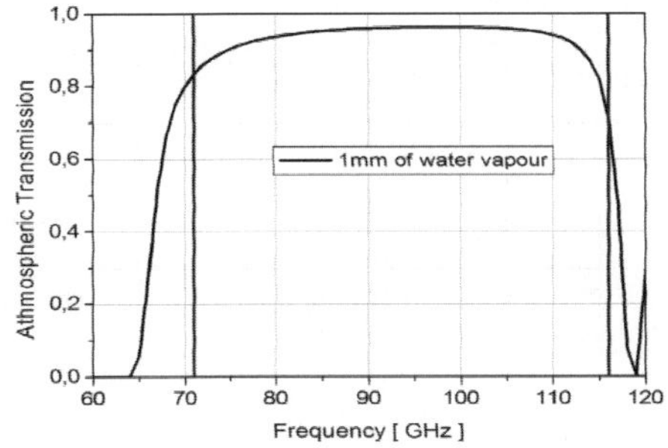

Fig.1: Transmission of the atmosphere in the 3mm band.

EMIR RECEIVER SPECIFICATIONS

The EMIR [1] receiver (see Fig. 2)is composed of four cryogenic dual polarization SIS modules covering the 3mm band, the 2mm band, the 1.3mm band and the 0.8mm band. The specifications of the different modules are the following:

-Band 1 (3 mm): 71-116 GHz (since dec-2015; 84-116 GHz before)
-Band 2 (2 mm): 127-179 GHz
-Band 3 (1.3 mm): 200-276 GHz
-Band 4 (0.8 mm): 276-365 GHz
-Technology: side band separating SIS receivers, delivering 4 x 4-12 GHz IF channels per receiver module

Fig.2: External (left) and internal (right) views of the EMIR receiver cryostat.

At room temperature, a set of three dichroic filters, provided by QMC [2], are mounted on a translating frame in front of the cryostat windows. These filterscan be inserted in the optical path to allow simultaneous observations of the same point of the skywith two different frequency bands. The available band combinations are:

- Band 1 (3 mm) + band 2 (2 mm)
- Band 1 (3 mm) + band 3 (1.3 mm)
- Band 2 (2 mm) + band 4 (0.8 mm)

OPTICAL MODULE DESIGN

The EMIR upgraded optical module is presented in Fig 3. It is composed of a pair of focusing mirrors (an elliptical mirror plus a parabolic mirror) which ensure a proper frequency independent illumination of the sub-reflector. Inside the cryostat, those mirrors are cooled at 15 K. Some epoxy supports allow to thermally disconnect those mirrors form the other parts of the module (SIS mixers, feed horn, Ortho Mode Transducer, cryogenic IF isolators and low noise amplifiers ...) which operate at 4 K.

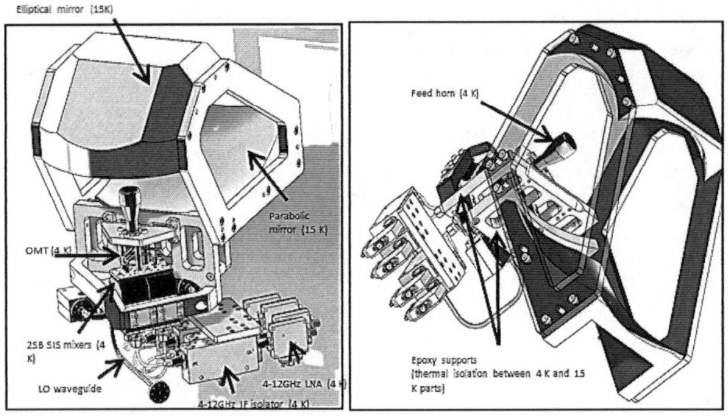

Fig 3: Mechanical 3D drawings of the 71-116GHz optics + RF module.

RF MODULE DESIGN

The upgraded 3mm RF module is composed of:

- a corrugated feed horn with circular waveguide output coupled to an Ortho Mode Transducer (OMT) which diplexes the two linear orthogonal polarizations of the receiver;
- two side band separating SIS mixers [3] with integrated IF couplers delivering four 4-12 GHz IF channels;
- A local oscillator waveguide splitter that distributes the LO power to the two polarizations;
- 4 cryogenic isolators and 4 cryogenic low noise amplifiers connected to the SIS mixers outputs

Some pictures of the assembled RF module are presented in Fig. 4.

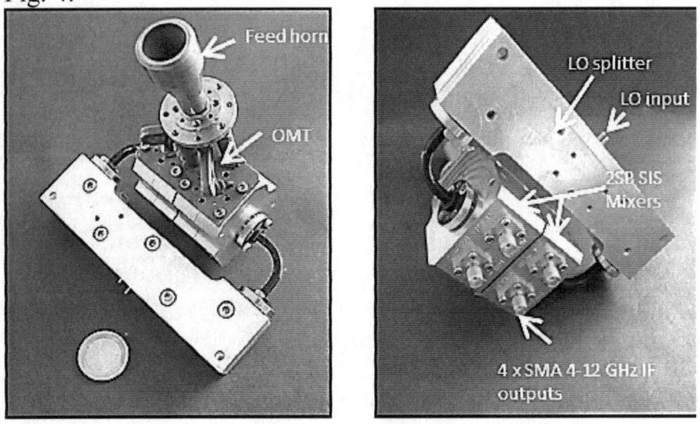

Fig.4: Views of the assembled 71-116GHz RF module.

INSTALLATION AT 30-M TELESCOPE, AND TESTS

Before it installation at the IRAM 30-m, the 71-116 GHz module has been first tested into a dedicated test cryostat. The integrated (in the 4-12GHz IF band) noise temperatures, the noise temperatures in the IF band and the image band rejection of the module have been characterized.Fig.5 shows the excellent noise performances reached by this module (~30K of noise temperature in the full 71-116GHz band)

The optical performances (co and cross-polarization patterns, co-alignment on the sky between the two polarizations) have also been measured.

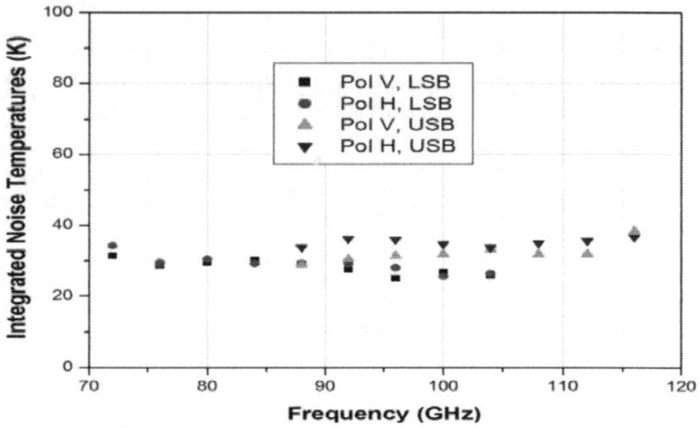

Fig.5: Integrated noise temperatures of the 71-116GHz RF module (including the OMT).

After these laboratory tests, the "old" 84-116GHz module has been replaced by this new module at the IRAM 30-m, into the EMIR cryostat. A picture of the 71-116 GHz module once installed into the receiver is presented Fig. 6.

Fig.6 : Integration of the 3mmupgraded module into the EMIR receiver.

CONCLUSIONS

The 71-116GHz EMIR module was successfullyinstalled at Pico Veleta in December 2015. State of the art performances are obtained in the whole very wide frequency range (~ 50% of bandwidth) of this receiver, which demonstrates the possibility of covering two ALMA bands in the same time (ALMA band 2 + band 3) with a single receiver. This new EMIR 3mm module is now widely used by astronomers: in particular, the astronomical observations which are made in the new part of the band covered (71-84GHz) represent now ~ 10% of the observing time at Pico Veleta.

ACKNOWLEDGMENT

The authors thank the IRAM mechanical group for the mechanical designs and manufacture of the RF and optical components and the IRAM SIS group for the fabrication of the SIS junctions.

REFERENCES

[1] M. Carter, B. Lazareff et al., "The EMIR Multi-band mm-Wave Receiver for the IRAM 30m Telescope", Astronomy &Astrophysics, vol. 538, pp. 13, 2012

[2] QMC Instruments

[3] D. Maier et al., "Fully integrated sideband-separating SIS mixers for the NOEMA receivers"in Proc. 25thISSTT, Moscow, Russia, 2014

Superconducting Local Oscillators; Development and Optimization.

P.N. Dmitriev[1], L.V. Filippenko[1], K.V. Kalashnikov[1,2], M.E. Paramonov[1], K.I. Rudakov[1,2], and V.P. Koshelets[1,*]
[1]*Kotel'nikov Institute of Radio Engineering and Electronics, Moscow, 125009, Russia*
[2]*Moscow Institute for Physics and Technology (State University), Dolgoprudny, Russia*
*Contact: valery@hitech.cplire.ru, phone +7-495-6293418

Abstract— **Different types of the superconductor local oscillators were considered for integration with a SIS-mixer to build fully superconducting integrated receivers (SIR). The Josephson Flux Flow Oscillators (FFO) based on Nb-AlOx-Nb and Nb-AlN-NbN junctions have proven to be the most developed for such integration. The continuous frequency tuning of the FFO over the 250 - 750 GHz frequency range and the possibility of FFO phase stabilization have been achieved. The output power of the FFO is sufficient to pump integrated on the same chip SIS mixer in a wide frequency range; the FFO power can be electronically adjusted. The FFO free-running linewidth has been measured between 0.3 and 5 MHz; resulting in the spectral ratio of the phase-locked FFO from 99 to 70% over the whole frequency range. The possibility of reaching the phase noise of the order of - 90 dBc at an offset from a carrier frequency of more than 100 kHz has been demonstrated experimentally.To improve further FFO parameters and to extend its frequency range a number of new FFO designs were developed and investigated. The goal is to simplify the FFO operation at lower frequencies (through Fiske steps suppression at frequencies below Josephson self-coupling boundary) as well as to extend the FFO operation frequency beyond 1 THz.In this report an overview ondevelopmentof superconducting integrated THz local oscillators is presented.**

INTRODUCTION

Josephson junctions have been considered as natural terahertz oscillators for more than half a century, ever since Josephson discovered the effects named after him [1], [2].Since that time, many quite different types of Josephson oscillators have been proposed and studied [3] – [13], but only a few of them were developed at level suitable for real applications. Let us consider one of the most attractive applications - the direct integration of a Josephson Local Oscillator (JLO) with the most sensitive heterodyne SIS mixer. There are a number of important requirements of the JLO's properties to make it suitable for application in the phase-locked Superconducting Integrated Receiver (SIR). The continuous frequency tuning of the JLO over a wide frequency range (usually more than 100 GHz) and a possibility of the JLO's phase stabilization at any frequency in the operation range are required for most applications. The output power of the JLO should be sufficient to pump the matched SIS mixer within a wide frequency range and it can be electronically adjusted. Obviously, the JLO should emit enough power to pump an SIS mixer (of about 1 μW), taking into account a specially designed mismatch of about 5–7 dB between the JLO and the SIS mixer, which should be introduced to avoid leakage of the input signal to the LO path. It is a challenge to realize the ultimate performance of the separate superconducting elements after their integration into a single-chip device. Another very important issue is the linewidth of the JLO. Even for wideband room-temperature PLL systems, the effective regulation bandwidth is limited by the length of the cables in the loop (about 10 MHz for a typical loop length of two meters). This means that the free-running JLO linewidth has to be well below 10 MHz to ensure stable JLO phase locking with a reasonably good spectral ratio (SR) — the ratio between the carrier and the total power emitted by the JLO.

NB-BASED FLUX-FLOW OSCILLATORS

The Josephson Flux Flow Oscillators (FFO) [14] – [19] based on Nb-AlOx-Nb and Nb-AlN-NbN junctions have proven [20] – [25] to be the most developed superconducting local oscillator for integration with an SIS mixer in a single-chip submm-wave Superconducting Integrated Receiver [26] – [29]. The FFO is a long Josephson tunnel junction of the overlapped geometry in which an applied DC magnetic field and a DC bias current, I_B, drive a unidirectional flow of fluxons, each containing one magnetic flux quantum, $\Phi_0 = h/2e \approx 2*10^{-15}$ Wb. Symbol h represents Planck's constant and e is the elementary charge. An integrated control line with the current $I_{CL}$is used to generate the DC magnetic field that is applied to the FFO. According to the Josephson relation, the junction oscillates with a frequency $f = (1/\Phi_0)*V$ (about 483.6 GHz/mV) if it is biased at voltage V. The fluxons repel each other and form a chain that moves along the junction. The velocity and density of the fluxon chain, and thus the power and frequency of the submm-wave signal emitted from the exit end of the junction due to the collision with the boundary, may be adjusted independently by the appropriate settings of I_B and I_{CL}. The FFO differs from the other members of the Josephson oscillator family by the need for these two control currents, which in turn provides the possibility of an independent frequency and power tuning.

We experimentally investigated a large number of the FFO designs. The length, L, and the width, W, of the FFO used in our study were 300–400 μm and 4–28 μm, respectively. The value of the critical current density, J_C, was in the range of 4–8 kA/cm^2, giving a Josephson penetration depth of $\lambda_J\sim$ 6–4 μm. The corresponding value of the specific resistance was $Rn*L*W$ is~ 50–25 Ohm*μm^2. For the numerical calculations we used a typical value of the London penetration depth, $\lambda_L\approx$ 90 nm for all-Nb junctions, and a junction-specific capacitance $Cs \approx 0.08$ pF/μm^2. The active area of the FFO (i.e., the AlO_x or the AlN tunnel barrier) is usually formed as a long window in the relatively thick (200–250 nm) SiO_2 insulation layer, sandwiched between the two superconducting films (the base and wiring electrodes). The so-called "idle" region consists of the thick SiO_2 layer adjacent to the junction (on both sides of the tunnel region) between the overlapping electrodes. It forms a transmission line parallel to the FFO. The width of the idle region (W_I = 2–14 μm) is comparable to the junction width. The idle region must be taken into account when designing an FFO with the desired properties. In our design, it is practical to use the flat-bottomed electrode of the FFO as a control line in which the current I_{CL} produces the magnetic field, which is mainly applied perpendicular to the long side of the junction.

Previously, the Nb-AlOx-Nb or Nb-AlN-Nbtrilayers were successfully used for the FFO's fabrication. Traditional all-Nb circuits are constantly being optimized but there seems to be a limit for linewidth optimizations at certain boundary frequencies due to the Josephson self-coupling (JSC) effect [21], as well as a high frequency limit, imposed by the Nb gap frequency (~700 GHz). This is the reason to develop novel types of junctions based on materials other than Nb. We reported on the development of the high-quality Nb-AlN-NbN junction-production technology [30]. The implementation of an AlN tunnel barrier in combination with an NbN top superconducting electrode provides a significant improvement in the quality of the SIS junction. The gap voltage of the junction Vg = 3.7 mV. From this value, and the gap voltage of the Nb film Δ_{Nb}/e = 1.4 mV, we have estimated the gap voltage of our NbN film as Δ_{NbN}/e = 2.3 mV [25]. The use of Nb for the top "wiring" layer is preferable due to smaller losses of Nb when compared to NbN below 720 GHz. Furthermore, the matching structures developed for the all-Nb SIRs can be used directly for the fabrication of receivers with Nb-AlN-NbN junctions. The general behavior of the new devices is similar to that of the all-Nb ones; even the control currents, necessary to provide magnetic bias for the FFO, were nearly the same for the FFOs of similar designs.

A family of the Nb-AlN-NbN FFO IVCs, measured at different magnetic fields produced by the integrated control line, is presented in Fig. 1 (L = 300 μm, W = 14 μm, W_I = 10 μm). A single SIS junction with an inductive tuning circuit was employed as a harmonic mixer (HM) for the linewidth measurements. The tuning and matching circuits were designed to provide "uniform" coupling in the frequency range of 400–700 GHz. Measured values of the HM current induced by the FFO oscillations (HM pumping) are shown in Fig. 1 by the color scale. The HM pumping for each FFO bias point was measured at a constant HM bias voltage of 3 mV (pumping is normalized on the current jump at the gap voltage, I_g = 140 μA). From Fig. 1 one can see that an FFO can provide a large enough power over the wide frequency range, which is limited at higher frequencies only by the Nb superconducting gap in transmission line electrodes (base and wiring layers) and below 400 GHz by the design of the matching circuits.

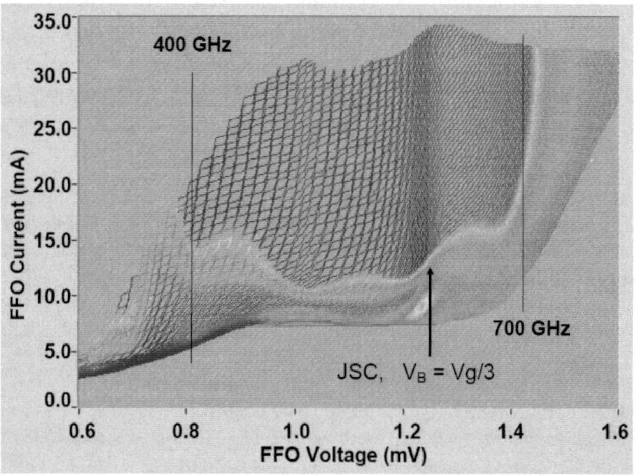

Fig.1IVCs of the Nb-AlN-NbN FFO, measured at different magnetic fields produced by the integrated control line. The color scale shows the level of the DC current's rise at the HM induced by the FFO. The red area marks the region of the FFO's parameters where the HM current induced by the FFO exceeds 25% of the Ig. This level is well above the optimal value for an SIS-mixer operation.

The feature at approximately 600 GHz where the curves get denser is a JSC (Josephson Self-Coupling) boundary voltage. It was initially observed for all-Nb FFOs [21]. The JSC effect is the absorption of the FFO-emitted radiation by the quasi-particles in the cavity of the long junction. It considerably modifies the FFO's properties at the voltages $V \approx V_{JSC} = 1/3*Vg$ (V_{JSC}corresponds to 620 GHz for the Nb-AlN-NbN FFO). Just above this voltage, the differential resistance increases considerably; that results in an FFO-linewidth broadening just above this point. This, in turn, makes it difficult or impossible to phase-lock the FFO in that region. For an Nb-AlOx-Nb FFO, the transition corresponding to $V_{JSC} = Vg/3$ occurs around 450 GHz. Therefore, by using the Nb-AlN-NbN FFOs we can cover the frequency gap from 450 to 550 GHz that is imposed by the gap value of all-Nb junctions. The feature in Fig. 1 around 1 mV is very likely due to a singularity in the difference between the superconducting gaps $\Delta_{NbN} - \Delta_{Nb}$.

Continuous frequency tuning at frequencies below 600 GHz for the Nb-AlN-NbN FFOs of moderate length is possible, although the damping is not sufficient to completely suppress the Fiske resonant structure at frequencies below Vg/3. For short junctions with a small α (wave attenuation factor), the distance between the steps in this resonant regime can be as large, so that it is only possible to tune the FFO within a certain set of frequencies. For a 300–400 μm long Nb-AlN-NbN junction, this is not the case — the quality factor of the resonator formed by a long Nb-AlN-NbN Josephson junction is not so high at frequencies > 350 GHz. Therefore, the resonance steps are slanting and the distance between them is not so large (see Fig. 1). This allows us to set any voltage (and any frequency) below V_{JSC}, but for each voltage, only a certain set of currents should be used. Therefore, in this case, we have the regions of forbidden bias-current values, which are specific for each voltage below V_{JSC}, instead of the forbidden voltage regions for the Fiske regime in Nb-AlOx-Nb FFO [25]. Special algorithms have been developed for automatic working-point selection in flight.

The typical current-voltage characteristics (IVCs) of an Nb-AlN-NbN SIS junction of an area approximately 1 μm^2 is given in Fig. 2, which represents both the unpumped IVC (the solid line) and the IVC when pumped by an Nb-AlN-NbN FFO at different frequencies (dotted lines). One can see that the FFO provides more than enough power for the mixer pumping. In this experiment, we used the test circuits with low-loss matching circuits tuned between 400 and 700 GHz. Even with the specially introduced 5 dB FFO/SIS mismatch (required for the SIR operation), the FFO delivered enough power for the SIS mixer's operation in the TELIS frequency range of 400–700 GHz. An important issue for the SIR's operation is a possibility to tune the FFO's power, while keeping the FFO frequency constant. This is demonstrated in Fig. 3, where the IVCs of an SIS mixer pumped at the FFO frequency of 500 GHz are shown, while they were being pumped at different FFO bias currents (different powers). Our measurements demonstrated [24], [28] that the FFO power can be adjusted in the range of 0–15 dB while keeping the same frequency, by proper adjustment of the FFO control line current.

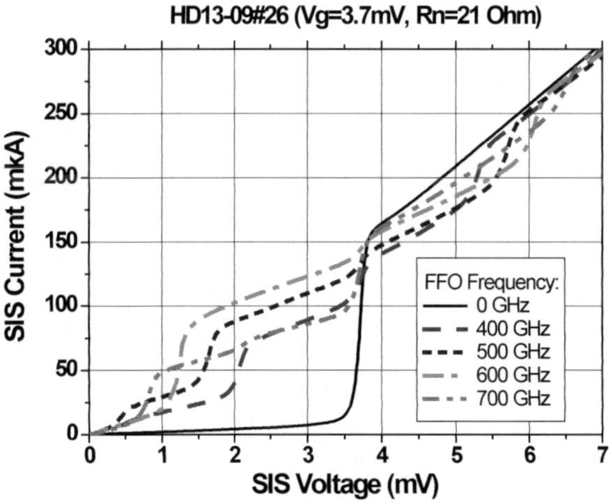

Fig. 2. The IVCs of the SIS mixer: unpumped = solid curve; pumped at different frequencies = dashed and dotted lines (color online).

LINEWIDTH OF THE FFO AND ITS PHASE-LOCKING

The FFO linewidth (LW) has been measured in a wide frequency range from 300 GHz up to 750 GHz by using a specially developed experimental technique [20] – [24]. A specially designed integrated circuit incorporates the FFO junction, the SIS harmonic mixer and the microwave matching circuits. Both junctions are fabricated from the same Nb/AlN/NbN or Nb/AlOx/Nbtrilayer. A block diagram of the set-up for the linewidth measurements is described in [22]. The FFO signal is fed to the harmonic mixer (a SIS mixer operated in Josephson or quasiparticle mode) together with a 17–20 GHz reference signal from a stable synthesizer. The required power level depends on the parameters of the HM; it is about of 1 μW for a typical junction area of 1 μm^2. The intermediate frequency (IF) mixer product ($f_{IF} = \pm (f_{FFO} - n \cdot f_{SYN})$ at ~ 400 MHz is first boosted by a cooled HEMT amplifier (T_n5 K, gain = 30 dB) and then by a high-gain room-temperature amplifier.

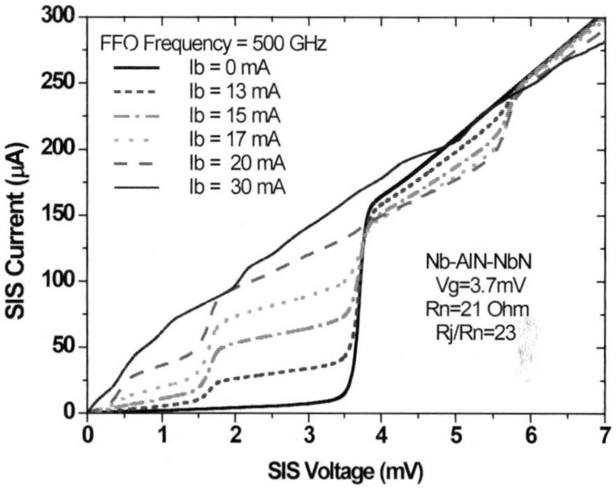

Fig. 3. The IVCs of the SIS mixer: unpumped = black solid curve; pumped at different FFO bias currents (different powers) = lines with symbols; FFO frequency = 500 GHz (color online).

In order to accurately measure the FFO line shape, the IF signal must be time-averaged by the spectrum analyzer. To remove low-frequency drift and interference from the bias supplies, temperature drift, etc., we used a narrow band (< 10 kHz) Frequency Discriminator (FD) system with a relatively low loop gain for the frequency locking of the FFO. With the FD narrow-band feedback system that stabilizes the mean frequency of the FFO (but which does not affect FFO's line shape), we can accurately measure the free-running FFO linewidth, which is determined by the much faster internal ("natural") fluctuations (see Fig. 4). The measured data are symmetrized relative to the center's frequency; these data are shown by diamonds. The profile of the FFO line recorded when biased at the steep Fiske step (FS), where the differential resistance is extremely small, can be different from the one measured on the smooth Flux Flow step. Theoretically [58], the shape is Lorentzian for wide-band fluctuations, while for narrow-band interference, at frequencies smaller than the autonomous FFO linewidth δf_{AUT}, the profile will be Gaussian; the theoretical curves are also shown in Fig. 4 for comparison. The theoretical lines providing the best fit near the peak are shown by the solid line and the dashed line for the Lorentzian and Gaussian profiles, respectively. The coincidence between the calculated curve and the symmetrized experimental data is excellent, and actually better than 5% in the emitted power, if a minor amplifier's nonlinearity of about 0.4 dB is taken into account.

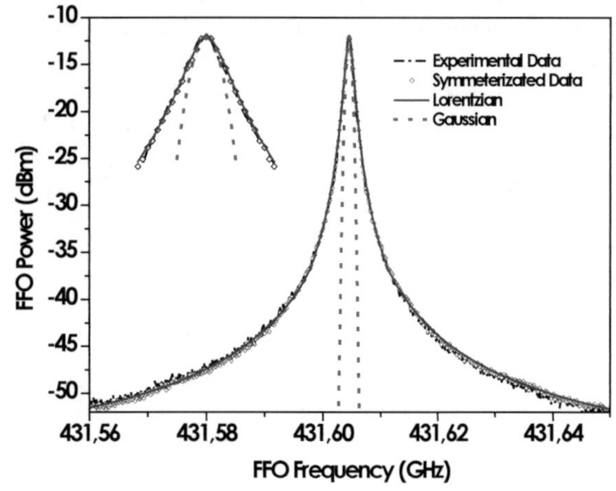

Fig. 4. The FFO spectrum measured when biased on the Fiske step (V_{FFO} = 893 μV, R_d = 0.0033 Ω, R_d^{CL}= 0.00422 Ω, δf_{AUT} = 1.2 MHz) is represented by the dash-dotted line. The symmetrized experimental data are shown by diamonds. The fitted theoretical Lorentzian and Gaussian profiles are shown by solid and dotted lines, respectively. The inset shows a close-up view of the central peak with the frequency axis multiplied five times [80].

The resulting IF signal is also supplied to the Phase-Locking Loop (PLL) system. The phase-difference signal of the PLL is fed to the FFO control-line current [15, 16, 75, 78–81]. Wideband operation of the PLL (10–15 MHz full width) is obtained by minimizing the cable loop's length. A part of the IF signal is delivered to the spectrum analyzer via a power splitter (see Fig. 5, 6). All instruments are synchronized to the harmonics of a common 10 MHz reference oscillator. Dependencies of the free-running FFO linewidth and the Spectral Ratio (SR) for the phase-locked FFO on frequency for two different FFO technologies (Nb-Ox-Nb and Nb-AlN-NbN) are presented in Fig. 7. One can see that SR > 70% can be realized for Nb-AlN-NbN FFO in the range of 250–750 GHz.

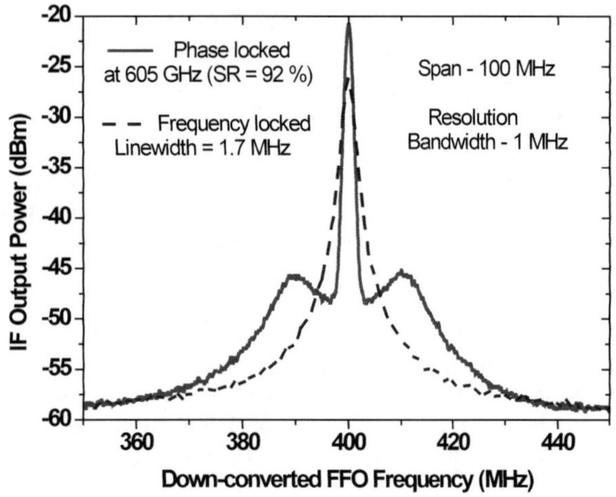

Fig. 5. Spectra of the Nb-AlN-NbN FFO operating at 605 GHz (blue dashed line = frequency locked by FD; red solid line = phase-locked). Linewidth = 1.7 MHz; Spectral Ratio = 92 % (color online).

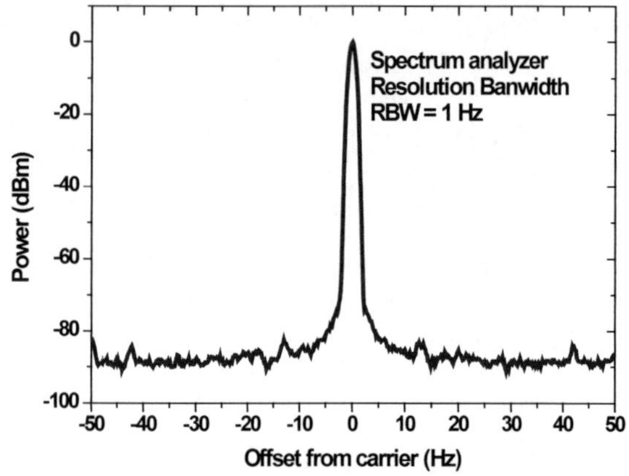

Fig. 6. Spectra of the phase-locked Nb-AlN-NbN FFO operating at 605 GHz. Span = 100 Hz, RBW = 1 Hz, signal-to-noise ratio = 87 dB as measured in a bandwidth of 1 Hz.

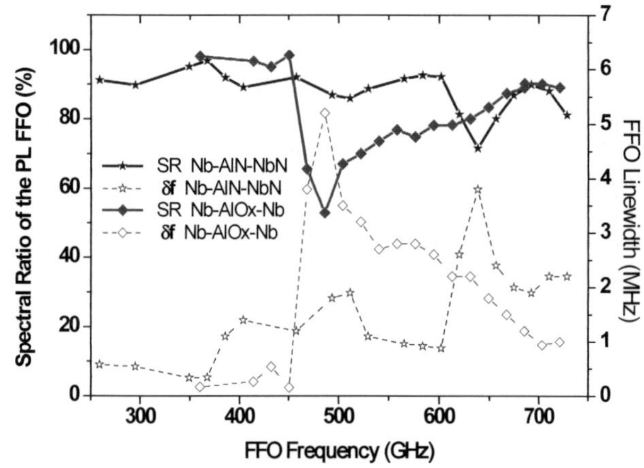

Fig. 7.Dependencies of the free-running FFO linewidth and the Spectral Ratio for the phase-locked FFO on frequency. Data are presented for two different FFO technologies: Nb-Ox-Nb (represented by diamonds) and Nb-AlN-NbN (asterisks).

CRYOGENIC PHASE DETECTOR

The local oscillator for the superconducting integrated receiver based on Flux flow oscillator has a tuning range of 250–750 GHz; in this range a free-run linewidth at half-height (–3 dB in power) may vary from hundreds kHz to several dozen of MHz (in some specific cases). When a conventional room-temperature (RT) PLL is used for the FFO phase locking, the FFO signal firstly down-converted at the harmonic mixer (HM) from hundreds of GHz to 400 MHz, and then the low frequency signal is amplified and sent from the cryostat to the semiconductor RT PLL. The phase of intermediate frequency signal is compared with the reference at the RT PLL, and the error signal is sent back to the cryostat, where it adjusts the FFO frequency. Long connecting cables are used to reduce the heat flow into the cryostat; however, it leads to a delay of the feedback signal of ~10 ns. The electronic block of the PLL adds another 7 ns, resulting in a limitation of about 15 MHz for the synchronization band of the RT PLL for FFO.

For efficient locking of wide Lorentzian lines emitted by FFO a PLL system with a very wide regulation bandwidth is required(due to slow decrease of the noise level with offset from the carrier). To overcome the limitations of the traditional RT PLL, we have developed the cryogenic high-harmonic phase detector (CHPD) [33], [34].Implementation ofthe SIS junction both for down-conversion of oscillator frequency and generation of feedback signal to control the FFO frequency allows us to place all PLL elements in close vicinity to the oscillator. In turn, this provides significant reduction of loop time delay (less than 4 ns) and extremely large regulation bandwidth (up to 70 MHz).Since cryogenic PLL system consists of only superconductive and low-consumption elements, it could be integrated on the single chip with locked oscillator. As it is shown in Fig. 8, the CHPD PLL system could efficiently synchronize highly broad emission lines[29], [34].

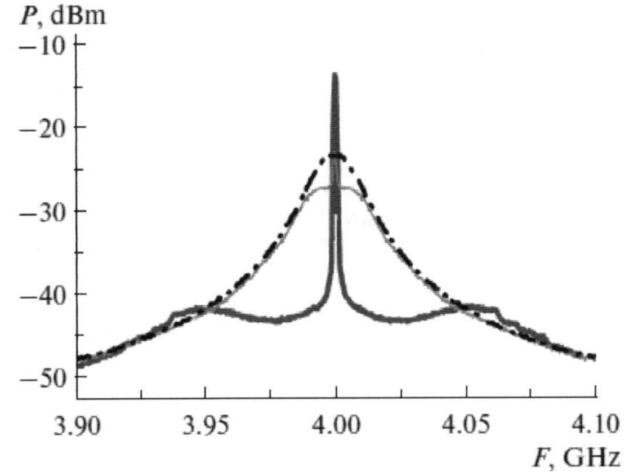

Fig. 8. Experimentally measured FFO emission spectra downconwerted from 634 GHz. The dashed and dotted line represents the autonomous FFO oscillation spectrum (line width, 16.8 MHz); the thin line represents the FFO spectrum synchronized using a room-temperature PLL (spectrum ratio 6%). The bold line shows the FFO spectrum synchronized by CHPD (spectrum ratio 84%).

Suppression of the Fiske Steps

As it was already mentioned, continuous tuning of the FFO frequency below Vg/3 is limited due to presence of the Fiske steps (see Fig. 1). To overcome this limitation and considerably supress the Fiske resonances we have developed new FFO design with additional resistive elements (Fig. 9). It makes possible to increase damping and completely suppress the reflected electromagnetic waves and the Fiske resonant structure at frequencies just below Vg/3, see Fig. 10 and 11.

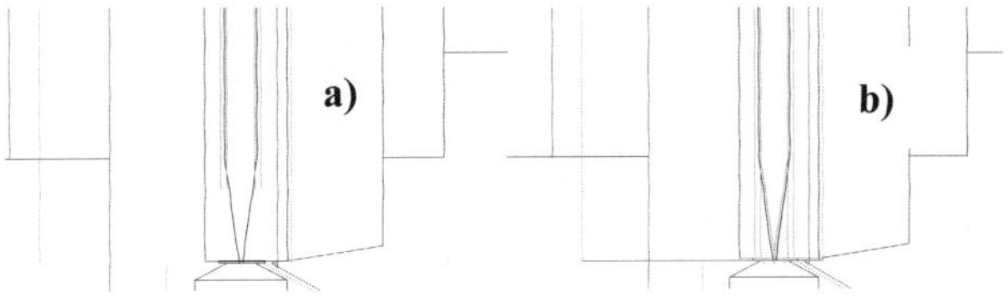

Fig. 9. Two layouts of the FFO with additional resistive elements (shown by green color)for suppression of the Fiske resonant structure at frequencies below Vg/3.

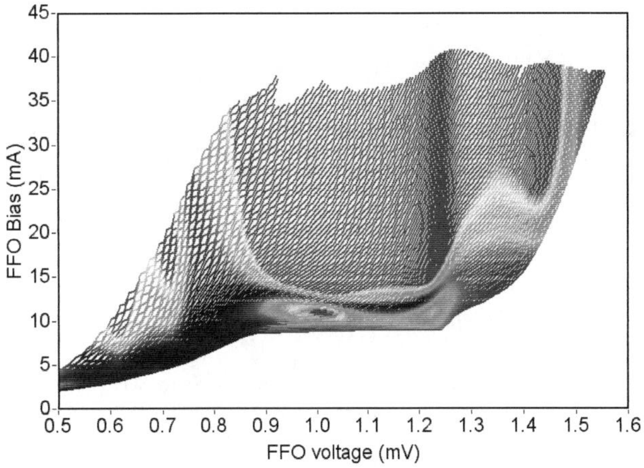

Fig. 10. IVCs of the Nb-AlN-NbN FFO with additional resistive elements for suppression of the Fiske resonant structure (design shown in Fig. 9 a);see Fig. 1 for comparison and detailed information)/

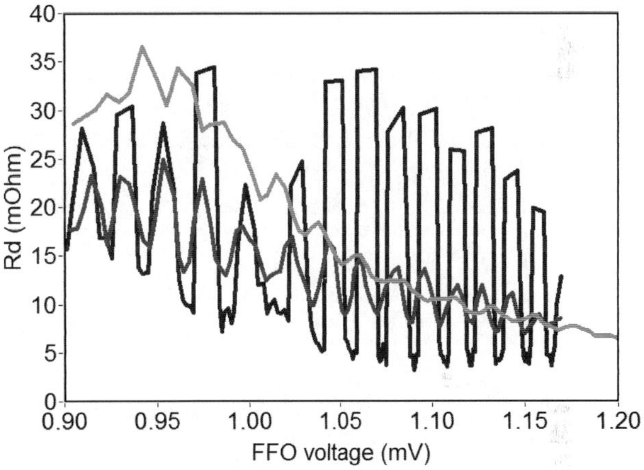

Fig. 11. The differential resistance of the FFO IVCs measured at FFO current of about 25 mA. Data for FFO without suppression (see Fig.1) – blue curve; data for designs a) and b) are shown by red and green curves correspondingly.

Conclusions

The Flux Flow Oscillators (FFO) based on Nb-AlOx-Nb and Nb-AlN-NbN junctions provide unique combination of parameters unreachable for any competing technique:
- 250 – 750 GHz tuning range (in practice might be limited by the SIS matching circuitry);
- FFO frequency and power can be electronically adjusted; furthermore FFO can be phase-locked at any frequency providing spectral ratio > 70 % with phase noise of the order of -90 dBc;
- FFO was integrated with an SIS mixer in a single-chip sub-THz Superconducting Integrated Receiver (SIR) for the atmospheric-research instrument TErahertz and submillimeter LImb Sounder (TELIS).

Acknowledgment

This work was supported by the Ministry of Education and Science of the Russian Federation (No. 14.607.21.0100; ID RFMEFI60714X0100).

REFERENCES

[1] B.D. Josephson(1962). Possible new effects in superconductive tunnelling.*Phys. Lett.*1, pp. 251-253.

[2] B. D. Josephson, (1964). Coupled Superconductors, *Rev. Mod. Phys.* 36, pp. 216-220.

[3] K. K. Likharev, Dynamics of Josephson Junctions and Circuits; Gordon and Breach, London, 1986.

[4] C. Varmazis, R. D. Sandell, A. K. Jain, and J. E. Lukens, “Generation of coherent tunable Josephson radiation at microwave frequencies with narrowed linewidth”, *Appl. Phys. Lett.* 33, 357 (1978).

[5] A. K. Jain, K. K. Likharev, J. E. Lukens, and J. E. Sauvageau (1984).Mutual phase-locking in Josephson junction arrays. Phys. Rep. **109**, 309-426.

[6] B. Dueholm, O. A. Levring, J. Mygind, N. F. Pedersen, O. H. Soerensen, and M. Cirillo (1981). Multisoliton Excitations in Long Josephson Junctions.*Phys. Rev. Lett.***46**, 1299.

[7] Joergensen E., Koshelets V.P., Monaco R., Mygind J., Samuelsen M.R., Salerno M., “Thermal Fluctuations in Resonant Motion of Fluxon on a Josephson Transmission Line: Theory and Experiment”, *Phys. Rev. Lett.*,**49**, pp. 1093-1096, (1982).

[8] M. Cirillo and F. Lloyd (1987).Phase lock of a long Josephson junction to an external microwave source. *J. Appl. Phys.***61**, 2581.

[9] P. A. A. Booi and S. P. Benz (1994).Emission linewidth measurements of two-dimensional array Josephson oscillators, *Appl. Phys. Lett.***64**, 2163.

[10] Barbara, P., Cawthorne, A. B., Shitov, S. V. and Lobb, C. J. (1999). Stimulated Emission and Amplification in Josephson Junction Arrays,*Phys. Rev. Lett.***82,** pp. 1963-1965.

[11] Morvan Salez and Faouzi Boussaha, Fluxon modes and phase-locking at 600 GHz in superconducting tunnel junction nonuniform arrays, Journal of Applied Physics 107, 013908 (2010).

[12] Song, F., Müller, F., Scheller, T., Semenov, A. D., He, M., Fang, L., Hübers, H.-W.andKlushin, A. M. (2011). Compact tunablesubterahertz oscillators based on Josephson junctions, *Appl. Phys. Lett.***98**, 142506.

[13] M A Galin, A M Klushin, V VKurin, S V Seliverstov, M I Finkel, G N Goltsman, F Müller, T Scheller and A D Semenov, “Towards local oscillators based on arrays of niobium Josephson junctions”, Supercond. Sci. Technol. 28 (2015) 055002 (7pp)

[14] Nagatsuma, T., Enpuku, K., Irie, F. and Yoshida, K. (1983). Flux-flow type Josephson oscillator for millimeter and submillimeter wave region, J. Appl. Phys. **54**, 3302.

[15] Nagatsuma, T., Enpuku K., Sueoka K., Yoshida, K. and Irie F., Flux-flow type Josephson oscillator for millimeter and submillimeter wave region. II. Modeling, J. Appl. Phys. **56**, 3284 (1984).

[16] Nagatsuma, T., Enpuku, K., Yoshida, K. and Irie F. (1985). Flux-flow type Josephson oscillator for millimeter and submillimeter wave region. III. Oscillation stability, J. Appl. Phys. **58**, 441.

[17] J. Qin, K. Enpuku and K. Yoshida, Flux-flow type Josephson oscillator for millimeter and submillimeter wave region. IV. Thin-film coupling, J. Appl. Phys. **63**, 1130 (1988).

[18] Y. M. Zhang and P. H. Wu, J. (1990).Numerical calculation of the height of velocity-matching step of flux-flow type Josephson oscillator, Appl. Phys. **68**, 4703-4709.

[19] Y. M. Zhang, .Winkler, T. Claeson (1993). Detection of mm and submm wave radiation from soliton and flux-flow modes in a long Josephson junction," IEEE Trans. Appl. Superconductivity, **3**, pp. 2520 – 2523

[20] Koshelets V P, Shitov S V, Shchukin A V, Filippenko L V, and Mygind J, "Linewidth of Submillimeter Wave Flux-Flow Oscillators". *Appl. Phys. Lett.***69** 699-701 (1996)

[21] Koshelets V P, Shitov S V, Shchukin A V, Filippenko L V, Mygind J, and Ustinov A V. "Self-Pumping Effects and Radiation Linewidth of Josephson Flux Flow Oscillators". *Phys Rev B***56** 5572-5577 (1997)

[22] V.P. Koshelets, S.V. Shitov, A.V. Shchukin, L.V. Filippenko, P.N. Dmitriev, V.L. Vaks, J. Mygind, A.M. Baryshev, W. Luinge, H. Golstein, "Flux Flow Oscillators for Sub-mm Wave Integrated Receivers", *IEEE Trans. on Appl. Supercond.* 9, 4133–4136 (1999).

[23] Koshelets V P and Mygind J, “Flux Flow Oscillators For Superconducting Integrated Submm Wave Receivers”, Studies of High Temperature Superconductors, edited by A.V. Narlikar, NOVA Science Publishers, New York, **39** 213-244 (2001)

[24] Koshelets V. P., Dmitriev P. N., Ermakov A. B., Sobolev A. S., Torgashin M. Yu., Kurin V. V., Pankratov A. L, Mygind J. (2005). Optimization of the Phase-Locked Flux-Flow Oscillator for the Submm Integrated Receiver, *“IEEE Trans. on Appl. Supercond.”,* **15,** pp.964-967

[25] M.Yu. Torgashin, V.P. Koshelets, P.N. Dmitriev, A.B. Ermakov, L.V.Filippenko, and P.A. Yagoubov, “Superconducting Integrated Receivers based on Nb-AlN-NbN circuits“ “IEEE Trans. on Appl. Supercond.”, vol. 17, pp.379- 382, 2007.

[26V. P. Koshelets, S. V. Shitov, L. V. Filippenko, A. M. Baryshev, H. Golstein, T. de Graauw, W. Luinge, H. Schaeffer, H. van de Stadt "First Implementation of a Superconducting Integrated Receiver at 450 GHz"; *Appl. Phys. Lett.*, 68, 1273 (1996).

[27] V.P. Koshelets, S.V. Shitov, “Integrated Superconducting Receivers” Superconductor Science and Technology, vol 13, pp. R53-R69, (2000).

[28] V.P. Koshelets, M. Birk, D. Boersma, J. Dercksen, P.N. Dmitriev, A.B. Ermakov, L.V. Filippenko, H. Golstein, R. Hoogeveen, L. de Jong, A.V. Khudchenko, N.V. Kinev, O.S. Kiselev, P.V. Kudryashov, B. van Kuik, A. de Lange, G. de Lange, I.L. Lapitsky, S.I. Pripolzin, J. van Rantwijk, A. Selig, A.S. Sobolev, M.Yu. Torgashin, V.L. Vaks, E.de Vries, G. Wagner, P.A. Yagoubov, “Integrated Submm Wave Receiver: Development and Applications”, - Chapter in the book “Nanoscience Frontiers - Fundamentals of Superconducting Electronics”, Springer Serie: Nanoscience and Technology 35372, pp. 263-296, Editor: AnatolieSidorenko (August 2011).

[29] V.P. Koshelets, P.N. Dmitriev, M.I. Faley, L.V. Filippenko, K.V. Kalashnikov, N.V. Kinev, O.S. Kiselev, A.A. Artanov, K.I. Rudakov, A. de Lange, G. de Lange, V.L. Vaks, M.Y. Li, H.B. Wang, “Superconducting Integrated Terahertz Spectrometers”, IEEE Transactions on Terahertz Science and Technology, vol. 5, pp 687- 694, (July 2015).

[30] P.N. Dmitriev,I.L. Lapitskaya, L.V. Filippenko, A.B. Ermakov, S.V. Shitov, G.V. Prokopenko, S.A. Kovtonyuk, and V.P. Koshelets. “High Quality Nb-based Integrated Circuits for High Frequency and Digital Applications”, *“IEEE Trans. on Appl. Supercond.”*, vol. 13, No 2, pp. 107-110, 2003.

[31] Koshelets V P, Ermakov A B, Dmitriev P N, Sobolev A S, Baryshev A M, Wesselius P R, Mygind J, “Radiation linewidth of flux flow oscillators”, *Superconductor Science and Technology* **14** 1040 – 1043 (2001)

[32] V.P. Koshelets, S.V. Shitov, L.V. Filippenko, V.L. Vaks, J. Mygind, A.B. Baryshev, W. Luinge, N. Whyborn, " Phase Locking of 270-440 GHz Josephson Flux Flow Oscillator", Rev. of Sci. Instr., v. 71, No 1, pp. 289-293, (2000).

[33] A. V. Khudchenko, V. P. Koshelets, P. N. Dmitriev, A. B. Ermakov, P. A. Yagoubov and O.M. Pylypenko, “Cryogenic Phase Locking Loop System for Superconducting Integrated Receiver”, Supercond.Sci. and Technol.,**22**,085012 (2009).

[34] Kalashnikov, K.V., Khudchenko, A.V. and Koshelets, V.P. (2013). Harmonic Phase Detector for Phase Locking of Cryogenic Terahertz Oscillator, *Appl. Phys. Lett.***103**, pp.102601-1-102601-4.

Improvement of the Planar Schottky Diode Capacity Model for the Implementation in the Non-linear Harmonic Balance ADS Simulator for Multipliers Design

D. Moro-Melgar1*, A. Maestrini1, *Member, IEEE*, J. Treuttel1, L. Gatilova1&2, F. Tamazouzt1, T. Vacelet1, J. Mateos3, *Member, IEEE*, T. González3, B. G. Vasallo3

[1]*LERMA, Observatoire de Paris, PSL Research University, CNRS, Sorbonne Universités, UPMC Paris06, F-75014, France*
[2]*Laboratoire de Photonique et de Nanostructure, CNRS, 91460 Marcoussis, France*
[3]*(USAL) Universidad de Salamanca, Salamanca 37008, Spain*
* Contact: diego.moro-melgar@obspm.fr

A 2-dimensional "*ensemble*" Monte Carlo (MC) physical simulator, previously used to study HEMTs in [1]-[2], has been used for studying planar GaAs-Schottky barrier diode structures (SBDs) where there is a two dimensional electron transport. By using this physical MC simulator, the analytical junction capacity model proposed in [3] has been extended for structures where the existence of surface charges placed in the semiconductor-dielectric interfaces has been considered. The influence of the substrate on the junction capacity when the epilayer thickness is strongly reduced has also been studied by including a degeneracy model in MC simulations [2]. This work has been carried out in parallel with the design of a 1.2 THz heterodyne receiver for the JUICE-SWI mission project and the experimental characterization of the 600 GHz heterodyne front-end receiver presented in [4]. This study has allowed us to define the junction capacity for real geometries of the Schottky anodes used in the experimental available devices fabricated with the LERMA-LPN process, presented in [5]. The improved capacity model has been implemented in the non-linear harmonic balance ADS simulator by defining a SDD model (Symbolically-Defined Device), which reproduces the electrical behavior of the standard Schottky diode model integrated in the software, but also includes the improved junction capacity model. Both models are based on a simplification of the Lumped Elements Circuit (LEC) [6], typically used in the design of these devices [4]-[5], where the series impedance of the Schottky junction circuit is approximated by a single constant resistance. The SDD model has finally been introduced in the same ADS test-benches, linked with HFSS simulations, with which the frequency doubler and mixer used in [4] were designed and optimized.

This work has experimentally demonstrated that the implementation of the improved junction capacity model in HFSS-ADS simulations when including the approximation of the series impedance by a single constant resistance, is able to accurately reproduce the experimental results of a frequency doubler at 280 GHz as long as it works in a pure varactor mode. In addition, the good reliability of the LERMA-LPN fabrication process has also allowed us to identify a lack of accuracy of the LEC Schottky diode model when the diodes are strongly pumped by the local oscillator input signal. A non-pure varactor operation mode of the diodes has been remarked in these cases in which the second harmonic generation efficiency is a combination of the varactor and varistor operation mode. This work lays the basis for the improvement of the Schottky diode model based in the LEC model, necessary for correctly simulating the diodes in varistor operation mode in ADS-HFSS simulations.

References

[1] J. Mateos, T. González, D. Pardo, V. Hoel, H. Happy and A. Cappy. "Design optimization of AlInAs-GaInAs HEMTs for high-frequency applications. *IEEE Trans. Electron Devices*, vol. 51, no 4, p. 521-528, 2004.

[2] J. Mateos, T. González, D. Pardo, S. Bollaert, T. Parenty and A. Cappy."Improved Monte Carlo algorithm for the simulation of δ-doped AlInAs/GaInAs HEMTs".*IEEE Trans. Electron Devices*, vol. 47, no 1, p. 250-253, 2000.

[3] J. T. Louhi. "The capacitance of a small circular Schottky diode for submillimeter wavelengths".*Microwave and Guided Wave Letters, IEEE Microwave Guided Wave Lett.*, vol. 4, no 4, p. 107-108m 1994.

Design of a Terahertz Wire-wrap Backward Wave Oscillator

Changpeng Xu[1*], Yong Yin[1], Liangjie Bi[1], Zhiwei Chang[1], Zhang Zhang[1], and Lin Meng[1]
[1]*University of Electronic Science and Technology of China, Chengdu, Sichuan610054, China*
*Contact: superxcp@163.com

Abstract— **In this paper, an innovative slow-wave circuit, applied the wire-wrap structure, is proposed for the design of backward-wave oscillator (BWO) operating in terahertz band. Compared with the conventional BWO, the novel BWO has the relatively low accelerating voltage around 1 kV and low beam current of 0.1 A (the current density is 188 A/cm^2). The Particle-in-cell (PIC) results predict that this BWO is capable of producing the output power over 400 mW at 324GHz and the range of 3dB voltage tuning bandwidth is from 311 GHz to 338 GHz. Moreover, the advantage that the wire-wrap SWS is constructed by wrapping the fine copper wire avoids the difficulty caused by conventional micro-fabrication technology. The simulation results presented by various parameters demonstrate that the novel wire-wrap structure can be applied as a promising slow-wave structure for terahertz radiation source application.**

INTRODUCTION

Recent years, the research of the terahertz (THz) wave has been considerable interest in radio astrophysics and astronomy, security inspection, radar applications, imaging, biology and materials science [1]-[3]. The availability of the terahertz wave is consequently become the precondition to terahertz development. The generation of THz wave is mostly depended on vacuum electron devices (VEDs) with slow-wave structure (SWS). Backward-wave oscillator (BWO), a typical vacuum electron device, is considered as the most promising device to product THz radiation signal and SWS is the pivotal component using to exchange energy between electron beam and the electromagnetic field

The conventional SWS, such as corrugated waveguide, sinusoidal waveguide and the folded waveguide were introduced as the popular structure for the realization of BWO [4]-[6]. But all the SWSs mentioned above used modern micro-fabrication technology, such as deep reactive ion etching, deep X-ray LIGA process and UV/SU-8 lithography to process [7]. Unfortunately, the machine accuracy including the accuracy on the dimensions and the surface roughness of the metal walls bring great adverse influence on the device performance. It is hard to process non-destructive SWS. So a novel wire-wrap SWS and fabrication method are put forward and apply to design BWO to get over this challenge.

MODEL AND DESIGN

A. *Slow-wave Structure Design*

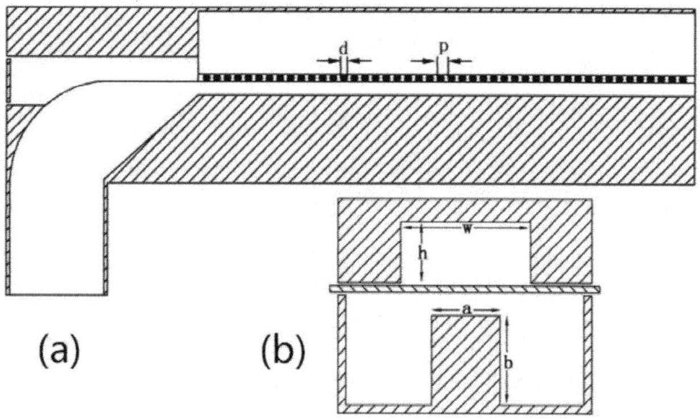

Fig. 1 Schematic configuration of the novel wire-wrap BWO: (a) the sectional view under the vertical axis direction and (b) the front view of the whole tube.

TABLE I
THE DIMENSIONS OF BACKWARD WAVE OSCILLATOR

G	Geometrical parameter	Quantitative Value and Units
d	Copper diameter	25 μm
p	Period length	50 μm
w	Cavity width	0.4 mm
h	Cavity height	0.3 mm
a	Ridge width	0.2 mm
b	Ridge height	0.37 mm

The specific structure of wire-wrap slow wave circuit is shown in Fig. 1 and the dimensions are listed in Table II. This structure features a rectangular cavity, periodic fine copper wire and a rectangular ridged waveguide. The ridge on the bottom of waveguide is used to increase the z-component electric field when electron beam interact with the z-component electric field. The physical realization of this structure was accomplished with two component parts. The fine copper wire twines around the upper cover plate to construct the slow wave circuit. Before that, the fine copper wire is firstly stretched to the diameter of 25 μm. In the progress of machining SWS, the fine copper wire keeps smooth. So the problem caused by the conventional micro-fabrication can be avoided because the fine copper wire is extremely uniform and smooth.

On the contrary of forward wave devices which output its energy at the end of collector, the backward wave devices output its energy at the end of electron gun. Based on this principle, a standard rectangular output waveguide is designed at the end of electron gun to output signal. In addition, a slope is added to match the coupling impedance of the interaction chamber and output waveguide.

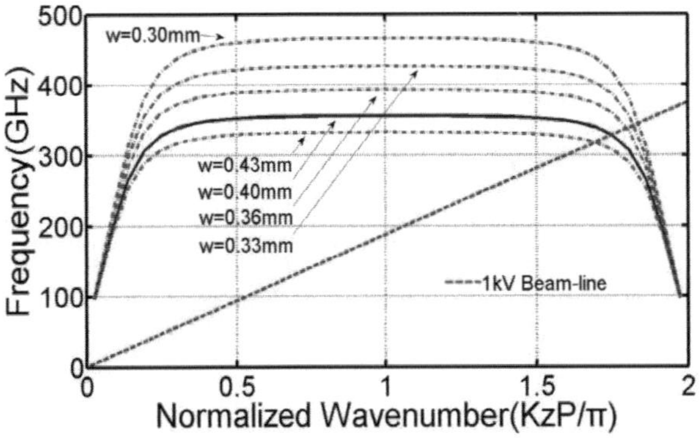

Fig. 2 Dispersion curves with different cavity widths *w*.

B. High Frequency Characteristics

The dispersion properties of the fundamental mode obtained by the eigenmode calculation with CST are plotted in Fig. 2. In this structure, the dispersion behaviour is mainly determined by the cavity width *w* on the upper cover plate. Except the dispersion for the nominal dimensions in Table III, the curves for different values of cavity width *w* are plotted for the purpose of providing flexibility in choosing dimension of device. The beam line of 1 kV intersects the dispersion curve at the operating frequency about 324 GHz. At this intersection, the velocity of electrons is almost synchronous with the phase velocity of the backward wave which is equal to the slope of the beam line.

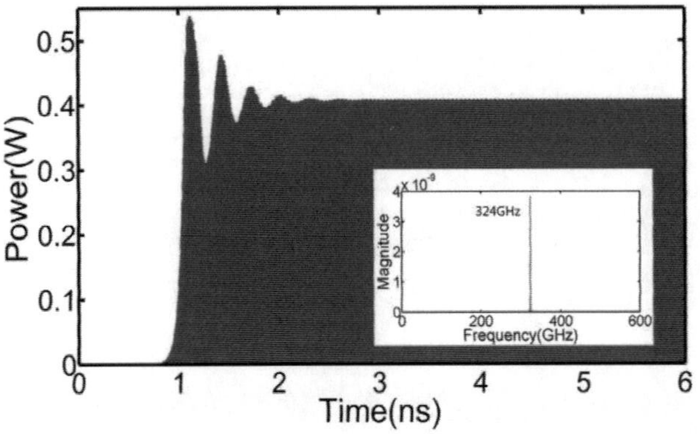

Fig. 3 Time plot of the output power with a inset of its frequency spectrum.

PERFORMANCE SIMULATION AND RESULT ANALYSIS

In order to test the performance of the wire-wrap BWO, the necessary simulations are performed by particle-in-cell (PIC). The initial pencil electron beam with a diameter of 0.26mm has a relatively low introducing voltage about 1 kV and low beam current of 0.1 A, thus the current density is 188 A/cm^2. The BWO structure has 46 periods and the longitudinal distance between two centres of copper wire is 50 μm. A series of typical parameters, such as power, frequency and 3dB voltage tuning bandwidth are obtained.

As is clear from Fig. 3, the wire-wrap backward wave oscillator is capable of producing stable output peak power of 400 mW with the start oscillator time about 1.2 ns. The inset shows a pure frequency spectrum operating at 324GHz. At this operating point, the interaction chamber has the strongest interaction impedance. When the electrons travel along the SWS, the electrons gradually bunch and give its kinetic energy to the electromagnetic field. Fig. 4 shows the variation trend of output power as a function of introduced voltage. As can be seen, this BWO can get a 3dB voltage tuning bandwidth of 27GHz range from 311 GHz to 338 GHz. There exists a fluctuant phenomenon between 1 and 1.1 kV due to mode competition. Given that the operation voltage is under the low condition and the current density is small, the BWO with wire-wrap SWS is considered as an effective solution for BWO application.

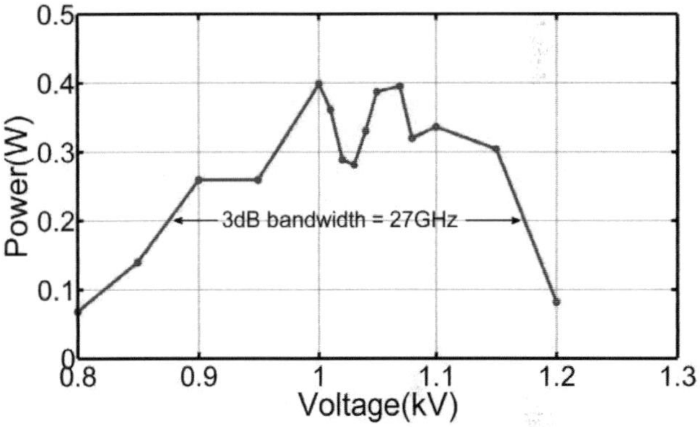

Fig. 4 Output power as a function of introducing voltage

CONCLUSIONS

In conclusion, a novel wire-wrap SWS for backward wave oscillator has been designed. The BWO has the relatively low operating voltage of 1kV and low current density of 188 A/cm^2. Based on these conditions, the device can produce the output peak power over 400 mW operating at THz band. Moreover, the processing method is another original advantage in the processing progress. The experiment of BWO with the proposed wire-wrap SWS is in plan.

ACKNOWLEDGMENT

This work was supported in part by the National Natural Science Foundation of China under Grant 61201011 and the Fundamental Research Funds for the Central Universities under Grants ZYGX2015J037 and ZYGX2015J039.

REFERENCES

T. C. Bowman, M. El-Shenawee, L. K. Campbell, "Terahertz Imaging of Excised Breast Tumor Tissue on Paraffin Sections," *IEEE Antennas and Propagation*, vol. 63, pp. 2088–2097, May. 2015.

H. Quast, T. Loffler, "3D-terahertz-tomography for material inspection and security," in *Proc. IEEE IRMMW-THz*, 2009, pp. 1-2.

S. S. Ponomarenko, S. A. Kishko, V. V. Zavertanniy, Ed., "400-GHz Continuous-Wave Clinotron Oscillator," *IEEE Plasma Science*, vol. 41, pp. 82–86, Jan. 2013.

M. Mineo, C. Paoloni, "Corrugated Rectangular Waveguide Tunable Backward Wave Oscillator for Terahertz Applications," *IEEE Electron Devices*, vol. 57, pp. 1481–1484, Jun. 2010.

X. Xu, Y. Wei, F. Shen, Ed., "A watt-class 1-THz backward-wave oscillator based on sine waveguide," *Phys. Plasma*, vol. 19, no. 4, Jan. 2012.

A. V. Aksenchyk, A. A. Kurayev, "Folded waveguide TWT and BWO for frequency range 37–3000 GHz," in *Proc. IEEE IVEC,* 2007, pp. 1–2.

Y. M. Shin, L. R. Barnett, D. Gamzina, Ed., "Terahertz vacuum electronic circuits fabricated by UV lithographic molding and deep reactive ion etching," Appl. Phys. Lett., vol. 95, pp. 181505-1–181505-3, Nov. 2009.

Design and Analysis of a Y-band Extended Interaction Oscillator with a Pseudospark-Sourced Electron Beam

Zhang Zhang[1]*, Yong Yin[1], Liangjie Bi[1], Zhiwei Chang[1], Changpeng Xu[1] and Lin Meng[1]
[1]University of Electronic Science and Technology of China, *Chengdu, Sichuan610000, China*
* Contact: sunshinezz6688@163.com

Abstract—**In this paper, a pseudospark(PS) discharge system instead of traditional electron gun is designed for the pencil beam Y-band extended interaction oscillator (EIO). The characteristics of EIO circuit and the PS discharging voltage of the electron beam are studied to optimize the performance of the Y-band EIO. The CST simulation results show that the averaged output power over 24 W with a main frequency of ~286 GHz can be achieved by using a 30.75 kV PS discharge voltage. The advantage of the newly proposed device is that high current density electron beam pulse is transported in the positive-ion focusing channel without a guiding magnetic field. Meanwhile, high impedance(R/Q) and high gain per unit length is another characteristic of the device.**

Index Terms—**Extended interaction oscillator (EIO), pseudospark-sourced electron beam, vacuum electronics.**

INTRODUCTION

At present, terahertz (0.1-20 THz) sources are a currently active research area and are important for applications to high data rate communications, biological spectroscopy, molecular spectroscopy and biomedical diagnostics etc [1-4]. EIO is a novel type of vacuum electronic high power terahertz sources. It has the advantages of high interaction impedance and high gain per unit length. For the development of Terahertz EIO, the high frequency characteristics and the parameters of EIO circuit are researched. Meanwhile, PS discharge is one of gas discharge, which operates in the hollow cathode, axially symmetric parallel electrodes and planar anode configuration [5], [6]. It can produce high current density, narrow beam diameter and axially symmetric electron beam pulse. The recent research about pseudospark-sourced electron beam show that several thousand Amperes electron beam pulses produced by PS source can be transported distances up to 20cm in 3mm diameter beam tunnel without a guiding magnetic field [7]. Using the pseudospark-sourced electron beam instead of traditional electron gun can become a novel way to generate terahertz wave.

In this article, we present research result that demonstrate the successful design and optimize of Y-band EIO based on a pseudospark-sourced electron beam for a 30.75 kV 0.5A beam with a 0.2mm diameter beam tunnel, which has produced over 24 W of peak output power at Y-band with the help of 3-D particle-in-cell simulation.

CIRCUIT DESIGN

A Y-band EIO circuit is presented in this paper. The photograph of EIO circuit is shown in Fig.1. The material of the device shell is the copper to ensure a better heat dissipation environment for circuit. The circuit consists of identical nine slots and two symmetrical coupling cavities located up and down respectively. A 0.2mm diameter beam tunnel passes through the centre of the nine-slot slow wave structure. The power is exported to the WR-3 standard waveguide attached to upside coupling cavity through a coupling hole. The EIO operates in the 2π mode. Contour of the electric field distribution component Ez at Y-Z and X-Y cross-section and electric field strength along the Z direction is shown in Fig.2. Based on the structure parameters of EIO circuit, the dispersion curve of Y-band EIO is shown in Fig.3. This result show that the operation frequency is ~286 GHz and the dispersion curve of the EIO circuit is in synchronism with a 30.75 kV electron beam. Under the circumstance, we can guarantee a more effective interaction between the interaction circuit and the electron beam.

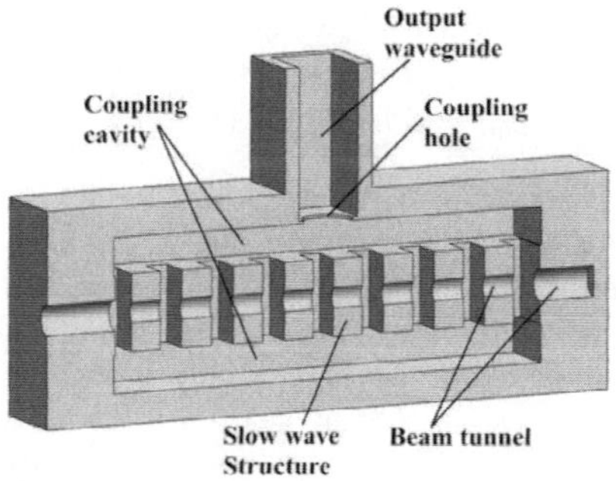

FIG.1. The photograph of EIO circuit.

With the help of simulations, the dimensions of EIO circuit are optimized, including coupling hole, slot, coupling cavity, and electron beam tunnel. The dc beam was designed for a voltage of 30.75 kV, an electron current of 0.5 A, and a guiding magnetic field of 0.5 Tesla is adopted. When the beam-wave synchronous condition was met, the oscillation of the circuit was established and power was exported stably after 17 ns. Fig.4 shows simulated output single and frequency spectrum. The averaged output power was about 24 W and the operation frequency was 286 GHz.

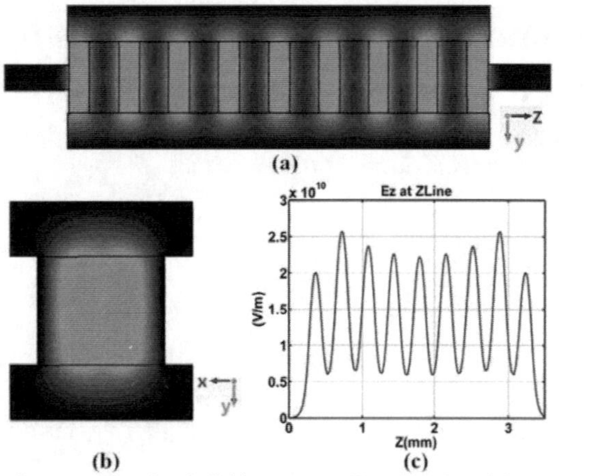

FIG.2. Contour of the electric field component Ez at (a)Y-Z and (b)X-Y cross-sections(from CST), and (c)the field strength along the Z direction.

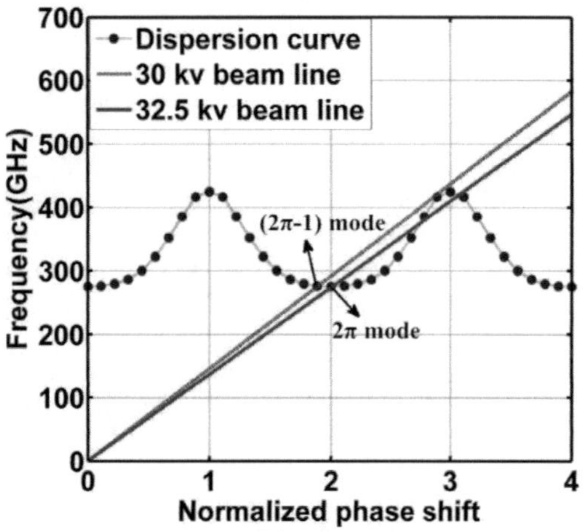

FIG.3. Dispersion curve of Y-band EIO circuit.

The operation voltage range and the oscillation startup time are important parameters of the EIO with pseudospark-sourced electron beam. Electron beam interacts with the standing wave of the EIO circuit. Fig.5 illustrates the relationship between the output power and the oscillation startup time versus different beam voltage at 0.5 A beam current. As shown in Fig.5, operation voltage ranges from 28.5 kV to 32 kV. The fast falling edges of the output power at 32 kV was due to the change of the operation mode. When beam voltage goes beyond operation voltage range, the electron velocity is too fast, which may motivate other mode of the cavity. The Q_o will decrease and Q_e will increase. The operating voltage region of the 2π mode in the EIO was 3.5 kV.

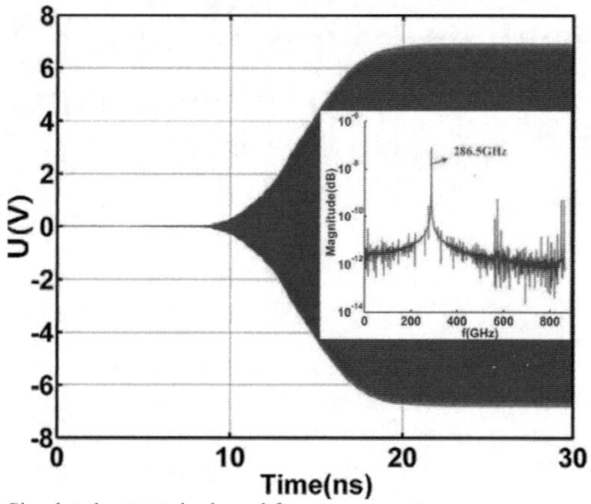

FIG.4. Simulated output single and frequency spectrum.

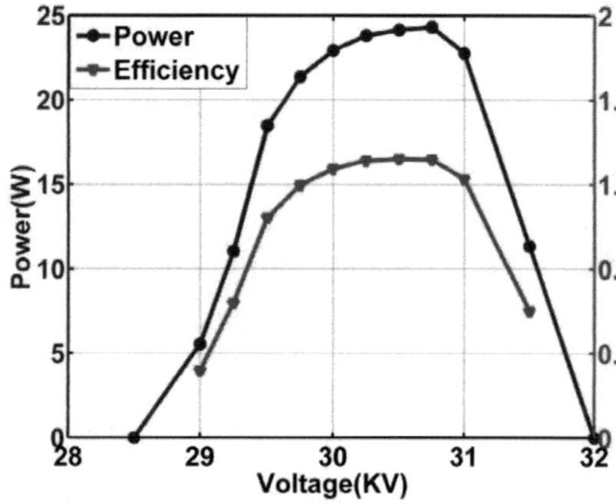

FIG.5. The averaged output power and efficiency versus voltage.

CONCLUSIONS

In this paper, a CST 3-D model of EIO circuit was established to study the correlation interaction properties of the cavity operating mode. Under the condition that the injected beam diameter is 0.2 mm, 30.75 kV beam voltage, 0.5 A beam current and the operating frequency is about 286 GHz, the averaged output power is achieved about 24 W. Meanwhile, using the pseudospark-sourced electron beam instead of a thermionic electron beam achieve a novel way to generate terahertz wave.

ACKNOWLEDGMENT

This work was supported by the National Natural Science Foundation of China under Grant 61201011 and the Fundamental Research Funds for the Central Universities under Grants ZYGX2015J037 and ZYGX2015J039.

REFERENCES

[1] J. B. Baxter, and G. W. Guglietta, "Terahertz spectroscopy," *Anal. Chem.*, vol.83, pp. 4342-4368, May 2011.

[2] J. H. Booske, R. J. Dobbs, C. D. Joye, C. L. Kory, G. R. Neil, G. S. Park, J. Park, and R. J. Temkin, "Vacuum electronic high power terahertz sources," *IEEE Trans. Terahertz Sci. Technol.*,vol.1, pp.54-74, Sep.2011.

[3] J. H. Booske, "Plasma physics and related challenges of millimeter-wave-to-terahertz and high power microwave generation," *Phys. Plasmas,* vol.15, p.055502,2008.

[4] P. H. Siegel, "Terahertz technology," *IEEE Trans. Microwave Theory Tech.* , vol.50,pp. 910-928, Mar.2002.
[5] K. Frank and J. Christiansen, "The fundamentals of the pseudospark and its applications," *IEEE Trans. Plasma Sci.*, vol.17, pp.748-753, Oct.1989.
[6] S. Muhl and A. Perez, "The use of hollow cathodes in deposition processes: A critical review," *Thin Solid Films*, vol. 579, pp. 174-198, Mar. 2015.
[7] H. Yin, A. W. Cross, A. D. R. Phelps, D. Zhu, W. He, and K. Ronald, "Propagation and post-acceleration of a pseudospark-sourced electron beam," J. Appl. Phys., vol.91, pp. 5419-5422, Apr. 2002.

340 GHz Frequency Multiplier with Unbalance Circuit Based on One Schottky Diodes Chip

Jiang Jun[12], Lu Bin[12], Tian Yaolin[12], Huang Wei[12], Miao Li[12], Zhang Jian[12] , Deng Xian-Jin[12] and Tang Jianxiang[3*]
1 Institute of Electronic Engineering, CAEP, Mianyang 621900, China
2 Microsystem and Terahertz Research Center, CAEP, Mianyang 621900, China
3 The Department of Power, CAEP , Mianyang 621900, China
* Contact: 000jiangjun@163.com

Abstract— **This Terahertz is a new across research field. This paper introduces a design of multiplier at 340 GHz with discrete GaAs planar Schottky diode. The 50um thick quartz circuit substrate is flip-chip mounted for diode thermal dissipation. A proposed Schottky diode model improves the accuracy of design which is considered behaviours of current voltage (I/V) and capacitance voltage (C/V), plasma resonance and skin effect. Diode embedding impedances were calculated by full-wave analysis and harmonic-balance (HB) simulation tools with an coaxial port to represent the nonlinear junction for circuit matching. In the circuit, one stage Compact Suspended Microstrip Resonators (CSMRs) are used for low-pass filter (LPF) at input port and a hammer-head filter is used for DC port. Two low-pass filters minimize the length/width ratio. The doubler is self-biasing and fix-tuned, the highest efficiency is 5.1% and output power is 1.1 mW @ 329.6 GHz with input power 20mW.**

INTRODUCTION

Frequency multipliers based on Schottky barrier diodes play a crucial role at frequencies ranging from 300 GHz to 3 THz. Over the years, these devices have remained critical to a variety of submillimeter-wave heterodyne-based instruments [1] including radiometers for space-borne applications, receivers for ground-based radio astronomy, and sources for vector network analyser frequency extenders. Planar Schottky technology made tremendous progress in the late 1990s, principally thanks to the astrophysics community that supported the construction of the heterodyne instrument on Herschel [2] [3]. Because of low output power of semiconductor three ports device, GaAs planar Schottky diode technology plays a crucial role in THz and sub-THz regions in two decades. Frequency multipliers based on Schottky diode are nonlinear devices that generate harmonics of an input sine signal. Through matching networks at the input and output frequencies, optimize the transfer of power from the fundamental frequency to the desired harmonic and get the wanted harmonic and suppress undesired ones, Fig. 1 shows the schematic diagram of the multiplier. Because the balanced multiplier has a high output power and efficiency, it becomes the popularity topology for frequency multiplication.

However, due to the limit of planar Schottky Diode and membrane technics, most of multiplier had a low frequency and output power: the frequency is less than 400 GHz [4] and the max output power is up to 20mW at 170GHz which is failing to drive a 340GHz balanced doubler. This paper focuses on 340GHz frequency multiplier including waveguide block, quartz and one Schottky Diodes Chip with three anodes in series.

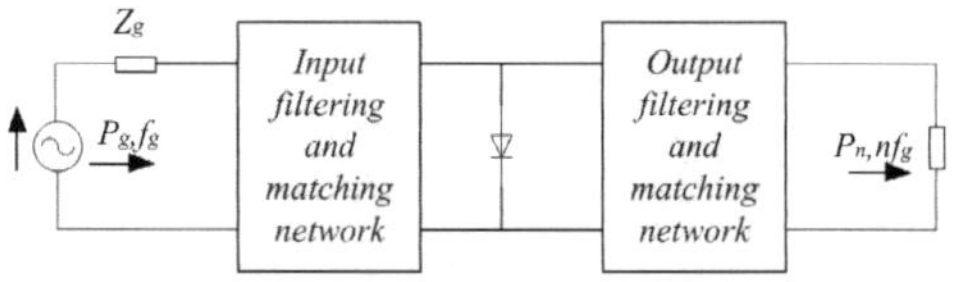

Fig. 1 The Schottky diode multiplier schematic diagram

DESIGN

The structure diagram of the 340GHz frequency multiplier is shown in Fig. 2. The quart circuit extend from the bias voltage port, input waveguide (WR5) to output port WR2.8. There are two low pass CSMRs (Compact Suspendered Microstrip Resonators) filter in the circuit beside input port respectively. The Schottky Diodes Chip placed between ground and Suspendered Microstrip.

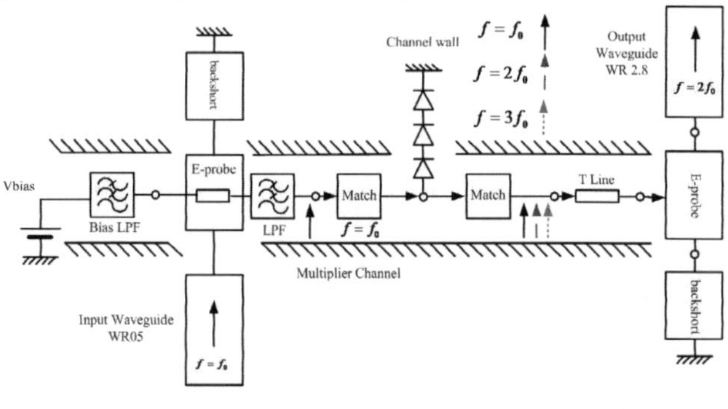

Fig. 2 Structure diagram of the 340GHz frequency doubler

Schottky diode model: Schottky Diodes Chip with three anodes in series is used in this doubler. The Schottky diode model includes two parts: diode spice model in ADS (Harmonic Simulation Software) and 3D structure model in HFSS (3D electromagnetic simulation software). The spice model can be got from I-V and C-V curves [5]. Each anode has a series resistance of 3 ohm, ideal factor 1.25, Junction capitation Cj(0) is 30fF, forward voltage is 0.65V and Cpar is about 6fF.

Circuit simulation: combine harmonic simulation software with 3D electromagnetic simulation software, an optimum result can be got. When the bias voltage is -3V and input power is 13dBm, impedances at input and output are Z_LO=7+12i, Z_RF=81+162i. Finally, the simulation result is shown in Fig. 3. The simulation includes five parts: input probe, output probe, Diode placement, improved CSMRs filters and match networks.

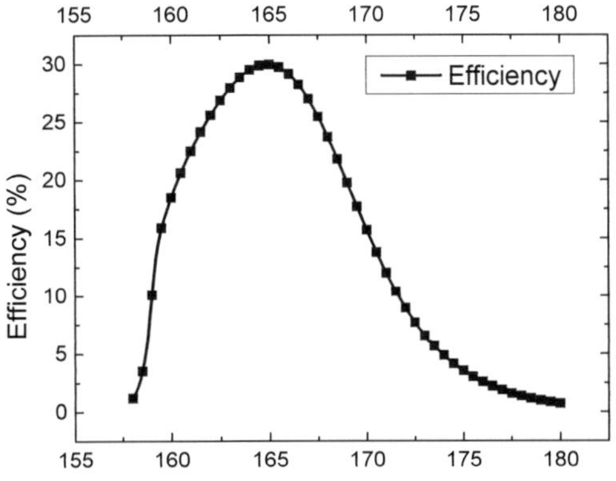

Fig. 3 Simulation results with ADS and HFSS, shows the efficiency vs drive frequency

In the quartz circuit, the CSMRs lowpass filter have a great performance with a passband from DC to 200GHz and stopband from 270GHz to more than 450GHz. The structure and simulation results of CSMRs filter between input waveguide and Schottky Diodes Chip are shown in Fig. 4.

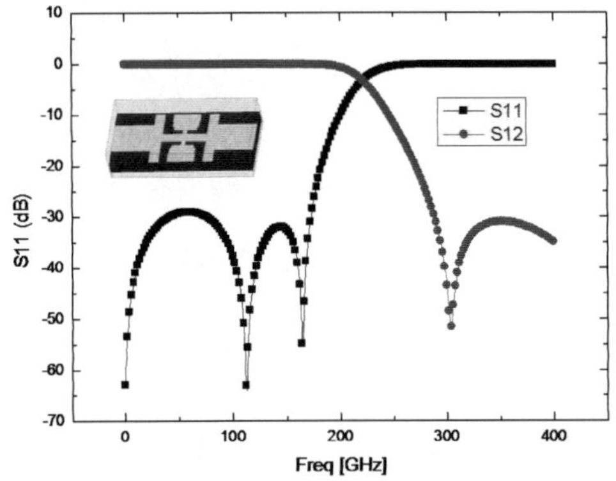

Fig. 4 simulation results and structure of CSMRs filter

Architecture And Measurements

The fabricated 340GHz frequency doubler is shown in Fig. 5. The multiplier is made up a waveguide block split in the E-plane. The planar diodes are flipped-chip mounted between a 50um quartz-based microstrip circuits and metal channel wall.

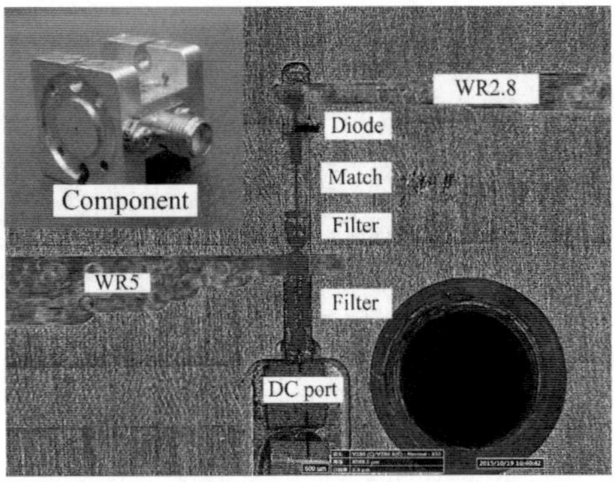

Fig. 5 Photograph and detail of 340GHz multiplier

For the measurements, a commercial signal generator was used to driver an MMIC based quadrupler with max output power of 40mW at 80GHz~86GHz [6], and all the measurements were done at room temperature. W-band amplifier has a power of 110mW to drive a 170GHz multiplier with 20mW output power. 340GHz multiplier was driven by 170GHz multiplier and detected by PM4 (Power Meter 4). The testing platform is shown in Fig. 6.

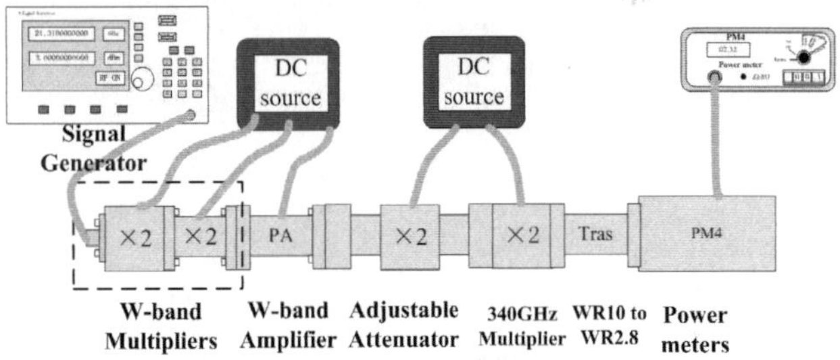

Fig. 6 340GHz multiplier testing platform

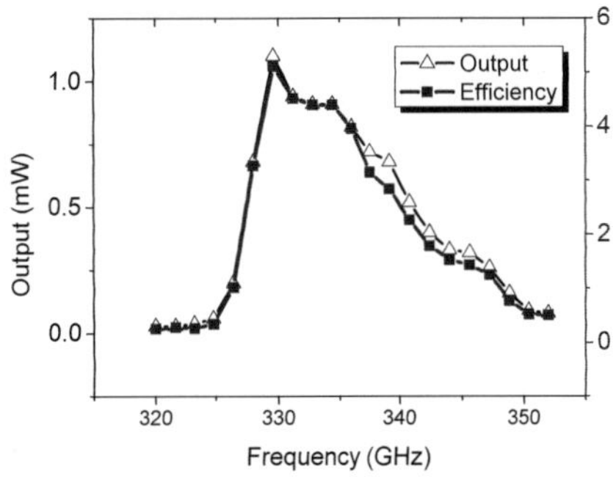

Fig. 7 340GHz multiplier output power and efficiency vs output frequency.

Conclusions

This paper focused on design and measurements of 340GHz multiplier. It presented one kind of unbalance multiplier structure based on one Schottky Diodes Chip and improved CSMRs filter. CSMRs filter can reduce the ratio of length/width and it is useful for fabrication. That one Chip used in this component reduced the cost of multiplier and it is suitable for low power driver. The details of measurements have been shown in Fig. 7. The highest efficiency is 5.1% and output power is 1.1 mW @ 329.6 GHz with input power

20mW. The typical tested power is 3% and output power above 0.5mW in 328GHz ~ 340.4GHz.

ACKNOWLEDGMENT

This work was supported by Institute of Electronic Engineering, Microsystem and Terahertz Research Center, China Academy of Engineering Physics (CAEP-IEE) and National Key Basic Research Program of China (973 Program) with Grant No.2015CB755406.

REFERENCES

A. V. Räisänen, “Frequency multipliers for millimeter and submillimeter wavelengths,” Proc. IEEE, vol. 80, no. 11, pp. 1842–1852,Nov. 1992.

H. van de Stadt, Detectors for the heterodyne spectrometer of FIRST, in: Proceedings of the ESA Symposium “The Far Infrared and Submillimeter Universe”, Grenoble, April 15–17, 1997, p. 457.

G.L. Pilbratt, The Herschel mission, scientific objectives, and this meeting, in: Proc. Eur. Space Agency Symp., ESA paper SP-460, December 2000, pp. 13–20.

Jun J, Jian Z, Xianjin D, et al. 340GHz frequency multipler without matching circuit based on Schottky diodes[J]. INFRARED AND LASER ENGINEERING. 2014, 43(12): 4028-4034.

Miao L, Jiang J, Wang C, et al. A 340GHz Sub-Harmonic Mixer Based on Planar Schottky Diodes: IRMMW-THz[Z]. 20141-4.

Yue He, Xianjin Deng, Cheng Wang and Bin Lu, "W-band Quadrupler based on Multi-Chip Module and Schottky Barrier Diodes", Key Engineering Materials Vols.645- 646, pp80- 85,2015

Experimental Investigation of a Twin-Bridges Superconducting Switch

Boon-Kok Tan*‡, Ghassan Yassin*, Faouzi Boussaha† and Christine Chaumont†
*Department of Physics (Astrophysics), University of Oxford, Keble Road, Oxford OX1 3RH, UK
†Observatoire de Paris, Pôle Instrumental-GEPI – 77 Avenue Denfert-Rochereau, 75014 Paris, France.
‡Contact: boonkok.tan@physics.ox.ac.uk, phone +44 1865 273 303

Abstract—**We present the design and some preliminary measured results of a planar superconducting on/off switch comprising two niobium nitride (NbN) bridges deposited across the slotline section of a unilateral finline. The two bridges are separated by a distance of λ/4, such that the superconducting impedance of the bridges could be cancelled out at the resonance frequency. Both the NbN bridges were switched from the superconducting state to the normal state via a bias current exceeding the critical current of the NbN film. A millimetre wave source calibrated with a terahertz power meter is used to illuminate the switch, and the response of the switch in each state was measured using a superconductor-insulator-superconductor (SIS) chip as a direct detector. Preliminary measured results agreed generally well with our simulations, especially when the multiple wave reflection effect is included in our model.**

I. Introduction

A static superconducting on/off switch that can be easily integrated as part of the planar detector circuit would simplify significantly the design of a millimetre and sub-millimetre large format receiver array. An important example is to form a planar phase-switching circuit using a pair of switches, replacing the rotating half-wave plate or rotating waveguide that often are bulky and consist of moving parts. This is particularly useful for the construction of a highly sensitive pseudo-correlation polarimeter used to measure the B-mode polarisation signal of the Cosmic Microwave Background [1], [2]. The superconducting on/off switch can also be used to form an on-chip modulator to reduce the $1/f$ noise of an astronomical receiver.

II. Resonantly-Tuned Twin-Bridges Superconducting Switch

In previous papers [3], [4], [5], we have presented the design and the measured performance of a 220 GHz superconducting on/off switch comprising a number of niobium nitride (NbN) bridges deposited across a slotline. Here, we exploited the strongly current-dependent complex impedance of the NbN film near its critical current value to operate the bridges as a switch. This binary change of impedance can be instigated by applying a DC bias current across the bridges. This way, any incoming RF signal would see two substantially different complex impedance states, hence will either pass through the transmission line with minimal loss or reflected back to the input port with high return loss [6].

The complex impedance of a superconducting bridge is composed of two parts: R_s, the resistive part of its surface impedance, and the inductance part $L = L_g + L_k$ where L_g is the geometric inductance and L_k is the kinetic inductance. The kinetic inductance arise only in the superconducting state where $R_s \approx 0$. In the normal state $R_s = R_N$, its thin film normal resistance. Mathematically,

$$Z_S = i\omega(L_k + L_g), \quad \text{and} \tag{1a}$$
$$Z_N = R_N + i\omega L_g, \tag{1b}$$

where $\omega = 2\pi f$ is the angular frequency of the incoming RF signal, and Z_S and Z_N are the impedance of the switch at the superconducting and the normal states respectively.

For an ideal switch, the bridges should has an impedance closed to zero in one state, and an impedance approaching infinity in another state. However, even with a highly resistive superconducting film such as niobium titanium nitride (NbTiN) or NbN films, the normal resistance is still finite for a strip with a certain realistic geometry. On the other hand, at very high angular frequency, the superconducting kinetic inductance could induce a very high complex impedance that is closed to this normal resistance value. Therefore, the switching ratio (i.e., the difference in power transmission between the two states) is limited primarily by the properties of the available superconducting material.

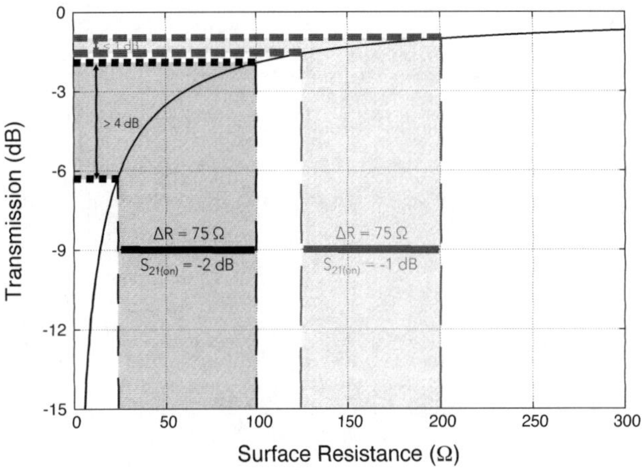

Fig. 1. The nonlinear relation between the surface resistance of a superconducting bridge shunting a slotline and its power transmission characteristic.

In previous attempts, we employed a number of short bridges across the slotline, instead of a single-shunting strip, to manoeuvre the total impedance of the switch at both states. This concept is illustrated in Figure 1, which plots the nonlinear relation between the surface impedance of a superconducting strip across a slotline and its power transmission characteristic. As can be seen, by halves the surface impedance value, the difference in power transmission between the superconducting and the normal state could be improved by several dB level. However, a fundamental disadvantage of this solution is that the employment of multiple bridges unavoidably decrease the RF transmission when the switch is closed, where the insertion loss S_{21} is now reduced from -1 dB to -2 dB.

A potentially much more effective solution to improve the switching ratio without affecting the power transmission characteristic of the switch is by creating a parallel resonant circuit using a twin-bridge tuning structure. Here, a second NbN bridge is placed a quarter-wavelength away from the first NbN bridge, so that the complex impedance of the first bridge is conjugated by the presence of the second bridge, and therefore cancelling out the unwanted superconducting inductance. In this case, the incoming RF signal would propagate through the transmission line unimpeded, since the bridges are virtually disappear now. On the other hand, when the bridges are in the normal state, the bridges become a low resistive load and the incoming wave will be reflected with minimal ohmic losses.

Figure 2 shows the predicted power transmission in both superconducting and normal state for a switch comprising two 50 nm thick NbN bridges (5 μm $\times$ 2 μm) separated by a 300 μm long slotline (5 μm width), simulated using Ansys High Frequency Structure Simulator (HFSS). As can be seen, the power transmission near 235 GHz is closed to 0 dB when the bridges are superconducting. At this central frequency, the switching ratio is closed to 8 dB, and decreases only gradually away from the resonance frequency. Although this resonance-bridges design is inherently narrow band at about 10–15 GHz, but it minimise the transmission losses significantly compared to the previous multi-bridges design. Hence, it would suit well with applications that do not require ultra-wide operational bandwidth, such as astronomical observations where the spectral line position is well-known.

III. Preliminary Results

Based on this idea, we fabricated a series of superconducting on/off switch chips with different resonance frequencies in the range of 200–260 GHz. They comprise two NbN bridges of 2 μm width deposited across a 5 μm wide slotline, supported by a 100 μm quartz substrate. The NbN bridges are modulated between the superconducting and the normal states via a DC current that alternate above and below the NbN critical current value. The RF signal is fed to the device via a unilateral finline taper, as shown in Figure 3, and transmitted through the switch via another similar taper. The chip is housed within a rectangular waveguide along the E-plane, with the front end of the waveguide block connected to a millimetre horns, as shown in Figure 4 and 5. The RF signal from a local oscillator (LO) is coupled to the horn with a pair of parabolic mirror and a beam splitter. A THz power meter is placed after the beam splitter to monitor the strength of the LO output signal to ensure the consistency of the power level across the measured bandwidth. The transmitted signal through the switch is measured at the other end by observing the pumped DC current-voltage (IV) curves of a superconductor-insulator-superconductor (SIS) chip, placed several millimetre away from the switch chip along the same rectangular waveguide

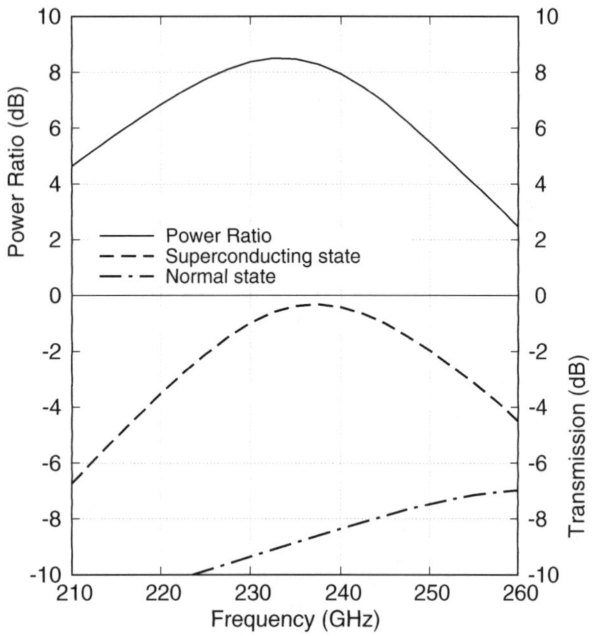

Fig. 2. The response of a superconducting on/off switch with two identical 50 nm thick NbN bridges separated by a quarter wavelength long slotline.

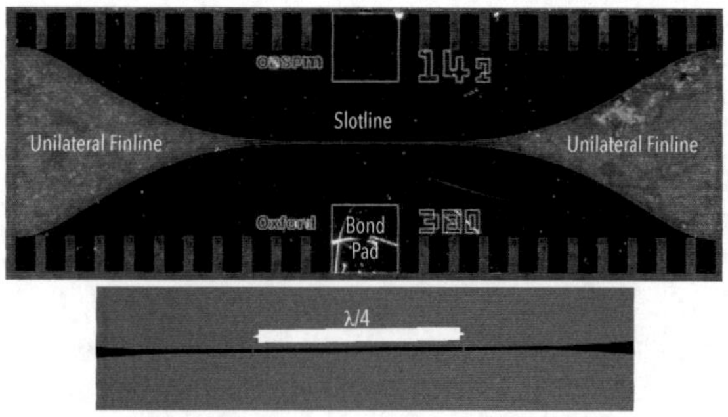

Fig. 3. A planar superconducting switch chip comprising two resonantly-tuned NbN bridges deposited across the slotline section of a back-to-back unilateral finline taper.

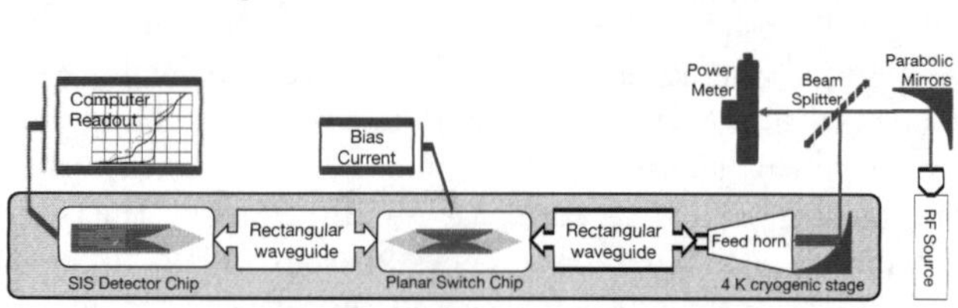

Fig. 4. Experimental setup for measuring the response of the superconducting on/off switch using an SIS device as a direct power detector.

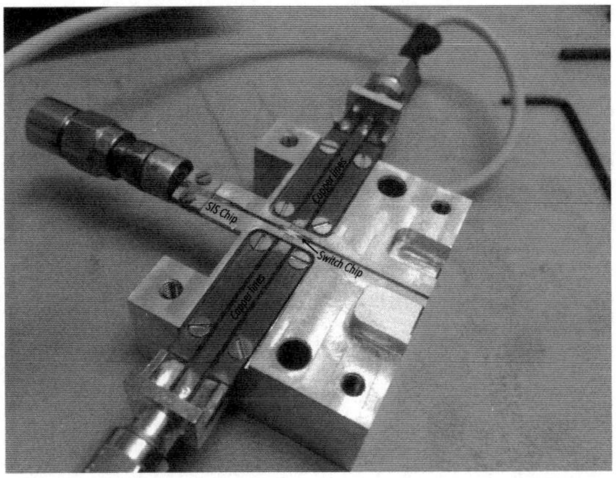

Fig. 5. Both the SIS detector chip and the switch chip are supported across the split-plane of a rectangular waveguide with a set of deep grooves on each side of the waveguide wall. The bias current for the switch is supplied through two copper line with bond wires connecting the copper strip to the switch's electrode. The SIS device is biased using a copper line and bond wires from the rear of the block.

channel. The detail of the SIS detector chip can be found in [9].

In Figure 6, we show the measured responses of a NbN switch with $\lambda/4 = 300\ \mu$m, in the frequency range of 210–260 GHz. It is clearly seen that the pumping levels of the SIS device changed by a few dB when the switch is alternated between the superconducting and the normal state. The highest switching ratio was measured at about 8 dB level, consistent with the HFSS prediction albeit a shift of frequency which appears to correspond to a twin-bridges design where the separation distance is 280 μm instead of 300 μm. However, it hard not to notice that the switching ratio varies periodically with frequency, although the general trend of the gradual decrease away from the central frequency is observed. We suspect that this periodically variation is caused by the existence of standing waves established by multiple wave reflections between the switch chip and the SIS detector chip. In our current setup, both the chips were fabricated on a relatively thick (100 μm) quartz substrate, and both the chips do not have matching notches to taper the impedance mismatch between the unloaded waveguide and the chips.

To further investigate the interaction between the switch chip and the SIS device, we used SuperMix, a quantum mixing software package developed at Caltech [10]. The SuperMix model was formed by cascading the HFSS calculated scattering parameters of both the switch (in either state) and the SIS detector chip, along with a section of empty waveguide 12 mm long (estimated for the distance between the chips) inserted between the two chips. This allow us to estimate the pumping level of the SIS device when the switch is alternated between the superconducting and the normal state, taking into account the frequency-dependent power coupling behaviour of the SIS device. Figure 6 (b) shows the result of this simulation, and it can be seen clearly now that a periodic dependence does exist if the empty waveguide is included in the model. The general behaviour of the predicted switching curve also matches very well to the measured one. This therefore reaffirm the effect of

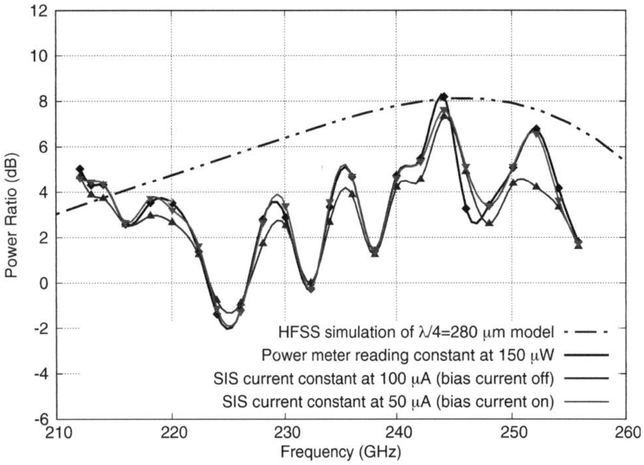

(a)

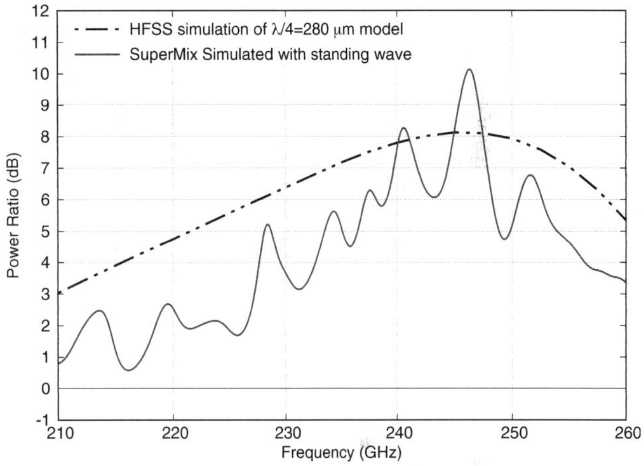

(b)

Fig. 6. (a) The power ratio measured from the SIS IV curves at a fixed bias point when the switch is alternated between superconducting and normal state. The solid black curve shows the power ratio variation with the LO input power fixed at 150 μW, overplotted with the HFSS predicted behaviour with the distance between the two NbN bridges at 280 μm. The blue and red curves show the similar power ratio variation, with the blue curve representing the case where the SIS tunnelling current (at the fixed bias point) is maintained at 100 μA level when the bias current to the switch is disconnected; while the red curve for the case where the tunnelling current is fixed at 50 μA when the biasing to the switch is on. All three measurements show an almost identical switching behaviour. (b) The power ratio variation of the switch estimated by the SuperMix model, including a section of empty waveguide between the two chips. The periodic variations and the peak of switching ratio near 245 GHz is highly resemblance of the behaviour measured experimentally.

standing waves that masked the actual behaviour of the switch.

Despite the issue with the standing wave, we have demonstrated experimentally that the resonantly-tuned superconducting bridge design can be used as an efficient on/off switch in millimetre circuits. The general behaviour, such as the power transmission when the bridges are superconducting is higher than the normal state, as well as the gradual roll-off away from the resonance frequency, follow closely the trend predicted by our calculation. Works are currently underway to minimise the multiple reflection effect by reducing significantly the distance between the two chips, and potentially fabricating both the switch and the detector circuit on the same chip.

IV. Conclusion

We have presented the design and the measured responses of a resonantly-tuned planar superconducting on/off switch comprising two NbN bridges deposited across the electrodes of a back-to-back finline chip. We simulated the performance of these resonantly-tuned superconducting switches using 3-D electromagnetic package, and the simulation model shows that at resonance, it has close to unity transmission while retaining relatively large switching ratio across the designated bandwidth. Using an SIS device as a direct power detector, we have measured a maximum switching ratio of $\sim$8 dB at 245 GHz, although the typical power ratio varies considerably across the 210−260 GHz due to the standing wave interaction between the switch and the SIS chip. Nevertheless, we managed to reproduce the measured results using SuperMix, taking into account the coupling behaviour of the SIS chip and the effect of an empty waveguide between the chips. The simulated result agrees very well with the measurements, establishing our understanding that the performance deviated from the intended design was caused by the introduction of unwanted standing waves. The measurement of the remaining devices within the same batch is still on going and works are underway to improve the measurement setup so that we can measure the response of the switch more prominently without the influence of the standing waves.

References

[1] P. K. Grimes, G. Yassin, L. S. Kuzmin, P. D. Mauskopf, E. Otto, M. E. Jones, and C. E. North, "Investigation of planar switches for large format CMB polarization instruments," in *Society of Photo-Optical Instrumentation Engineers (SPIE) Conference Series*, vol. 6275 of *Society of Photo-Optical Instrumentation Engineers (SPIE) Conference Series*, July 2006.

[2] G. Yassin, L. S. Kuzmin, P. K. Grimes, M. Tarasov, E. Otto, and P. D. Mauskopf, "An integrated superconducting phase switch for cosmology instruments," *Physica C Superconductivity*, vol. 466, pp. 115–123, Nov. 2007.

[3] B.-K. Tan, G. Yassin, E. Otto, and L. Kuzmin, "Experimental investigation of a 220 ghz superconducting switch with multiple nano-bridges," in *Millimeter Waves and THz Technology Workshop (UCMMT), 2015 8th UK, Europe, China*, pp. 1–4, 2015.

[4] B.-K. Tan, G. Yassin, E. Otto, and L. Kuzmin, "Investigation of the Dynamic Range of Superconducting Nano-Bridge Switches," in *Twenty-six International Symposium on Space Terahertz Technology*, 2015. .

[5] B.-K. Tan, G. Yassin, E. Otto, and L. Kuzmin, "Experimental Investigation of a Superconducting Switch at Millimetre Wavelengths," *IEEE Transaction on Terahertz Science and Technology*, pp. 121–126, jan 2016. .

[6] B.-K. Tan, G. Yassin, L. Kuzmin, E. Otto, H. Merabet, and C. North, "A Superconducting Millimetre Switch with Multiple Nano-Bridges," in *Twenty Fifth International Symposium on Space Terahertz Technology*, Apr. 2014.

[7] V. I. Belitskii, M. A. Tarasov, S. A. Kovtoniuk, L. V. Filippenko, and O. V. Kaplunenko, "Low noise completely quasioptical SIS receiver for radioastronomy at 115 GHz," *International Journal of Infrared and Millimeter Waves*, vol. 13, pp. 389–396, Apr. 1992.

[8] P. Grimes, *Design and analysis of 700 GHz Finline Mixers*. PhD thesis, University of Cambridge, United Kingdom, 2006.

[9] B.-K. Tan, F. Boussaha, C. Chaumont, and G. Yassin, "A 220 GHz Finline Mixer with Ultra-Wide Instantaneous Bandwidth," in *Twenty Sixth International Symposium on Space Terahertz Technology*, Mar. 2015.

[10] J. Ward, F. Rice, G. Chattopadhyay, and J. Zmuidzinas, "Supermix: A flexible software library for high-frequency circuit simulation, including sis mixers and superconducting elements," in *Proceedings, Tenth International Symposium on Space Terahertz Technology*, pp. 269–281, 1999.

Broadband Antireflective Subwavelength Structures for Large Diameter Silicon Lenses

T. Nitta[1*], Y.Sekimoto[2], K. Noda[3], S. Sekiguchi[4], M. Sekine[4], S. Shu[4], Y. Yamada[1], S. Hisamatsu[1], H.Matsuo[2], A.Dominjon[2], M.Naruse[5], N.Kuno[1], and N.Nakai[1]

[1]*Institute of Physics, University of Tsukuba, 1-1-1 Tennodai, Tsukuba, Ibaraki 305-8577, Japan*
[2]*Advanced Technology Center, National Astronomical Observatory of Japan, 2-21-1 Osawa, Mitaka, Tokyo 181-8588, Japan*
[3]*Oshima Prototype Engineering Co., 3-10-28 Nishikubo, Musashino-shi, Tokyo, Japan*
[4]*Department of Astronomy, The University of Tokyo, 7-3-1 Hongo, Bunkyo-ku, Tokyo 113-0033, Japan*
[5]*Graduate School of Science and Engineering, Saitama University, 255 Shimo-Okubo, Sakura-ku, Saitama 338-8570, Japan*
* Contact: nitta.tomu.fw@u.tsukuba.ac.jp

We have been developing a millimeter-wave camera with microwave kinetic inductance detector (MKID) for ground based wide field-of-view observations. In our cold optics design,two silicon lenses with diameter of 200-mm and 300-mm are used as focusing elements. Silicon is ideal because of their low dielectric loss and high refractive index, but an anti-reflection coating is required to reduce reflections at the surfaces. To solve this problem, we introduce an antireflective subwavelength structure. The antireflective subwavelength structures have some merits compared with dielectric material coatings because the structure acts as an antireflective layer with only one material.

To get a larger bandwidth, taper structures with period of 265- μ m and depth of 700- μ m were fabricated on a silicon flat sample by a special dicing blade. The transmittance of structure was measured with a Fourier Transform Spectrometer at cryogenic temperatures. In the result, the measured average transmittance between 170- and 350-GHz was about 94%, and the measured result is in good agreement with the simulation. We present fabrication, simulation and measurement results.

Beam Pattern Measurements of a Picket-Potter Feed Hornat 1.9 THz

J. L. Kloosterman1*, K. K. Davis2, C.E. Groppi2, J. H. Kawamura1, M. Underhill2

[1]Jet Propulsion Laboratory *, Pasadena, CA, 91109 USA
[2]Arizona State University, Tempe, AZ, 85287 USA
* Contact: Jenna.L.Kloosterman@jpl.nasa.gov

Pickett-Potter feed horns with a circular to rectangular waveguide transition have been designed,fabricated, and tested to advance the development of large format heterodyne arrays. These feed horns have severalnotable advantages over the standard diagonal feed horns because they have lower cross-polarization properties andare easier to design, machine, and assemble. We present beam pattern and cross-polarization measurements toverify performance of the new feed horn and waveguide transition design using a waveguide-coupled hot electronbolometer (HEB) mixer and a 1.9 THz multiplier chain local oscillator (LO) source. We employ an electronicallychopped LO source and a nitrogen cold load on an XY stage for measurements. Theoretical and measured beamwaists and cross polarization are in close agreement.

Transmission and Reflection Properties of Dielectric Materials for THz Instrumentation

A. O. Pienkina[1,2], M. C. Wiedner[1*] , F. Defrance, Y. Delorme[1]
[1]LERMA, Observatoire de Paris, PSL Research University, CNRS, Sorbonne Universités, UPMC Univ. Paris 06, F-75014, Paris, France
[2] PhLAM, UMR8523, CNRS - Université Lille 1, Villeneuve d'Ascq, 59650, France
*Contact: martina.wiedner@obspm.fr, phone +33-1-50 41 20 32

Abstract— **The objective of this work is to present the results of measuring the refractive index and absorption coefficient of different materials (HDPE, Mylar, Zitex, Teflon, and Silicon) at 600 GHz and 1.4 THz at a temperature of 293K. The knowledge of these material properties at THz frequencies is essential in order to design low loss optics for astronomical receivers. This work draws particular attention to the polarization dependence of the reflection and transmission, which we carefully measured and compared to theory.**

INTRODUCTION

In radio astronomy, high density polyethylene (HDPE) is one of the most widely used materials for lens systems and cryostat windows. Mylar is commonly used in beam splitters (such as diplexers in heterodyne receiver or in Fourier-transform spectrometers (FTS)) and also as cryostat windows. Zitex is often used as a cryogenic IR blocking filter. Silicon has found extensive use in quasi-optic lenses and also as cryostat windows. The angular and polarization dependent transmission of these materials is investigated.

MEASUREMENTS

We developed instrumental set-ups (Fig. 1) using a solid state LO source for these measurements.

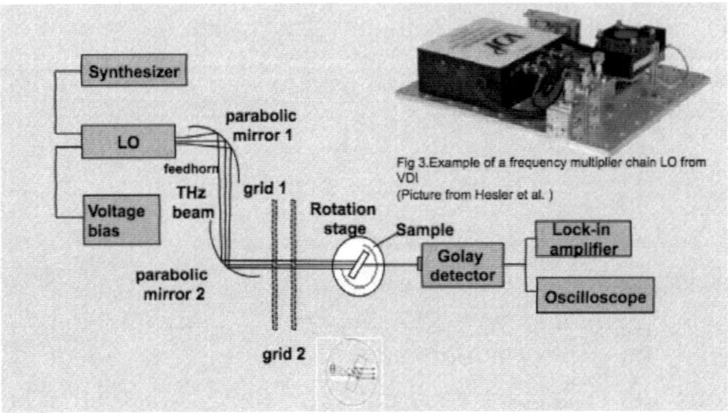

Fig. 1 Experimental set-up to measure the transmission of different materials as a function of angle and polarization.

In a first step we showed that the angular dependence of the transmission can be well described by formulas, see example fit to data at 1.4 THz.

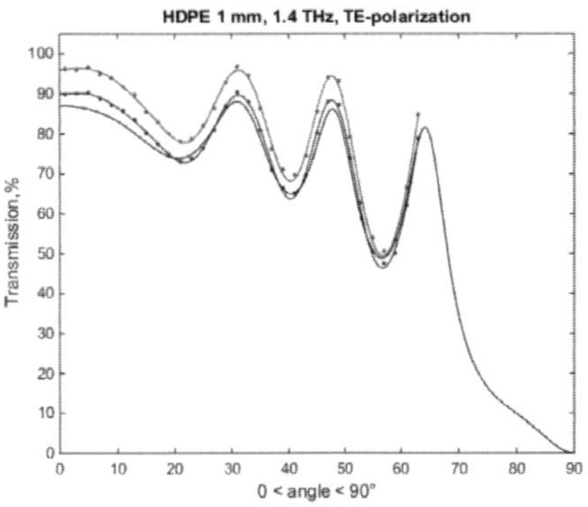

Fig. 2 Transmission of 1mm thick HDPE at 1.4 THz as a function of incident angle. The red curve shows the maximum, the blue curve the minimum of the measurements (due to standing waves, noise of the detector etc.) and the black line is the theoretical curve using the best fitting parameters for the refractive index n and the absorption coefficient alpha.

Once these were established and our measurement accuracy improved we used the formulas to derive the thickness of the sample, the refractive index and the absorption coefficient.

TABLE I
BEST FIT VALUES FOR MATERIAL PROPERTIES FOR MEASURMENTS AT 1.4 THZ

Material	**Properties**		
	Thickness	**n Refrac. index**	**α (mm^{-1}) absorption coeff**
HDPE	1mm	1.524	0.01-0.03
	2mm	1.523 - 1.524	0.01-0.05
Mylar	100 μm	1.7-2.0	0.8-1.5
	50 μm	1.7-2.0	0.7-2.0
	25 μm	1.70-1.75	1-3
	9 μm	1.7-1.8	1-4

Zitex	100 μm	1.2-1.25	1-4
Silicon	300 μm	3.415	0

Conclusions

We have carried out transmission measurements of different material commonly used in astronomical receivers at THz frequencies. The measurements were taken for different incident angles and different polarization and allowed us to derive the material properties.

We stress the importance of taking the polarization of the beam and its incident angle into account when building THz instrumentation.

Acknowledgment

The present work was funded by the French ANR under the Contract No. ANR-13-BS05-0008-02 IMOLABS.

Corrugated Horns for ALMA band 11 (1.25-1.57 THz)

A. Gonzalez[*1], K. Kaneko[1], T. Kojima[1], S. Asayama[1], and Y. Uzawa[2]

[1]National Astronomical Observatory of Japan*, Mitaka, Tokyo 181-8588, Japan
[2]National Institute of Information and Communications Technology, Koganei,Tokyo, 184-8795, Japan
* Contact: Alvaro.Gonzalez@nao.ac.jp

During the ALMA band 10 receiver production at NAOJ, up to 200 corrugated horns were fabricated and tested in the 787-950 GHz band. Of those, around 160 showed good performance in terms of co- and cross-polarization patterns. Many of these horns were fabricated in Japan by direct machining of aluminum. The acceptable yields suggest the possibility of using direct machined corrugated horns at even higher frequencies using the experience acquired from ALMA band 10 corrugated horn production. The main difficulty in the fabrication process was to mill the first corrugations in the horn throat, which are always the deepest. A way to simplify fabrication and presumably increase the yield is to shorten the horn by using a profile different from conical. A simple profile composed of connected conical sections has been used in the design of a horn for a future ALMA band 11(1.25-1.57 THz). This horn shows goodsimulated performance but cannot be modeled by conical corrugated horn quasi-optical models. For this, a new modeling method based on numerical near-field Gaussian beam fitting of the simulated fields has been developed. This model has been used to perform quasi-optical designs of relay optics, which show good simulated performance using Physical Optics.

The designed profiled corrugated horn is being fabricated and will be delivered in late January 2016. It will then be characterized at NAOJ, using an already established near-field beam measurement system. For comparison, a traditional long conical corrugated horn based on a re-optimized scaled version of the ALMA band 10 horn is also being manufactured.

This paper will present the proposed design method for THz profiled corrugated horns, together with the newly developed quasi-optical modeling technique. The results of simulations will be compared with measurements to validate the horn design. If available in time, measurements of the long conical corrugated horn will also be included and compared with those of the profiled corrugated horn. In short, this paper will assess the possibility of fabrication of corrugated horns for a future ALMA band 11 by state-of-the-art direct machining.

ACKNOWLEDGMENTS

Part of this work was made possible by the Visiting Scholar Program grant supported by the Research Coordination Committee, National Astronomical Observatory of Japan (NAOJ).

Fast On-the-Fly Near-field Antenna Measurement at 500GHz

J. Hu[1, 2, 3], Z. Lou[1, 2], J. P. Yang[1, 2], W. Miao[1, 2], Z. H. Lin[1, 2], S. C. Shi[1, 2]
1. Purple Mountain Observatory, Nanjing, China
2. Key Lab of Radio Astronomy, CAS, Nanjing, China
3. University of Chinese Academy of Sciences, China
Contact: jiehu@pmo.ac.cn

***Abstract*—A near-field antenna measurement system using On-the-Fly (OTF) method is described here. Compared with the step-and-integrate technique, which samples discrete positions at certain step, the OTF scan obtains the near-field data while the scanner is still moving, thus greatly reducing the measurement time. Measurement done at 500GHz is presented here, showing the OTF method can improve the efficiency by a factor of nearly 2.5 and the accuracy of the data is comparable with that get from the step-and-integrate method. The scan effect of OTF is also discussed**

I. Introduction

NEAR-FIELD measurements widely used to characterize electromagnetic radiation properties of a variety of antenna and quasi-optical systems at terahertz frequency [1]. It is often difficult to meet the criterion of far-field distance of $2D^2/l$ at short wavelength [2]. Besides, the feed antennas, especially those integrated with superconducting receivers in the cryostat, are preferred to bemeasured in the near-field range, as it is difficult to rotate the cryostat [3]. It is also important to fully characterize the feed horns in the near-field because the beam coupling usually takes place in the near-field region [4].

The usual method used to sample near-field data is step-and-integrate technique. It samples the near-field at fixed discrete positions, which means the scanner has to stop to sample the data. Such method is more vulnerable to the changes in the system, as it usually takes hours to map the entire 2-D data [5-7].

In order to reduce the measurement time, we apply the On-the-Fly (OTF) technique [8] to the near-field measurement system. The OTF technique is a commonly method used in astronomical observing. Contrasted to the step-and-integrate mapping, the OTF records the near-field data and scanner position information in a continuous way while the scanner is moving smoothly and rapidly across the near field. There are several advantages using the OTF mapping. First of all, the sample time is significant reduced, as there is no need to start and stop the scanner. Second, the system change will be reduced as the entire measurement time is less.

OTF measurement of the spiral antenna integrated with HEB at 500GHz has beentaken based on the near-field antenna measurement developed at Purple Mountain Observatory (PMO) [9]. This paper will first describe the measurement system, then the scanning effect of OTF will be discussed. Finally, the result of OTF scan will be given and it will be compared with that from step-and-integrate technique.

II. System Setup

The measurement system is shown in Fig.1. The detector of the system is a superconducting hot electron bolometer (HEB) mixer [10] integrated with the spiral antenna, which is mounted on the back of an elliptical lens with a diameter 10mm.

The RF and LO is generated by microwave synthesizer followed by an AMC with a multiplication factor of 36 separately. The RF and LO are injected into the HEB mixer by a beam splitter made of a 25 micron-thick Mylar film. The 360 MHz IF signal is amplified by a cryogenic HEMT low-noise amplifier and then by two stages of room-temperature amplifiers. The Probe of RF signal is a diagonal horn, which is mounted on a motor controlled XY scanner, which covers a span of 300 mm in each axis and has a resolution of 10 micron.

The phase reference for the test signal recorded by mixing the output of RF and LO base frequency generated by synthesizers, which produces the 10 MHz IF signal.

Data acquisition is done by a dual-channel high-speed ADC with a sampling rate of 800 MHz and a resolution of 12 bit/sample. The 360 MHz IF signal and 10 MHz reference are recorded simultaneously. Digital FFT is applied to obtain the spectrum of both channels.

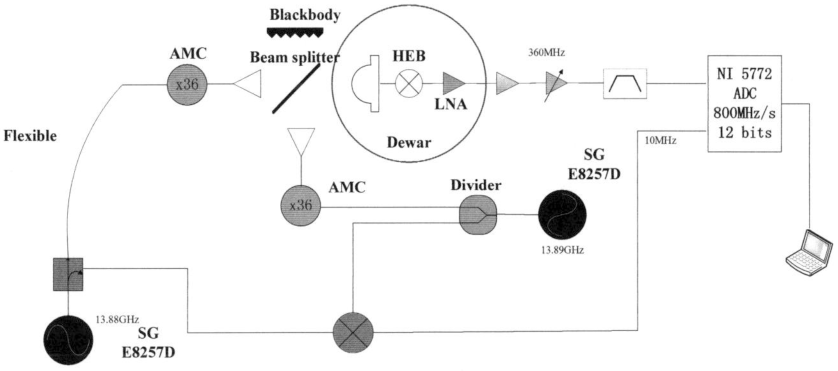

Fig. 1 Schematic of system layout

III. OTF IMPLEMENTATION

The measured co-polarized near-field data of the spiral antenna that is integrated with HEB mixer is shown in Fig. 2. The near-field of the antenna has been scanned on a regular rectangular grid of 80´ 80 mm. The used scan pattern is meandric along the vertical axis, meaning that horizontal scans (x direction) from left to right and vice-versa are taken, each time increasing the vertical position (y direction). The calibration is done by sampling a fixed point in the scanning area when a vertical scan is over [3]. The total scan time of OTF is about 900 s at the scan speed at 20 mm/s, which is about 2.5 times faster than the step-and-integrate scan.

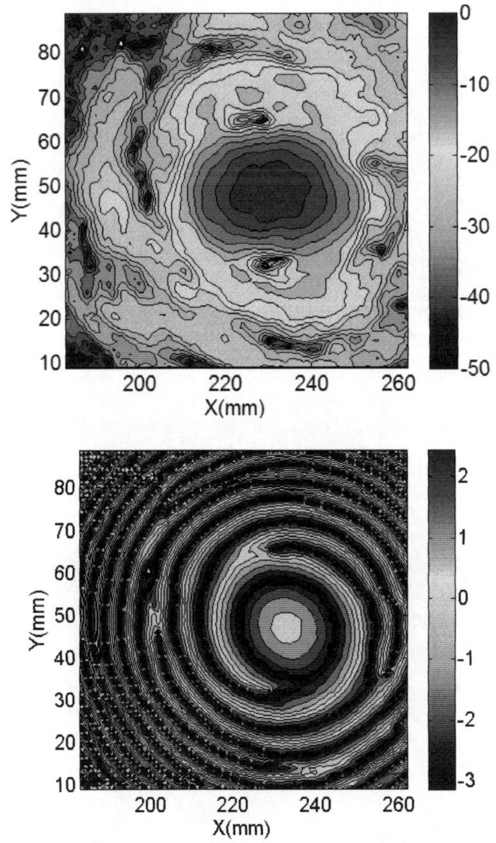

Fig. 2 Measured near-field amplitude and phase.

The near-field to far-field (NTF)transformationis done to compare the result of the OTF scan and the step-and integrate scan. The H-plane and E-plane result are shown in Fig. 3. No probe compensation is taken into consideration. The E-plane is exactly the same with that get from step-and-integrate. The H-plane shows a bit difference. This is because the scan direction is horizontal (x-direction), the receiver gain and phase instability affects the measurement data more across the y-direction [9].

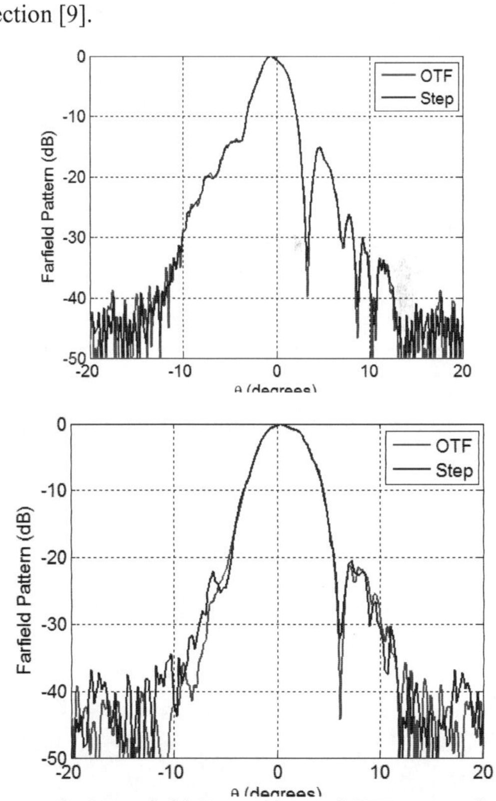

Fig.3 Far-field Comparison of OTF scan and Step-and-integrate scan. a) E-plane b) H-plane

IV. CONCLUSION

A successful OTF scan of the spiral antenna integrated with HEB is done at 500 GHz, which shows about 2.5 times faster than the step-and-integrate technique. Also, the far-fields get from the two technique are nearly the same.The OTF scan probably shows better accuracy as the scan time is much shorter and the system amplitude and phase drift are much smaller.

REFERENCES

[1] A. D. Yaghjw, "An overview of near-field antenna measurements", *IEEE Trans. Antennas Propagat.,* Vol. 34, No. 1, pp. 30-45, 1986.

[2] C. A. Balanis, Antenna Theory: Analysis and Design, 2nd ed. John Wiley &Sons, 1997.

[3] C. E. Tong, S. Paine and R. Blundell, "Near-field characterization of a 2-D beam pattern of submillimeter superconducting receivers," in *Proc. 5th Int. Symp. On Space Terahertz Technol.*, Michigan, USA, May 10-12, 1994, pp. 660-673.

[4] P. F. Goldsmith, Quasioptical Systems: Gaussian Beam Quasioptical Propagation and Applications. IEEE Press, 1998.

[5] P. Fuerholz and A. Murk, "Phase-corrected near-field measurements of the TELIS telescope at 637 GHz," *IEEE Tran. Antennas and Propagat.*vol. 57, no. 9, 2009.

[6] Y. Fujii, A. Gonzalez, M. Kroug, K. Kaneko, *et al.*, "The first six ALMA band 10 receivers," *IEEE Trans. THz Sci. Technol.*, vol. 3, no. 1, pp. 39-49, Jan. 2013.

[7] A. Gonzalez, Y. Fujii, T. Kojima, and S. Asayama, "Reconfigurable near-field beam pattern measurement system from 0.03 to 1.6 THz," *IEEE Tran. THz Sci. Technol.*, vol. 3, no. 2, pp 300-305, 2016.

[8] J. G. Mangum, D. T. Emerson, and E. W. Greisen, "The on the fly imaging technique," *Astronomy &Astrophysics*, vol. 474, no. 2, pp 679-687, 2007.

[9] Z. Lou, J. Hu, K. M. Zhou, W. Miao *et al.*, "A quasi-optical vector near-field measurement system at terahertz band," Rev. Sci. Instr., vol. 85, no. 6, 2014.

[10] W. Zhang, L. Jiang, Z. H. Lin, Q. J. Yao, *et al.*, "Development of a quasi-optical NbN superconducting HEB mixer," in *Proc. 16th Int. Symp. On Space Terahertz Technol.*, Goteborg, Sweden, 2005, pp. 209-213.

[11] J. Tuovinen, A. Lehto, and A. Raisanen, "A new method for correcting phase errors caused by flexing of cables in antenna measurements," IEEE Tran. Antennas and Propgat., vol. 39, no. 6, pp. 859-861, 1991.

A Three-disc Window based on Triangular Lattice of Dielectric Rods for High Power Gyro Amplifiers

Yelei Yao, Guo Liu, Fang Li, Liya Yang, Yong Luo

University of Electronic Science and Technology of China, Chengdu, Sichuan610000, China

A three-disc microwave window has a matching dielectric disc at each side of the central ceramic disc for W-band gyro amplifiers is studied. The two discs are made of two dimensional triangular lattice of BeO rods to obtain a lower effective relative permittivity as well as keep a high thermal conductivity. Numerical simulation results show a better than -15 dB reflection over the operating frequency band of 0.09~0.11THz. The propose structure can meet the strict requirements of the high power broadband gyro amplifiers.

A WR-4 Optically-Tunable Waveguide Attenuator with 50 dB Tuning Range and Low Insertion Loss

Zhenguo Jiang[1], Md I. B. Shams[1], Syed M. Rahman[1], Patrick Fay[1], Cheuk-yuE. Tong[2] and Lei Liu[1*]
[1]*University of Notre Dame, Notre Dame, IN 46556, USA*
[2]*Harvard-Smithsonian Center for Astrophysics, Cambridge, MA, USA*
*Contact: lliu3@nd.edu

Abstract—**We report the development of a compact WR-4 (170-260 GHz) optically-tunable waveguide attenuator based on the interaction between electromagnetic waves and photo-induced free carriers in semiconductors. This approach is promising for achieving superior performance, including a ~50 dB tuning range, low insertion loss, lowreturn loss (VSWR), and high tuning speed. Based on full-wave simulation results, an average attenuation level of ~50 dB can be achieved using a 1 mm long Ge absorber at a light intensity of 1 W/cm^2. By employing an E-plane taper design and energy absorption mechanism, return loss lower than -13 dB have been achieved. A prototype attenuator with Si absorber has been implemented using an E-plane split waveguide design. Initial measurement results show that ~14 dB attenuation can be achieved using a single illumination spot. The attenuation level can be improved by increasing the illumination region length employing multiple fibers.**

I. Introduction

Tunable attenuators/modulators are important components in millimeter wave and terahertz (THz) wave systems because many applications require the capability of controlling and varying the power of THz signals. Mechanically tunable waveguide attenuators have been widely used at microwave frequencies[1]. However, mechanically tuning approaches have the disadvantages of low tuning speed and high insertion loss. Although free space optically- and electronically-tunable attenuators have been demonstrated at THz range [2], [3], they have not yet been applied in waveguide configurations due to the challenges associated with feeding light and integrating circuits and components into waveguide structures.

In this paper, we report the development of optically-tunable terahertz waveguide attenuators with superior performance including high tuning speed, low insertion loss, low return loss and large tunable range. This is achieved using E-plane tapered high-resistivity semiconductor chips (absorbers) aligned in the wave propagation direction. The tunability is realized by illuminating the semiconductor chip using fiber-guided IR laser diodefor providing different levels of light intensity. For a prototype demonstration, a WR-4 waveguide attenuator has been designed and fully simulated, presenting lower than 1 dB insertion loss, lower than -15 dB reflection and a tuning range of ~20 dB using a Si absorber. A prototype tunable waveguide attenuator employing Si absorber has been implemented. Initial measurement results show that ~14 dB attenuation can be achieved using a single illuminationspot of ~0.5 mm × 0.5 mm. This tuning range can be further increased to larger than 50 dB usinggermanium (Ge) absorberand/orlonger illumination region based on simulation.

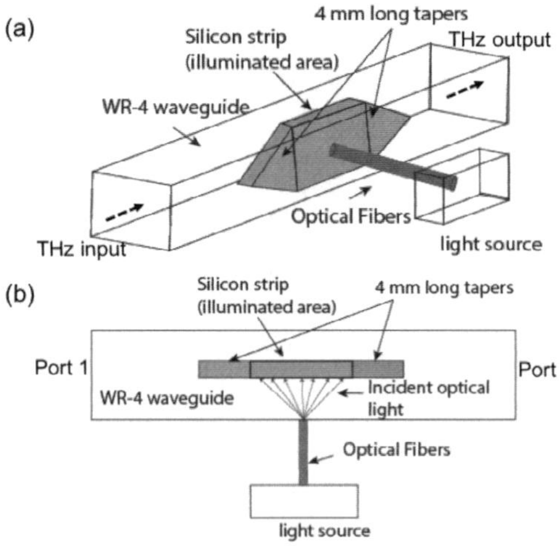

Fig. 1. The structure of a WR-4 tunable waveguide attenuator using photo-induced Si as absorber: (a) 3-D, (b) top view.

II. Optically-TunableWaveguide Attenuator

Photo illuminated/induced semiconductor(e.g. Si, Ge) can be employed for effective attenuation/modulation of THz waves [4]-[6]. In order to design a high performance tunable WR-4 waveguide attenuator, a physics-based modelfor a photo induced Si wafer was established at the frequency range of 170-260 GHz. Calculation resultsusingthis physics-based model[7] indicate that at low light intensity, energy absorption dominates the attenuation process. When the light intensity is higher (e.g., larger than 2 W/cm^2) the semiconductor becomes very conductive and reflection dominates. The absorption-dominated attenuation with a lower light intensity can be adopted for developing tunable THz waveguide attenuators with superior performance.Based on this mechanism, a WR-4 tunable waveguide attenuator has been designed and simulated. In order to minimize the reflection, the proposed tunable waveguide attenuator employs an E-plane high resistivity semiconductor chip (e.g. Si or Ge) aligned in the wave propagation direction as an absorber, as shown in fig. 1(a). To further reduce reflection for optimum matching, the silicon absorber is chosen to have a thickness of only ~70 μm with both ends trimmed to be 4 mm long taper structures. As shown in fig. 1(b), the center region of the absorber is designed to be approximately 1 mm long for a maximum attenuation of~20 dB (for Si absorber). This region will be illuminated by optical fiber guided laser diode through the narrow wall of the waveguide.

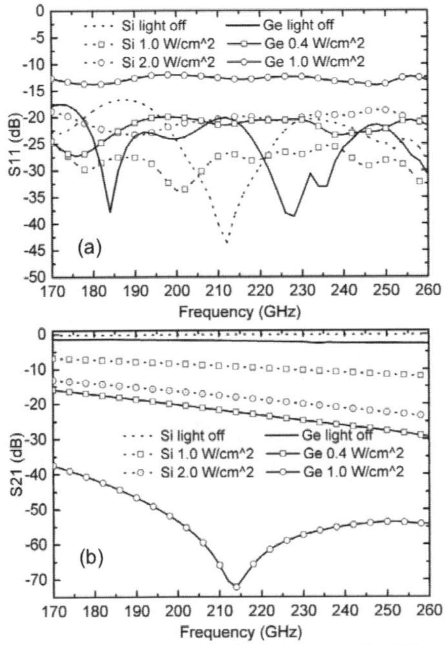

Fig. 2. Simulated s-parameters of the proposed WR-4 waveguide attenuator at different levels of illuminating light intensity.

The s-parameters of tunableattenuators implemented with Si and Ge absorbers have been simulated in HFSS at different levels of light intensity, as shown in fig. 2. Over the entire WR-4 frequency band, the attenuation can be continuously tuned with a maximum averaged value of ~20 dBfor Si based attenuator. More than 50 dB attenuation level can be achieved using Ge absorber at a lower illuminatingintensity due to Ge material's longer free carrier lifetime and higher carrier mobility. Owing to the E-plane tapered structure, THz reflection at the interface can be minimized resulting in a low simulated insertion loss.Taking into account the ohmic loss introduced by the waveguide walls, the overall device insertion loss is estimated to be lower than 1 dB for Si based absorber.Additionally, the return loss without light is lower than -13 dB owning to the absorption-dominated mechanism. The proposed attenuator can be easily implemented using E-plane split waveguide design and scaled to higher frequencies.

III. Prototype and Initial Measurement

A prototype Si based WR-4 tunable waveguide attenuator based on the above design has been implemented as shown in fig. 3 (a). A Si chip with taper structure and 1 mm long absorbing region was fabricated and installed to the block. The absorbing region was designed to be illuminated by two optical fibers through the holes. Initial measurements of tunable attenuation performance with only one illuminating spot have been performed and the results are shown in fig. 3 (b).During this testing, the Si chip was illuminated by a fibercoupled 808 nm laser diode under various biasing currents (light power levels). The quick initial test was performed using a WR-5 VDI source and a WR-10 Erickson powermeter. Therefore, the measured transmission curves show strong standing wave effect due to the reflection at both ends of the attenuator. It can be seen that tunable attenuation with a tuning range of up to ~14 dB can be achieved using a single illuminating spot of ~0.5 mm × 0.5 mm. We expect that ~ 20 dB tuning range can be achieved with a device VSWR smaller than 1.2 by using lower levels of light intensityand two illuminating spots. This result demonstrated that the proposed approach was promising for realizing THz tunable waveguide attenuator with high performance.

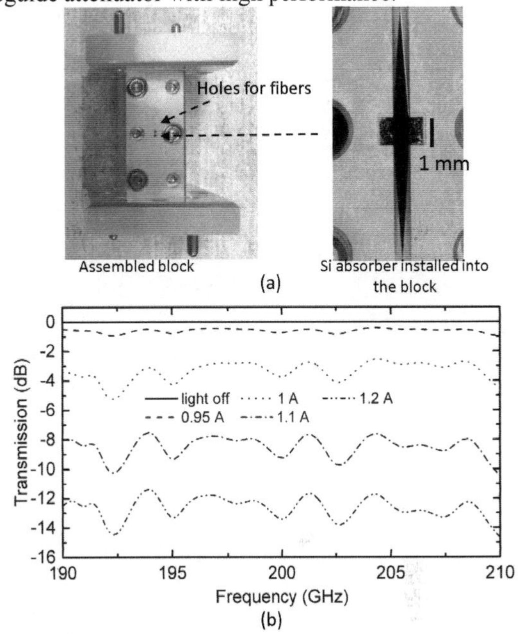

Fig. 3. (a) Pictures of the prototype WR-4 tunable waveguide attenuator. (b) Initial measurement results of the tunable attenuation operation (THz transmission at various driving currents) for the prototype attenuator.

IV. Conclusions

A novel approach for developing optically-tunable terahertz waveguide attenuators has been reported. A prototype high-speed tunable waveguide attenuator at WR-4 band has been designed and simulated. Based on the full-wave simulation results, an average of 20 dB attenuation level with an insertion loss lower than 1 dB can be achieved using Si absorber. Higher attenuation level can be obtained using longer absorber or semiconductor materials (e.g. Ge) with longer free carrier life-time and higher carrier mobility. A tuning speed of ~30 kHz has been estimated based on the linear recombination coefficient in silicon. Initial measurement results show that ~14 dB attenuation can be achieved using a single spot (~0.5 mm × 0.5 mm area)for illumination.Improved attenuator with longer absorber region employing 4 illuminating fibers (2 mm absorber region length) that can achieve ~50 dB attenuation will be fabricated and fully characterized soon.

Acknowledgment

This project is partially supported by NSF grant ECCS 1202452. The authors would like to thank NDnano at

University of Notre Dame and Harvard-Smithsonian Center for Astrophysics.

REFERENCES

[1] R. Sundararajn, E. Peterson, and R. Nowlin, "Attenuators." *Wiley Encyclopedia of Electrical and Electronics Engineering*, 1999.

[2] L-J. Cheng, and L. Liu, "Optical modulation of continuous terahertz waves towards cost-effective reconfigurable quasi-optical terahertz components." *Optics express*, vol. 21, no. 23, pp. 28657-28667, 2013.

[3] B. Sensale-Rodriguez et al., "Broadband graphene terahertz modulators enabled by intraband transitions." *Nature communications* 3, pp. 780, 2012.

[4] M. I. B. Shams et al., "Approaching real-time terahertz imaging with photo-induced coded apertures and compressed sensing." *Electronics Letters*, vo. 50, no. 11, pp. 801-803, 2014.

[5] M. Itrat B. Shams et al., "A Terahertz Reconfigurable Photo-Induced Fresnel-Zone-Plate Antenna for Dynamic Two-Dimensional Beam Steering and Forming," *International Microwave Symposium*, Phoenix, Arizona, 2015.

[6] A. Kannegulla et al., "Photo-induced spatial modulation of THz waves: opportunities and limitations," *Optics express*, vol. 23, no. 25, pp. 32098-32112, 2015.

[7] N. V. Smith, "Classical generalization of the Drude formula for the optical conductivity." *Physical Review B*, vol. 64, no. 15, pp. 155106, 2001.

Development of submicron high precision CFRP reflector

Liang Xu*, Yongjie Xie, Jiaoteng Ding, Yongjie Wang, Zhen Ma, Xuewu Fan
Xi'an institute of optics and precision mechanics, Chinese Academy of Sciences, Xi'an 710119, China
*Contact: liangx@opt.ac.cn, phone +86-029-88887674

Abstract—**Antenna gain affected by reflector surface figure accuracy and dimension stability directly, so one of the most important tasks is how to ensure the surface precision and dimensional stability. It is hard to control surface precision for springback of metal, so carbon fibre reinforced polymer (CFRP) usually be adopted to fabricate high precision reflector. With the rapid development of electronic technology, especially millimetre and terahertz wave technology, the precision of reflector needed increasingly. A Φ300mm CFRP flat reflector is developed for process study. In order to improve the thermal stability, a special "all CFRP" structure adopted. Optical replica process used to realize surface modification of CFRP reflector blank, final surface figure accuracy RMS reaching 0.1 μ m, and roughness Ra reaching 2nm. Further thermal stability tests show that the thermal stability reaching 13nm/℃. A Φ500mm CFRP aspherical reflector also fabricated, and surface accuracy reaching 0.4 μ m. The study is of certain reference value for the development of CFRP reflector in millimetre wave and terahertz wave band.**

INTRODUCTION

Due to the low density, high stiffness, low thermal expansion coefficient, etc., carbon fibre reinforced polymer (CFRP) material usually used to fabricate space antenna reflector. With the rapid development of electronic technology, especially millimeter wave and terahertz technology, there are challenges and opportunities for carbon fibre composites industry to develop antenna reflector with higher figure accuracy [1].

90s of last century, Composite Optics Incorporated (COI) conducted a large number of sub-millimeter and even infrared reflector, including SAO, JPL, MLS, and FIRST verification mirror [2] [3]. EADS-Astrium GmbH developed a 1.1m all CFRP reflector in order to accumulate technology for PLANCK [4], and surface accuracy RMS reaches 4μm, heat distortion only 2μm within a temperature range of 140 degrees.

A Φ300mm flat and a Φ500mm aspherical CFRP reflector developed in this study. The figure accuracy RMS is superior to 0.1 μ m, the thermal stability reaches 13nm/℃.The work is of certain reference value for high precision reflector fabrication.

PROCESS FLOW OF CFRP REFLECTOR

CFRP reflector comprises a front panel, back panel and meshgrid reinforced structure, and the same CFRP material system choice for the three parts. Taking into account the in-plane isotropic and process convenience, triangular meshgrid structure used.

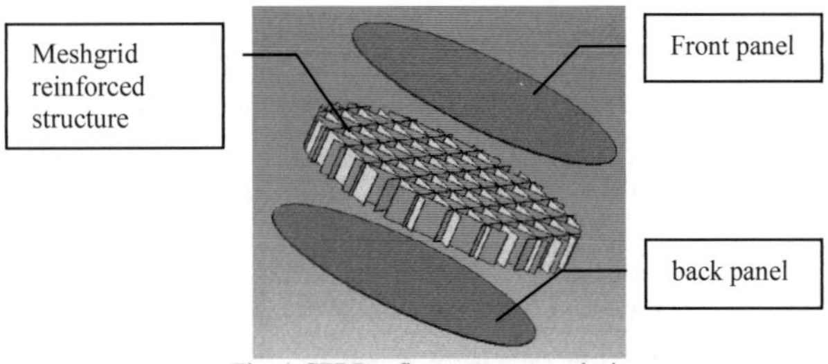

Fig. 1 CFRP reflector structure design

The process flow of high precision CFRP reflector shown in Figure 2:

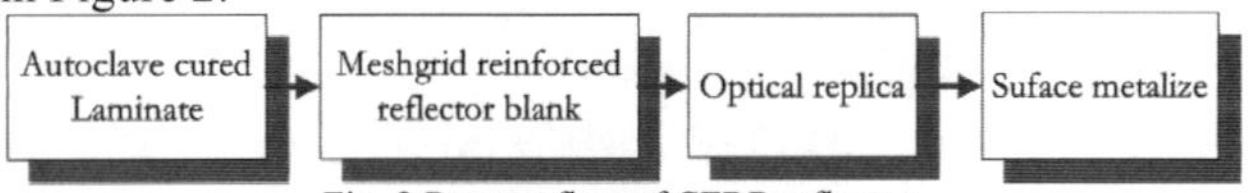

Fig. 2 Process flow of CFRP reflector

Material choice

In this study, epoxy matrix HM4J/3236 material selected to fabricate reflector, cured at 120 ℃temperature.

Autoclave moulding

In order to improve thermal stability, it requires the use of quasi-isotropic laminate. However, due to manufacture deviation of each ply, even quasi-isotropic laminate, still produces thermal deformation. So it needs to reduce thermal deformation from both ply design and manufacture deviation [5].Angle error with hand layup is about ± 2°, laminates with poor figure accuracy after autoclave curing. One automated fibre placement machine used and angle error can be controlled to ± 0.1°.

Reflector Blank

Taking into account the structure character of the CFRP laminate, the honeycomb sandwich structure is one of the most effective means to increase rigidity and thermal stability. Here a patented core structure designed with excellent thermal stability performance. The process flow of CFRP reflector blank shown in Figure 3.

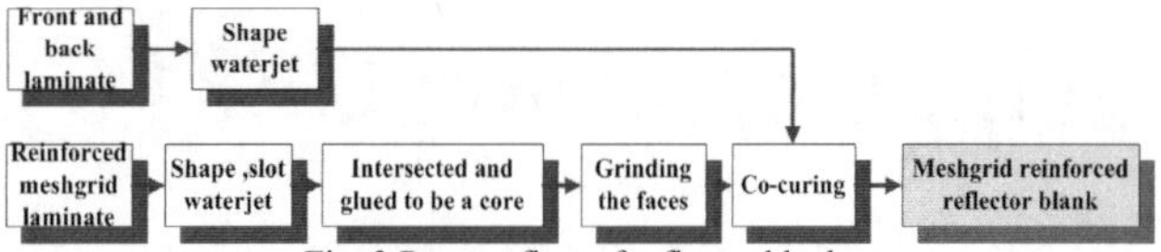

Fig. 3 Process flow of reflector blank

Optical Replica Process

Since the two-phase material properties of carbon CFRP, it can't be used as reflective surface, here the optical replication process selected for surface modification.

The main process flow of optical replication shown in Figure 4:

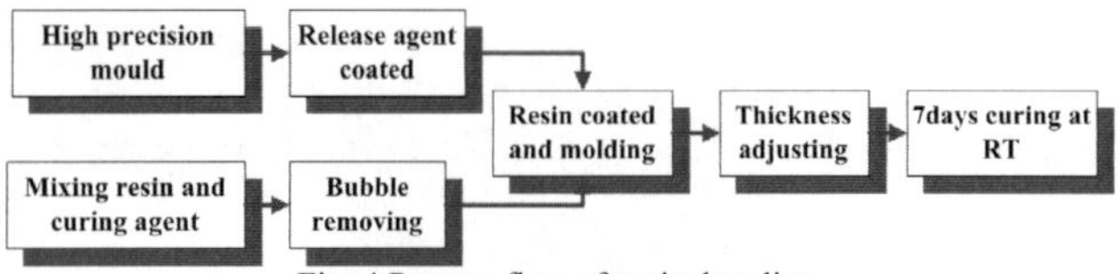

Fig. 4 Process flow of optical replica

Optical replica finished after 7 days curing at RT. Completed Φ300mm CFRP flat reflector shown in Figure 5. Figure accuracy is about 0.1μm, as shown in Figure 6, at 20.2 ℃ room temperature environment. Roughness Ra reached 2nm, completely copying the roughness of the mould.

Fig. 5 Φ300mm CFRP reflector completed

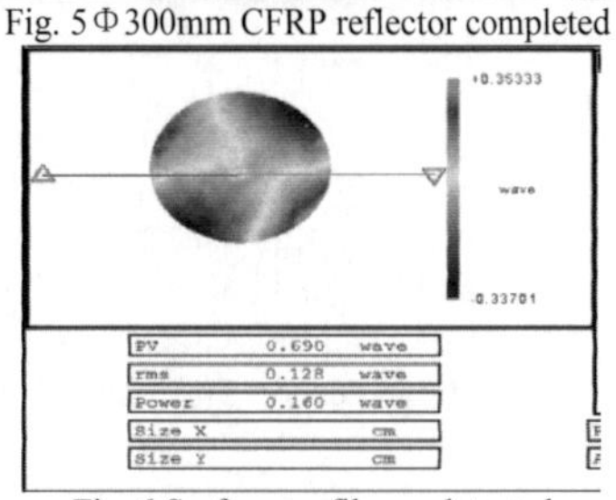

Fig. 6 Surface profile result tested

THERMAL STABILITY TEST

In order to verify thermal stability of the Φ300mm carbon fibre reflector, the laboratory temperature reduced from 20.2℃ to 11.4 ℃, surface accuracy variation shown in Figure 7.

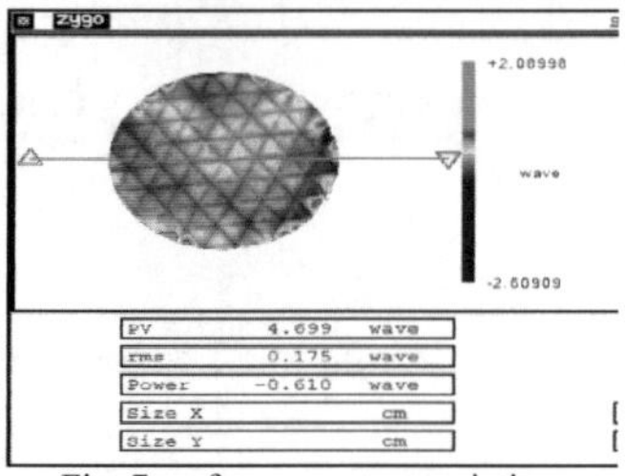

Fig. 7 surface accuracy variation

Figure 7 shows the deformation is mainly "grid effect", RMS variation 0.11 μm for 8.8 ℃ temperature difference, so its thermal stability is about 13nm/℃.

Φ500MM PARABOLIC REFLECTOR

Fig. 8 Φ500mm CFRP reflector completed

A Φ500mm CFRP parabolic reflector fabricated as shown in figure 8, and the figure accuracy RMS can achieve 0.4 μ m. The figure accuracy is enough for millimetre wave and terahertz application.

CONCLUSIONS

1) A Φ300mm CFRP flat reflector fabricated, Surface accuracy RMS eventually reached 0.22 μ m, roughness 3nm. Thermal stability reached 13nm/℃.

2) Meshgrid reinforced reflector blank adopted. The patented cell structure improves thermal stability.

3) The optical replication process can be used to modify surface, figure accuracy downgrade existed in current replica technology, and roughness completely copied.

4) Thermal deformation of "all CFRP" reflector is mainly about "grid effect".

5) Optical replication layer is not conductive, the metalize processing needed to improve the electrical properties of the reflector.

REFERENCES

[1]Liaohui Ao. "Application of CFRP in antennas," Telecommunication Engineering , 38[2]:41-44,(1998). (in Chinese)

[2]James P.Gormican, Shel Kulick and Eldon P.Kasl. Dimensionally Stable Graphite-fiber reinforced composite mirror technology. SPIE,1993, V1945.

[3]Abdel Abusafieh Dan Federico and others.Dimensional Stability of CFRP Composite for Space Based Reflectors.Optomechanical Design and Engineering, 2001, V4444.

[4]G.Helwig. Highly Dimensional Stable Composite Structures.Workshop on advanced Materials for High precision Detectors. 1994.

[5] XU Liang, XIE Yong-jie, DING Jiao-teng, etc, "The design and optimization of carbon fiber laminates based on thermal stability and bending rigidity," Fiber Reinforced Plastics/Composites, 2, 57-61, (2016).(in Chinese)

Development of Broadband Planar Ortho-mode Transducer with MKID for LiteBIRD Satellite

Shibo Shu* †, Shigeyuki Sekiguchi* †, Masakazu Sekine* †, Yutaro Sekimoto* †, Tom Nitta† ‡, Agnes Dominjon†, Takashi Noguchi†, Masato Naruse§, and Wenlei Shan †

* Department of Astronomy, Graduate School of Science, University of Tokyo, Japan

† Advanced Technology Center, National Astronomical Observatory of Japan, 2-21-1 Osawa, Mitaka, Tokyo, Japan

‡ Research Fellow of Japan Society for the Promotion of Science

§ Graduate School of Science and Engineering, Saitama University, Saitama, Japan

***Abstract*—We report on a design of broadband circular waveguide coupled planar ortho-mode transducer (OMT) with Microwave Kinetic Inductance Detector (MKID) for LiteBIRD mission, a small-size satellite for cosmic microwave background (CMB) polarization signal full-sky mapping at large angular scale by JAXA. In our 4-pixel prototype design, each single pixel is sensitive to two frequency bands (90 GHz and 150 GHz) corresponding to atmospheric window for testing at Nobeyama 45-m telescope. Silicon on insulator (SOI) has been selected for OMT structure and a broadband coplanar waveguide (CPW) 180-degree hybrid is designed to cancel higher modes of a circular waveguide and add two signals from the fundamental mode together. After a distributed microstrip bandpass diplexer, a microstrip line to coplanar waveguide transition structure couples signal to MKID and MKIDs are read out with frequency domain multiplexing. MKIDs are designed with Nb ground plane and Al/Ti bilayer central strip to achieve low frequency response, high sensitivity and also adjustable transition temperature. A 4-pixel module is under test and we plan to deploy these multi-chroic polarimeters on Nobeyama 45-m telescope.**

I. Introduction

WITH the successful scientific results of CMB from space missions of COBE [1], WMAP [2] and Planck [3], the temperature anisotropies and E-mode polarization signal of CMB have been well studied in the last two decades[4], [5]. However, for the B-mode polarization at large angular scale, which is believed to be generated by primordial gravitational wave after the Big Bang, is still not detected [6]. For future full sky B-mode polarization mission, background-limit multi-chroic detector array is necessary to achieve high sensitivity. LiteBIRD [7] is a next-generation satellite mission to measure the primordial B-modes polarization signals of CMB. The goal of LiteBIRD is to measure the tensor-to-scale ratio r to an uncertainty of $r = 0.001$ during 3 years observation. The LiteBIRD working group is currently considering two technologies for detection: Transition Edge Sensor [8] or MKID [9]. We are developing corrugated horn coupled OMT-MKID focal-plane for LiteBIRD[10]. For space mission, the radiation experiment has been done with aluminium MKID and we found no significant changes on resonator quality factor, responsivity, recombination time of quasi-particles and noise level[11]. This paper presents a prototype corrugated horn coupled octave-band planar OMT design with MKID for LiteBIRD mission.

In this design the incident wave from a corrugated horn is coupled to four planar OMT probes and coupled to high impedance CPW. An octave band corrugated horn array from 80-160 GHz has been fabricated with direct-machining at Advanced Technology Center of NAOJ and their beam pattern has been measure in [12]. After a CPW 180-degree hybrid structure which is based on [13], each polarization signal is added together and higher mode signals are sent to an absorber, which has been demonstrated by [14]. A CPW to microstrip transition structure is applied for following MS diplexer. This diplexer consists of 90 GHz and 150 GHz passbands with two 5-element Chebyshev MS shorted stub bandpass filters [15]. After the band separation, a CPW and MS combination structure dissipates the signal on a CPW central line [16], which is the central strip of MKID and is made of Al/Ti bilayer with gap frequency 70 GHz [17]. All other circuits and ground plane are made of Nb for lossless transmission line (<660 GHz).

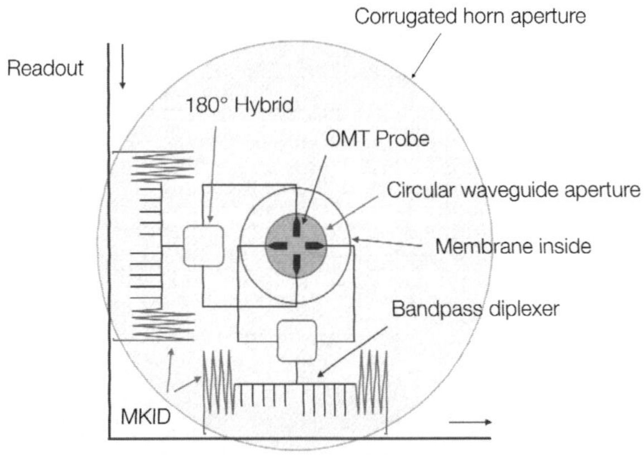

Fig. 1. Schematic figure of a single pixel design (not scaled). Inside the red circle there is only silicon membrane and outside is SOI wafer

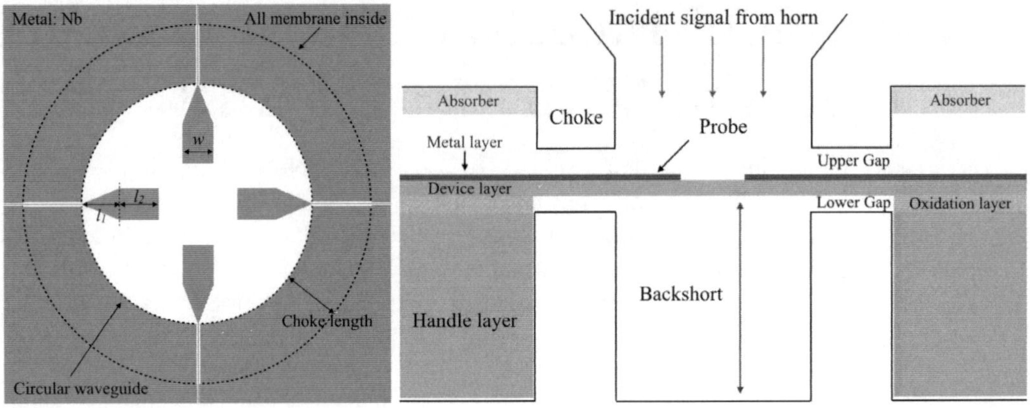

Fig. 2. Schematic figure of an OMT structure (not scaled). Left panel shows a top view of the OMT. Right panel shows a cross-sectional view of the OMT design. After choke ring, an absorber is attached to the top of circuits for absorbing leaky radiation from the gap.

II. Planar OMT Design

A planar OMT is designed with frequency range 80 - 160 GHz coupled after a circular waveguide. We note that this work follows the planar OMT design by McMahon et al. [18]. Figure 2 shows the planar OMT design with a silicon-on-insulator (SOI) wafer. Four OMT probes are suspended on a silicon membrane with a quarter wavelength backshort at the end of waveguide. Each probe is connected to a CPW with an impedance of $125\,\Omega$. After a short high impedance transmission line, the impedance of CPW on membrane is changed to $96\,\Omega$ on the SOI part with $20\,\mu\text{m}$: $3\,\mu\text{m}$: $20\,\mu\text{m}$ geometry. Since the length of choke structure is quarter wavelength, this $125\,\Omega$ transmission line acts as a quarter wavelength impedance transformer. The impedance of probe is calculated with $(125\,\Omega)^2/96\,\Omega \approx 160\,\Omega$.

TABLE I. Design parameters for the planar OMT

Waveguide diameter	2.4 mm
Choke length	500 μm
Backshort distance	500 μm
Probe width w	270 μm
Probe length l_1	300 μm
Probe length l_2	600 μm
125 Ω CPW geometry	26 μm : 3 μm : 26 μm
96 Ω CPW geometry	20 μm : 3 μm : 20 μm
Device layer thickness	6 μm
Insulator layer thickness	1 μm
Handle layer thickness	400 μm

Table I shows the design detail of the SOI wafer and the OMT structure. Careful simulations have been taken with HFSS [19] for optimizing the probe size, backshort distance and choke length. A simulation of the performance of the planar OMT is shown in Figure 3 with straight line profile for impedance transition part. The TE11 mode of the circular waveguide co-polarization coupling rate is 88.3% averaged for entire frequency range and cross-polarization is smaller than -60 dB. For $90\,\text{GHz}$ band and $150\,\text{GHz}$ bands, the average coupling rate is 91.3% and 91.6% respectively. Choke structure is defined by a device holder with metal boundary, which is made of aluminium. As mentioned by McMahon et al.[20], a $100\,\mu\text{m}$ gap of choke gives radiation loss less than 1.5% and here we keep both upper and lower gaps $35\,\mu\text{m}$ distance. This loss radiates from upper and lower gaps and may cause strong resonance with other metal boundaries. Therefore, an absorber is attached to the top part of the device holder.

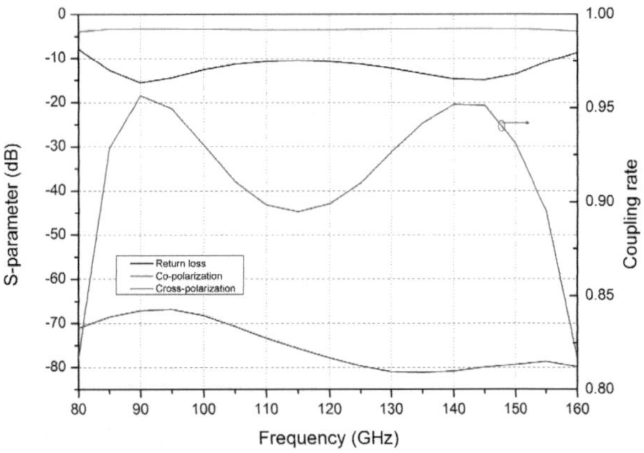

Fig. 3. Simulation result of an OMT. Co-polarization is the energy coupled to one side probe. Coupling rate shows co-polarization power of $TE11$ mode coupled to two probes with 88.3 % from 80 GHz to 160 GHz. Cross-polarization level is smaller than $-60\,\text{dB}$ in this frequency range.

III. Conclusions

We reported the design of an octave-band planar OMT coupled with MKID for a prototype MKID solution for the LiteBIRD mission. Two frequency bands of 90 GHz and 150 GHz are designed for a single pixel. An OMT design is realized with 6 m silicon membrane of an SOI wafer. The simulation result of the planar OMT shows co-polarization coupling rates are 91.3% and 91.6% averaged for 90 GHz band and 150 GHz bands respectively and cross-polarization is smaller than -60 dB.

ACKNOWLEDGMENT

The authors would like to thank Takafumi Kojima and Alvaro Gonzalez for their technical support. This work was supported by MEXT/JSPS KAKENHI Grant Numbers 25247022, 15H05743, 15H05891. This work was partially supported by JSPS Core-to-Core Program, A. Advanced Research Networks.

REFERENCES

[1] J. C. Mather *et al.*, "A preliminary measurement of the cosmic microwave background spectrum by the Cosmic Background Explorer (COBE) satellite," *Astrophysical Journal*, vol. 354, pp. L37–L40, May 1990.

[2] C. L. Bennett *et al.*, "The Microwave Anisotropy Probe Mission," *Astrophysical Journal*, vol. 583, pp. 1–23, Jan. 2003.

[3] J. M. Lamarre *et al.*, "The Planck High Frequency Instrument, a third generation CMB experiment, and a full sky submillimeter survey," *New Astronomy Review*, vol. 47, pp. 1017–1024, Dec. 2003.

[4] G. Hinshaw *et al.*, "Nine-year Wilkinson Microwave Anisotropy Probe (WMAP) Observations: Cosmological Parameter Results," *The Astrophysical Journal Supplement*, vol. 208, p. 19, Oct. 2013.

[5] Planck Collaboration *et al.*, "Planck intermediate results. XLVI. Reduction of large-scale systematic effects in HFI polarization maps and estimation of the reionization optical depth," *ArXiv e-prints*, May 2016.

[6] BICEP2/Keck and Planck Collaborations *et al.*, "Joint Analysis of BICEP2/Keck Array and Planck Data," *Physical Review Letters*, vol. 114, no. 10, p. 101301, Mar. 2015.

[7] M. Hazumi *et al.*, "LiteBIRD: a small satellite for the study of B-mode polarization and inflation from cosmic background radiation detection," in *SPIE Astronomical Telescopes + Instrumentation*, 2012, p. 844219.

[8] K. Irwin and G. Hilton, *Cryogenic Particle Detection*. Berlin, Heidelberg: Springer Berlin Heidelberg, 2005, ch. Transition-Edge Sensors, pp. 63–150. [Online]. Available: http://dx.doi.org/10.1007/10933596_3

[9] P. K. Day, H. G. LeDuc, B. A. Mazin, A. Vayonakis, and J. Zmuidzinas, "A broadband superconducting detector suitable for use in large arrays," *Nature*, vol. 425, no. 6960, pp. 817–821, 2003.

[10] S. S. Y. Sekimoto *et al.*, "Design of corrugated-horn-coupled MKID focal plane for LiteBIRD," in *SPIE Astronomical Telescopes + Instrumentation*, 2016.

[11] K. Karatsu, A. Dominjon, T. Fujino, T. Funaki, M. Hazumi, F. Irie, H. Ishino, Y. Kida, T. Matsumura, K. Mizukami, M. Naruse, T. Nitta, T. Noguchi, N. Oka, S. Sekiguchi, Y. Sekimoto, M. Sekine, S. Shu, Y. Yamada, and T. Yamashita, "Radiation Tolerance of Aluminum Microwave Kinetic Inductance Detector," *Journal of Low Temperature Physics*, feb 2016. [Online]. Available: http://link.springer.com/10.1007/s10909-016-1523-y

[12] M. S. S. Sekiguchi *et al.*, "Direct machined broadband corrugated horn array for millimeter observations," in *ISSTT 2016*, 2016.

[13] C.-H. Ho, L. Fan, and K. Chang, "New uniplanar coplanar waveguide hybrid-ring couplers and magic-T's," *IEEE Transactions on Microwave Theory and Techniques*, vol. 42, no. 12, pp. 2440–2448, Dec. 1994.

[14] G. Engargiola and R. L. Plambeck, "Tests of a planar L-band orthomode transducer in circular waveguide," *Review of Scientific Instruments*, vol. 74, no. 3, p. 1380, 2003.

[15] D. M. Pozar, *Microwave engineering*. Wiley. com, 2009.

[16] P. Day, H. Leduc, A. Goldin, T. Vayonakis, B. Mazin, S. Kumar, J. Gao, and J. Zmuidzinas, "Antenna-coupled microwave kinetic inductance detectors," *Nuclear Instruments and Methods in Physics Research Section A: Accelerators, Spectrometers, Detectors and Associated Equipment*, vol. 559, no. 2, pp. 561–563, 2006.

[17] A. Catalano, J. Goupy, H. le Sueur, A. Benoit, O. Bourrion, M. Calvo, L. Dumoulin, F. Levy-Bertrand, J. Macìas-Pérez, S. Marnieros, N. Ponthieu, and A. Monfardini, "Bi-layer Kinetic Inductance Detectors for space observations between 80-120 GHz," *arXiv:*, vol. 1511.02652, p. 7, Apr. 2015. [Online]. Available: http://arxiv.org/abs/1504.00281

[18] J. McMahon *et al.*, "Multi-chroic Feed-Horn Coupled TES Polarimeters," *Journal of Low Temperature Physics*, vol. 167, no. 5-6, pp. 879–884, Mar. 2012.

[19] ANSYS HFSS HP: http://www.ansys.com/.

[20] J. McMahon *et al.*, "Planar Orthomode Transducers for Feedhorn-coupled TES Polarimeters," in *American Institute of Physics Conference Series*, ser. American Institute of Physics Conference Series, B. Young, B. Cabrera, and A. Miller, Eds., vol. 1185, Dec. 2009, pp. 490–493.

Metamaterials-based terahertz filter

Zhenyu Zhao1,*, Wangzhou Shi1, and Wei Peng2
1 Department of Physics, Shanghai Normal University, Shanghai 200234, China
2 State Key Laboratory of Functional Materials for Informatics, Shanghai Institute of Microsystem and Information Technology, Chinese Academy of Sciences, Shanghai 200050, China

Abstract- **We investigate the terahertz (THz) responses of fractal concentric rectangular square resonators (CRS) induced by different mode coupling mechanisms. Near-field coupling results in the resonant mode redshift, while conductive coupling cause the results in the resonant mode blueshift. One can achieve a high Q mode with strong modulation depth by switching the capacitive coupling to conductive coupling in a fractal meta-atom.**

I. Introduction

The THz responses of metamaterials (MMs) are attributed to the shape, size, orientation, layout, and period of meta-atom (MA), which is a unit cell composed of resonators. Normally, the scale of single MAs needs to be ten times smaller than the operation wavelength, which restricts the miniaturization of device volume. A fractal structure-based MA pattern become a promising approach, which diverge the surface area of MAs within a finite volume[1]. It is found that one can achieve multi-frequency operation or broadband tenability in a MM composed of fractal MAs[1,2]. Actually, a fractal MA is a composite of many resonators at different scales. Therefore, it is significant to understanding the interaction between resonators of different scales in a fractal MA. The earlier works indicate that the interplay of resonating modes leads to unexpected THz response in coupled MAs based on a composite of wires and split-ring resonators, such as plasmon-induced transparency[3,4] and ultrasharp mode coming out[5]. The aforementioned two intrigue effects offer advantages for tunable filter and biosensor application. Therefore, it is significant to explore the THz response and mode coupling effect between the resonators of different fractal levels in one MA. we propose two types of fractal MAs made of concentric rectangular square resonators (CRS).In the first type of MA, each generated CRS is a simple duplication of the initiator CRS with a reduced length of boundary at the square root of 0.5to the adjacent CRS. The shape and orientation of generated CRS is totally the same as the initiator. This type of MA is termed as independent concentric rectangular square (I-CRS). In the second type of MA, the shape and reduction ratio of resonators are the same as I-CRS, however, the orientation of the generated one is rotated $\pi/2$ radius to the initiator. Since the reduction ratio is the square root of 0.5, the 4 vertices of smaller CRS resonators contact exactly the mid-point of the quadrilateral of adjacent larger CRS. This type of MAs is termed as junctional concentric rectangular squares (J-CRS). According to the fractal geometry [1,2], the fractal level refers to the number of generated smaller layer of CRS. As a consequence, the THz response of I-CRS and J-CRS can be compared at the same fractal scale. Due to the centro-symmetry of CRS, the THz polarization sensitive effect is excluded in our experiment. The THz transmittance of above fractal MAs are calibrated using a standard THz time-domain spectroscopy (THz-TDS) setup. The surface currents and electric field strengths of resonance modes are simulated. Finally, the origin of THz response of above two types of fractal MAs is discussed.

II. Results

A. Figures and Tables

The geometric parameters of fractal MAs are described as belows: The patterns of MAs are transferred onto 625 μm-thick <100>-oriented semi-insulating gallium arsenide (SI-GaAs) substrates by photolithography.The incident THz polarization and the pattern direction of I-CRS and J-CRS samples are presented in Fig. 1(a). Owing to the dielectric isotropy of <100>-oriented crystal, the normal line to the metal pattern layer is along with the crystallographic orientation of SI-GaAs. The relation between the surface

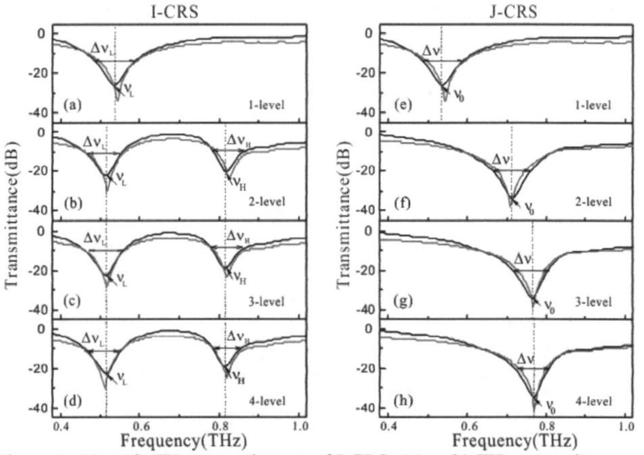

Figure 1. (a) ~ (d) THz transmittance of I-CRS; (e) ~ (h) THz transmittance of J-CRS. Black solid line: Experimental data. Red solid line: Simulation data. Dashed line: the central position of resonance modes

orientation of metal patterns and SI-GaAs substrate is illustrated in Fig.1 (b). The effective area of fractal MAs is 10 mm □10 mm. The MAs are metallized by a layer of 120 nm thick gold (Au) and 5 nm thick titanium (Ti). The lattice period is 100 μm, the quadrilateral-length of the first level CRS resonator is 65 μm, of which the width is 4 μm, respectively.

B. THz Response

The THz transmittance of both types of fractal MAs are presented in Fig 2. There are only two visible resonance modes in I-CRS. The high-order resonance frequencies of I-CRS are invisible even though the fractal is above 2. To J-CRS, however, the multiple resonance modes disappear while a single resonance mode occurs in the transmission spectrum of J-CRS. It is evident that ν_L and ν_H modes of I-CRS occurs redshift behavior with increasing the fractal level. Correspondingly, the ν_0 mode of J-CRS performs an obvious frequency blueshift. Meanwhile, the linewidth of ν_L and ν_H modes become narrower while that of ν_0 become broader. The results indicate that the Q factors of resonance modes of both MAs increase monotonically with the fractal levels. The surface current of resonance modes of both types of MAs are simulated to reveal the origin of variation of THz response.

C. Analysis

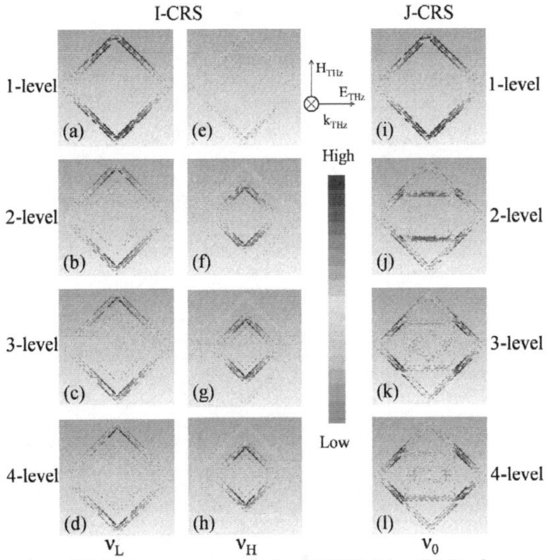

Fig. 2. (a) ~ (d) Surface currents of vL of I-CRS; (e) ~ (h) Surface currents of ν_H of I-CRS. (i) ~ (l); Surface currents of ν_0 of J-CRS. E_{THz}: the electric field of incident THz radiation. H_{THz}: the magnetic field of incident THz radiation. K_{THz}: the wave vector of incident THz radiation. Color bars: the relative strength of surface current.

It is evident that anti-parallel currents are produced in adjacent CRS resonators. The intensity of surface current indicates that single CRS experiences a strong coupling to the incident terahertz wave. When the fractal level increases from 2 to 4, however, the generated smaller CRS is weakly coupled to the incident THz radiation through capacitive interaction between the quadrilaterals of adjacent CRS. Therefore, a destructive interference of the scattered fields between the two adjacent resonators leads to the frequency redshift of ν_L, which is in agreement with the results in Fig.2. The similar phenomena are also observed in simulated surface current at ν_H, as shown in Fig. 3(e)~(h). The earlier works indicate that the capacitive mode coupling between two immediately adjacent resonators is proposed to be the origin of modes redshift. When the fractal levels increase up to 3 or 4, the outermost CRS and the innermost CRS undergoes much weaker coupling to the incident THz radiation field, hence the THz response for I-CRS mainly derives from the mode coupling of the biggest two adjacent resonators. On the other hand, there is no opposite surface current in J-CRS of different fractal levels. At the mode of ν_0, the current is strongly accumulated at the connected vertices of the adjacent CRS. The connected vertices play the role as the divergence points of the surface currents, which induce a current leakage from the outer large CRS to the inner small CRS. Since a single CRS works as a dipole oscillator, the effective length of the oscillator is reduced by the connection between the adjacent of CRS. The smaller the size of dipole oscillator, the higher the resonance frequency[5]. Therefore, the mode blueshift is attributed to the conductive coupling between the adjacent resonators in J-CRS.

III. Summary

In summary, the THz electromagnetic responses in fractal meta-atoms based on two-types of CRS resonators are investigated. In I-CRS, the capacitive coupling induces the redshift of multiple modes and reduces the MD. In J-CRS, the multiple modes are coupled conductively into a single resonance mode. The resonance modes appear to be blueshift in frequency spectra and the resonance strengths are increased when the fractal level increases. One can achieve a high Q mode with strong MD by switching the capacitive coupling to conductive coupling in a fractal MAs.

Acknowledgment

This work is financially supported by the Natioanl Natural Science Foundation of China (Grant No. 61307130).

References

[1]. F. Miyamaru, Y. Saito, M. W. Takeda, B. Hou, L. Liu, W. Wen, and P. Sheng, "Terahertz electric response of fractal metamaterial structures," Phys. Rev. B 77, 045124-1-045124-6 (2008).

[2].Q. Du, H.Yang, T.Lv, and X. Wang, "Multiband and polarization-independent left-handed metamaterial with cross fractal structure," Opt. Commun.301,74-77 (2013).

[3]. Y.Zhu, X. Hu, Y. Fu, H. Yang, and Q. Gong, "Ultralow-power and ultrafast all-optical tunable plasmon-induced transparency in metamaterials at optical communication range," Sci. Rep. 3, 02338-1-02338-7(2013).

[4]. X. Zhang, N. Xu, K. Qu, Z.Tian, R. Singh, J. Han, G.S. Agarwal, and W. Zhang, "Electromagnetically induced absorption in a three-resonator metasurface system," Sci. Rep. 5, 10737-1-10737-9 (2015).

[5]. I. Al-Naib, E. Hebestreit, C. Rockstuhl, F. Lederer, D. Christodoulides, T. Ozaki, and R. Morandotti, "Conductive Coupling of Split Resonators: A path to THz metamaterials with ultrasharp resonances," Phys. Rev. Lett. 112, 183903-1-183903-5 (2014).

Investigation of Temperature Dependence of Terahertz Spectra of Amino Acids

Ling Jiang*, Jiangping Yu, Yutian Xu, and Yunfei Liu

Nanjing Forestry University, Nanjing, JiangSu210037, China*
* Contact: jiangling@njfu.edu.cn

In this paper, we employ terahertz Time Domain Spectroscopy (THz-TDS) combined with temperature-tuned system to measurethe temperature dependence of terahertz spectroscopic features of various amino acids, such as β-Alanine, L-Tyrosine, L-Glycine, L-Asparagine and so on. We investigate the absorption frequencies of amino acids at different environmental temperatures, in order to study the temperature dependence of THz spectra.The measured absorption spectra arecompared with the results simulated by Density Function Theory (DFT) using hybrid functional B3LYP with basis set of 6-31 G (d).In the theoretical simulations, we vary the ambient temperatures of samples to observe the shift of absorption peak.The discrepancy between simulation and experiment results originates from that the molecular structure used in the simulation is gas phase molecular model, which is different from crystal structure.

Measurement of 461 GHz Atmospheric Opacity at Delingha

Sheng Li[1,2,3,*], Wen-ying Duan[1,2], Kun Zhang[1,2],Shao-liang Li[1,2,3], Qing Shi[1,2,3], and Sheng-cai Shi[1,2]
[1]*Purple Mountain Observatory, CAS, Nanjing, 210008, China*
[2]*Key Lab of Radio Astronomy, CAS, Nanjing, 210008, China*
[3]*University of Chinese Academy of Sciences, Beijing, 100049, China*
*Contact: shengli@pmo.ac.cn

Abstract—**POST (POrtable Submillimeter Telescope), mostly used for astronomical site evaluation and experimental observation, is a 30-cm transportable submillimeter telescope worked around 500 GHz. After an upgrade in 2014, POST was deployed at Delingha, a western city in China. Atmospheric opacity at 461 GHz was mainly measured there during the winters of 2015 and 2016. Total time of the measurements was more than 1000 hours. Statistical results show the quartiles of atmospheric opacity was 1.07, 1.51 and 1.95 during the observing season. ^{12}CO (J = 4-3) line of Orion A and M17SW were successfully detected and we also made a full mapping towards the Moon at some transparent days. Detailed results will be presented in this paper.**

I. Introduction

The 30-cm POST telescope was an old product of the age of 2000s [1]. Now it is mostly used for astronomical site evaluation and experimental observation at submillimeter wave lengths. Measurements of atmospheric opacity between 460 and 500 GHz have been made at some sites in western China [2][3]. Also, The first astronomical observation ever made with NbN superconducting tunnel junctions was realized on this telescope [4]. We upgraded its servo control system of antenna and some other modules of the receiver in 2014 [5]. Then, POST was shipped to Delingha (3200 m), a western city in China, for atmospheric opacity testing at 461 GHz. Non-consecutive measurements were made from Jan. to Apr. 2015 and in Jan. 2016 due to POST is not an unattended telescope yet and could not work under bad weather like rainy or snowy days. During the atmospheric opacity measurements, we took some time and successfully observed the ^{12}CO (J = 4-3) spectral line of two different source. A full mapping of the Moon at 461 GHz was also made while atmospheric transmission at this frequency was good.

II. Instrument and Measurement Methods

Before the newly upgrade in 2014, two other times of great upgrade was made on POST [6][7]. This time the upgrade involved SIS junction, detector bias, low temperature LNA, backend and antenna servo control system. Receiver noise temperature at 461 GHz was 230 K measured in field of Delingha.

The telescope operated as a tipper while measuring the atmospheric opacity. We digitalized the outputs of IF total power corresponding 14 different angles from zenith to near horizontal. Calibrator temperature was acquired twice at the start and end of each tip cycle. Then average was used as ambient temperature. When assuming a plane parallel uniform atmosphere, the zenith atmospheric opacity τ_0 could be fitted by the following equation. Here an approximation of $T_{atm} = T_{amb}$ was used. It took nearly four minutes of a measurement cycle to fit one τ_0.

$$T_{obs}(z) = T_{rx} + T_{amb}\left(1 - e^{-\tau_0 \sec(z)}\right)$$

Fig. 1 The 30-cm antenna of POST on site

III. Statistical Results

The measurements lasted more than 1000 hours in total and, removing one-third untrusted data, we got 10684 valid τ_0 data. These untrusted data almost obtained in bad weather with poor dynamic range which resulting in an incredible fitted τ_0 with large uncertainty. It could be determined that these time periods were unable to carry out astronomical observations. Consequently, the below statistical results should be better than the actual situation. More should be noticed is that the measurements were carried out mostly in the winter, so they could only reflect atmospheric transmission of the site to a certain extent.

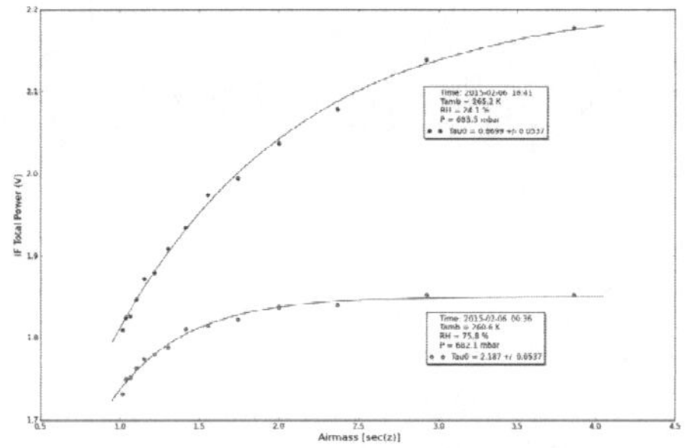

Fig. 2 Two fitted τ_0 at different time periods

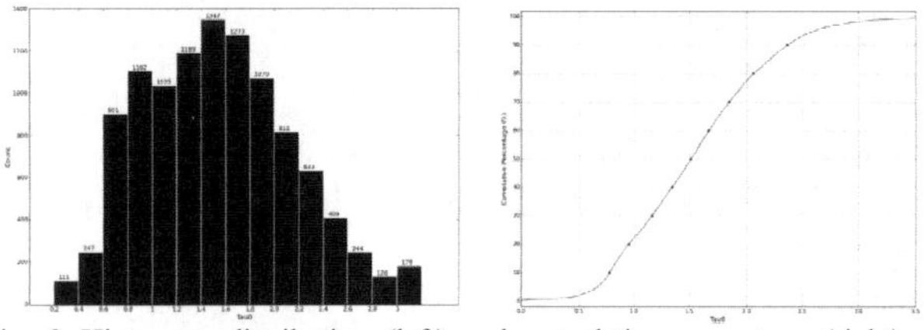

Fig. 3 Histogram distribution (left) and cumulative percentage (right) of measured τ_0 in the winter time

TABLE I
QUARTILES OF THE MEASURED τ_0

Quartile	Zenith pacity
25 %	1.07
50 %	1.51
75 %	1.95

IV. ^{12}CO (J = 4-3) LINE OBSERVATIONS

During the long time atmospheric opacity measurements, we made spectral line observations towards the standard sources of Orion A and M17SW when atmospheric transmission was tested good at 461 GHz (τ_0 around 1).

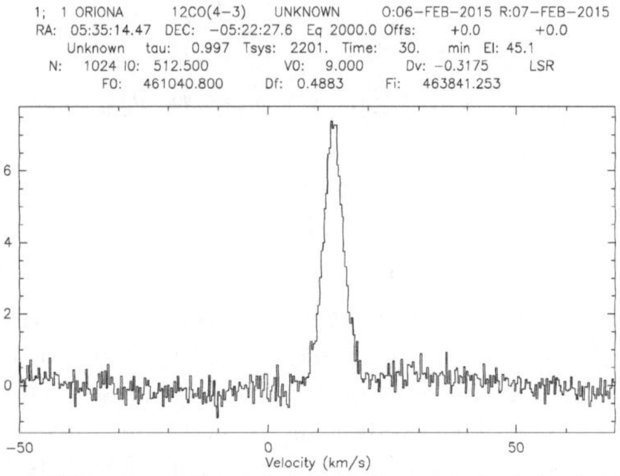

Fig. 4 ^{12}CO (J = 4-3) of Orion A observed in Feb. 6, 2015 (Total integration time was 30 mintues)

Also, we made a full mapping of the Moon by using the total power observation mode in Jan. 24, 2016. The observation was a 51 x 51 points 2D mapping with space interval of 1' in both AZ and EL directions. During the mapping, atmospheric opacity was around 0.7 indicating an transmission rate about 50 %. Fig. 5 shows the diameter of the Moon is about 30'.

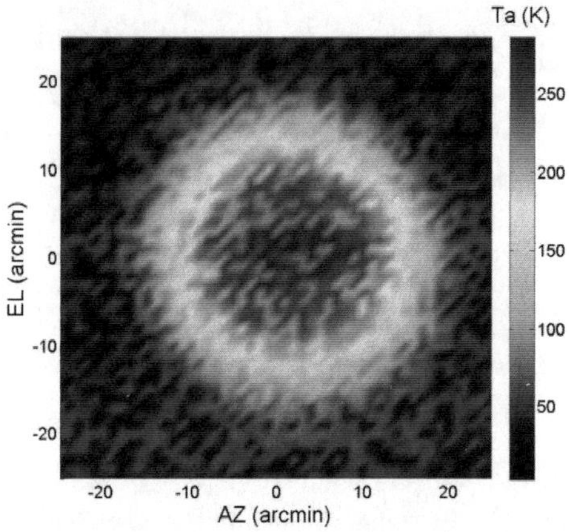

Fig. 5 The Moon mapping at 461 GHz, τ_0 was about 0.7 while observing

V. SUMMARY

POST telescope was transported to Delingha for atmospheric opacity evaluation at 461 GHz after a system upgrade. Inconsecutive measurements were made from Jan. to Apr. 2015 and in Jan. 2016. Statistical results show that quartiles are 1.07, 1.51 and 1.95 of the observing season. The real atmospheric transmission at this frequency should be worse for some bad data have not been counted in when standard errors of the fitting parameters are large. Moreover, POST was not working under some severe weather. These times were not suitable for astronomical observation and were not included in the statistics.

ACKNOWLEDGMENT

The authors would like to thank Xu-guo Zhang and other colleagues at Delingha Station for their help. This work was supported in part by National Natural Science Foundation of China (Grant No. 11127903, 41127901 and 11190012).

REFERENCES

[1] J. Yang, M.Ohishi, S. Shi, T. Noguchi, K. Miyazawa, J. Inatani, Y.Sekimoto, S. Yamamoto, H. Ozeki and S. Saito, "The POrtable SubmillimeterTelescope (POST) Project," *Ground-Based Astronomy in Asia*, (NAO:Japan),ed. N. Kaifu, pp. 565-568, 1996.

[2] Ji Yang, Cheng-hua Sun, Shu-ping Huang, Ohishi Masatoshi and Miyazawa Keisuke. "Measurement of Submillimeter Zenith Opacity at 492 GHz, " *Chinese Astronomy and Astrophysics*, vol. 28, pp. 367-376, 2004.

[3] Ming-feng Zhou, Qi-jun Yao, Sheng Li, Zhi-quan Luo and Ji Yang, "Measurements of 460 GHz Atmospheric Opacity at Yangbajing Observational Station, " *Chinese Astronomy and Astrophysics*, vol. 35, pp. 327-338, 2011.

[4] Jing Li, Masanori Takeda, Zhen Wang, Sheng-Cai Shi, and Ji Yang, "Low-noise 0.5THz0.5THz all-NbN superconductor-insulator-superconductor mixer for submillimeter wave astronomy," *Appl. Phys. Lett.*, vol. 92, 222504, 2008.

[5] Sheng Li, Wen-ying Duan, Kun Zhang, Dong Liu, Qi-jun Yao and Sheng-cai Shi, "System Performance of the Upgraded POST," *IRMMW-THz2015*, Hong Kong, 23-28 Aug. 2015.

[6] S.P. Huang, J. Li, J. Xu, A.Q. Cao, S.H. Chen, J. Huang, Z.H. Lin, S.C. Shi and J. Yang, *IRMMW-THz2006*, "A 500-GHz Superconducting SIS Receiver for The Portable Submillimeter Telescope, " Shanghai, 18-22 Sept. 2006.

[7] Qi-Jun Yao, Xue-Song Cao, Xu-Guo Zhang, Jing Li, Sheng-Cai Shi and Ji Yang, "A Compact 500GHz SIS Receiver System for POST," *APMC2009*, pp. 551-554, Singapore, 7-10 Dec. 2009.

Registered participants

Name	Institute/Company	E-mail
Pete Barry	Cardiff University	peter.barry@astro.cf.ac.uk
Harvey Beere	University of Cambridge	heb1000@cam.ac.uk
Faouzi Boussaha	Observatoire de Paris	faouzi.boussaha@obspm.fr
Jiajun Chen	University of Cambridge	j.chen@mrao.cam.ac.uk
Zhe Chen	University of Electronic Science and Technology of China	zhechen_uestc@126.com
Sergey Cherednichenko	Chalmers University of Technology	serguei@chalmers.se
Thomas Crowe	Virginia Diodes Inc	neumann@vadiodes.com
Wei Cui	Tsinghua University	cui@tsinghua.edu.cn
Gerhard De Lange	Netherlands Institute for Space Research	gert@sron.nl
Fabien Defrance	Observatoire de Paris	fabien.defrance@obspm.fr
Yan Delorme	Observatoire de Paris	yan.delorme@obspm.fr
Jiangqiao Ding	Nanjing University of Science & Technology	dingjiangqiao@126.com
Kun Dong	University of Electronic Science and Technology of China	dkun1990@126.com
Wenying Duan	Purple Mountain Observatory	wyduan@pmo.ac.cn
Andrey Ermakov	Chalmers University of Technology	andreye@chalmers.se
Bo Fang	University of Electronic Science and Technology of China	18200516559@163.com
Shang Fen	University of Electronic Science and Technology of China	1264186011@qq.com
Alexandre Féret	Observatoire de Paris	alexferet@gmail.com
Anne-Laure Fontana	Institut de Radioastronomie Rillimétrique	fontana@iram.fr
Jianrong Gao	Netherlands Institute for Space Research/Delft University of Technology	j.r.gao@tudelft.nl
John Garrett	University of Oxford	john.garrett@astro.ox.ac.uk
Lina Gatilova	Observatoire de Paris	lina.gatilova@obspm.fr
Chong Ge	Purple Mountain Observatory	chongge@pmo.ac.cn
Hugh Gibson	Gibson Microwave Design EURL	hughrpg@gmail.com
Gregory Goltsman	Moscow State Pedagogical University	goltsman@rplab.ru
Alvaro Gonzalez	National Astronomical Observatory of Japan	alvaro.gonzalez@nao.ac.jp
Matt Griffin	Cardiff University	matt.griffin@astro.cf.ac.uk
Frederic Gueth	Institut de Radioastronomie Rillimétrique	gueth@iram.fr
Peter Hargrave	Cardiff University	hargravepc@cardiff.ac.uk
Tetsuo Hasegawa	National Astronomical Observatory of Japan Chile Observatory	tetsuo.hasegawa@nao.ac.jp

Darren Hayton	Netherlands Institute for Space Research	d.j.hayton@sron.nl
Jeffrey Hesler	Virginia Diodes Inc	hesler@vadiodes.com
Ronald Hesper	Kapteyn Astronomical Institute, University of Groningen	r.hesper@sron.nl
Richard Hijmering	Netherlands Institute for Space Research	r.hijmering@sron.nl
Netty Honingh	University of Cologne	honingh@ph1.uni-koeln.de
Jie Hu	Purple Mountain Observatory	790575332@qq.com
Nan Huang	University of Electronic Science and Technology of China	294985115@qq.com
Yenru Huang	Academia Sinica Institute of Astronomy and Astrophysics	yenruhuang@asiaa.sinica.edu.tw
Heinz-Wilhelm Huebers	German Aerospace Center, Institute of Optical Sensor Systems	heinz-wilhelm.huebers@dlr.de
Yoon Dae Hwan	National Radio Research Agency	shwanta@msip.go.kr
Heo Hye Young	Korea Astronomy and Space Science Institute	hy1230@kasi.re.kr
Yoshihisa Irimajiri	National Institute of Information and Communications Technology	irimaji@nict.go.jp
Li Ji	Purple Mountain Observatory	ji@pmo.ac.cn
Xiaoqing Jia	Nanjing University	xqjia@nju.edu.cn
Ling Jiang	Nanjing Forestry University	jiangling@njfu.edu.cn
Shibo Jiang	MCEA	andrewjiangworking@gmail.com
Zhenguo Jiang	University of Notre Dame	zjiang@nd.edu
Jiang Jun	Institute of Electronic Engineering of Chinese Academy of Engineering Physics	000jiangjun@163.com
Matthias Justen	University of Cologne	justen@ph1.uni-koeln.de
Haoran Kang	Purple Mountain Observatory	kanghr@pmo.ac.cn
Hyunwoo Kang	Korea Astronomy and Space Science Institute	orionkhw@kasi.re.kr
Lin Kang	Nanjing University	kanglin@nju.edu.cn
Boris Karasik	Jet Propulsion Laboratory	bscrucian@gmail.com
Jonathan Kawamura	Jet Propulsion Laboratory	kawamuraj@charter.net
Pourya Khosropanah	Netherlands Institute for Space Research	p.khosropanah@sron.nl
Andrey Khudchenko	Netherlands Institute for Space Research	a.khudchenko@sron.nl
Jenna Kloosterman	Jet Propulsion Laboratory	jenna.l.kloosterman@jpl.nasa.gov
Takafumi Kojima	National Astronomical Observatory of Japan	t.kojima@nao.ac.jp
Valery Koshelets	Kotelnikov Institute of Radio Engineering and Electronics	valery@hitech.cplire.ru
Yuto Kozuki	Osaka Prefecture University	s_y.kozuki@p.s.osakafu-u.ac.jp
Sascha Krause	Chalmers University of Technology	sascha.krause@chalmers.se
Huang Kun	Institute of Electronic Engineering of Chinese Academy of Engineering Physics	2687515@qq.com
Bangwon Lee	Korea Astronomy and Space Science Institute	bwlee@kasi.re.kr
Jungwon Lee	Korea Astronomy and Space Science Institute	jwl@kasi.re.kr
Chaote Li	Academia Sinica Institute of Astronomy and Astrophysics	ctli@asiaa.sinica.edu.tw

Fang Li	University of Electronic Science and Technology of China	1311335082@qq.com
Hao Li	University of Electronic Science and Technology of China	lihao_uestc@126.com
Hua Li	Shanghai Institute of Microsystem and Information Technology	hua.li@mail.sim.ac.cn
Jibin Li	Purple Mountain Observatory	lijibin@pmo.ac.cn
Jing Li	Purple Mountain Observatory	lijing@pmo.ac.cn
Qin Li	Southeast University	liqin_iroi@seu.edu.cn
Sheng Li	Purple Mountain Observatory	shengli@pmo.ac.cn
Zhenqiang Li	Purple Mountain Observatory	zqli@pmo.ac.cn
Xu Liang	Xi'an institute of optics and precision mechanics of CAS	liangx@opt.ac.cn
Weimin Liao	Institute of Applied Physics and Computational Mathematics	lweim@vip.sina.com
Zhenhui Lin	Purple Mountain Observatory	lzh@mwlab.pmo.ac.cn
Dong Liu	Purple Mountain Observatory	dliu@pmo.ac.cn
Jie Liu	Purple Mountain Observatory	liujie@pmo.ac.cn
Lei Liu	University of Notre Dame	lliu3@nd.edu
Na Liu	University of Electronic Science and Technology of China	635733312@qq.com
Zheng Lou	Purple Mountain Observatory	zhenglou@pmo.ac.cn
Alain Maestrini	Observatoire de Paris	alain.maestrini@obspm.fr
Doris Maier	Institut de Radioastronomie Rillimétrique	maier@iram.fr
Ruiqing Mao	Purple Mountain Observatory	rqmao@pmo.ac.cn
Hiroshi Matsuo	National Astronomical Observatory of Japan	h.matsuo@nao.ac.jp
Imran Mehdi	Jet Propulsion Laboratory	imranmehdi@yahoo.com
Wei Miao	Purple Mountain Observatory	wmiao@mwlab.pmo.ac.cn
Behnam Mirzaei	Delft University of Technology	b.mirzaei@tudelft.nl
Diego Moro-Melgar	Observatoire de Paris	diego.moro-melgar@obspm.fr
Alessandro Navarrini	National Institute for Astrophysics	navarrin@oa-cagliari.inaf.it
Tom Nitta	University of Tsukuba	nitta.tomu.fw@u.tsukuba.ac.jp
Evgenii Novoselov	Chalmers University of Technology	eugene.novoselov@gmail.com
Chiko Otani	RIKEN	otani@riken.jp
Scott Paine	Smithsonian Astrophysical Observatory	spaine@cfa.harvard.edu
Francois Pajot	Institut de Recherche en Astrophysique et Planetologie	francois.pajot@irap.omp.eu
Songchon Park	National Radio Research Agency	pschon1000@msip.go.kr
Patrick Puetz	University of Cologne	puetz@ph1.uni-koeln.de
Yuan Qian	Purple Mountain Observatory	yuanqian@pmo.ac.cn
Hua Qin	Suzhou Institute of Nano-Tech and Nano-Bionics	hqin2007@sinano.ac.cn
Yuan Ren	University of Cambridge	yr235@cam.ac.uk

Shigeyuki Sekiguchi	University of Tokyo / National Astronomical Observatory of Japan	shigeyuki.sekiguchi@nao.ac.jp
Yutaro Sekimoto	National Astronomical Observatory of Japan	sekimoto.yutaro@nao.ac.jp
Wenlei Shan	National Astronomical Observatory of Japan	wenlei.shan@nao.ac.jp
Michael Shcherbatenko	Moscow State Pedagogical University	scherbatenko@prlab.ru
Qing Shi	Purple Mountain Observatory	shiqing@pmo.ac.cn
Shengcai Shi	Purple Mountain Observatory	scshi@pmo.ac.cn
Shibo Shu	The University of Tokyo	shibo.shu@nao.ac.jp
Andrey Smirnov	Astro Space Center, Lebedev Physical Institute of the Russian Academy of Sciences	asmirn@asc.rssi.ru
Peter Sobis	Omnisys Instruments AB	ps@omnisys.se
Boon Kok Tan	University of Oxford	tanbk@astro.ox.ac.uk
Hong Tang	University of Electronic Science and Technology of China	906578759@qq.com
Yaoling Tian	Institute of Electronic Engineering of Chinese Academy of Engineering Physics	272036657@qq.com
Edward Tong	Harvard-Smithsonian Center for Astrophysics	etong@cfa.harvard.edu
Jeanne Treuttel	Jet Propulsion Laboratory	jeanne.treuttel@obspm.fr
Xuecou Tu	Nanjing University	tuxuecou@nju.edu.cn
Yoshinori Uzawa	National Institute of Information and Communications Technology	uzawa@nict.go.jp
Hairen Wang	Purple Mountain Observatory	hairenwang@pmo.ac.cn
Hongchi Wang	Purple Mountain Observatory	hcwang@pmo.ac.cn
Junxian Wang	University of Science and Technology of China	jxw@ustc.edu.cn
Yan Wang	University of Electronic Science and Technology of China	wangyanuestc@163.com
Zheng Wang	Purple Mountain Observatory	zhwang@pmo.ac.cn
Martina Wiedner	Observatoire de Paris	martina.wiedner@obspm.fr
Ken Wood	QMC Instruments Ltd.	k.wood@terahertz.co.uk
Wentao Wu	Purple Mountain Observatory	wentaowu@pmo.ac.cn
Changpeng Xu	University of Electronic Science and Technology of China	superxcp@163.com
Lei Xu	Capital Normal University	2140602047@cnu.edu.cn
Leijun Xu	Jiangsu university	1503863@qq.com
Shuping Yan	Purple Mountain Observatory	yanshuping@pmo.ac.cn
Fei Yang	Southeast University	yangfei@seu.edu.cn
Ji Yang	Purple Mountain Observatory	jiyang@pmo.ac.cn
Jinping Yang	Purple Mountain Observatory	jpyang@pmo.ac.cn
Liusha Yang	University of Electronic Science and Technology of China	458298172@qq.com
Liya Yang	University of Electronic Science and Technology of China	yly1990@126.com
Yelei Yao	University of Electronic Science and Technology of China	yeleiyao@gmail.com
Ghassan Yassin	University of Oxford	g.yassin1@physics.ox.ac.uk

Lingzhen Zeng	Smithsonian Astrophysical Observatory	lingzhen.zeng@cfa.harvard.edu
Bo Zhang	University of Electronic Science and Technology of China	zhangbo.uestc@163.com
Hailong Zhang	Purple Mountain Observatory	hlzhang@pmo.ac.cn
Labao Zhang	Nanjing University	lzhang@nju.edu.cn
Shuinai Zhang	Purple Mountain Observatory	snzhang@pmo.ac.cn
Wen Zhang	Purple Mountain Observatory	wzhang@pmo.ac.cn
Xuguo Zhang	Purple Mountain Observatory	xuguo@pmo.ac.cn
Zhang Zhang	University of Electronic Science and Technology of China	sunshinezz6688@163.com
Guozhong Zhao	Capital Normal University	guozhong-zhao@126.com
Zhenyu Zhao	Shanghai Normal University	zyzhao@shnu.edu.cn
Jiaqiang Zhong	Purple Mountain Observatory	jqzhong@pmo.ac.cn
Kangmin Zhou	Purple Mountain Observatory	kmzhou@pmo.ac.cn
Shuai Zong	University of Electronic Science and Technology of China	paradise_zs112@126.com
Yingxi Zuo	Purple Mountain Observatory	yxzuo@pmo.ac.cn